Molecular Plant Biology
Volume One

The Practical Approach Series

Related **Practical Approach** Series Titles

Essential Molecular Biology V2 2/e
Radioisotopes 2/e
Plant Cell Biology 2/e
Bioinformatics: sequence, structure and databanks
Functional Genomics
Essential Molecular Biology V1 2/e
Differential Display
Gene Targeting 2/e
DNA Microarray Technology
In Situ Hybridization 2/e
Mutation Detection
Molecular Genetic Analysis of Populations 2/e
PCR3: PCR In Situ Hybridization
Antisense Technology
Genome Mapping
DNA and Protein Sequence Analysis
DNA Cloning 3: Complex Genomes
Gene Probes 1
Gene Probes 2
Non-isotopic Methods in Molecular Biology
DNA Cloning 1: Core Techniques
DNA Cloning 2: Expression Systems
Plant Cell Culture 2/e
PCR 1

Please see the **Practical Approach** series website at

http://www.oup.com/pas

for full contents lists of all Practical Approach titles.

No. 258

Molecular Plant Biology

Volume One

A Practical Approach

Edited by

Philip M. Gilmartin

Centre for Plant Sciences,
Faculty of Biological Sciences,
University of Leeds, UK.

and

Chris Bowler

Laboratory of Molecular Plant Biology
Stazione Zoologica, Naples, Italy.

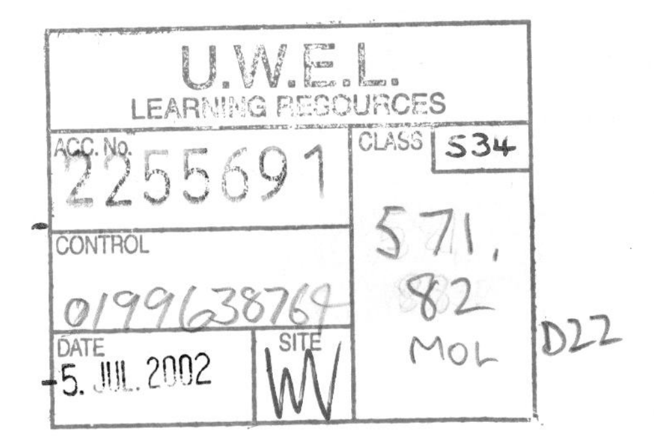

OXFORD
UNIVERSITY PRESS

OXFORD
UNIVERSITY PRESS

Great Clarendon Street, Oxford OX2 6DP

Oxford University Press is a department of the University of Oxford.
It furthers the University's objective of excellence in research, scholarship,
and education by publishing worldwide in

Oxford New York

Auckland Bangkok Buenos Aires Cape Town Chennai Dar es Salaam
Delhi Hong Kong Istanbul Karachi Kolkata Kuala Lumpur Madrid
Melbourne Mexico City Mumbai Nairobi São Paulo Shanghai Taipei
Tokyo Toronto

with an associated company in Berlin

Oxford is a registered trade mark of Oxford University Press in the UK and
in certain other countries

Published in the United States by Oxford University Press Inc., New York

First published 2002

A catalogue record for this title is available from
the British Library

Library of Congress Cataloguing-in-Publication Data
(Data available)

ISBN 0 19 963876 4 (Hbk)
ISBN 0 19 963875 6 (Pbk)

Typeset in Swift by Footnote Graphics, Warminster, Wilts
Printed in Great Britain on acid-free paper
by The Bath Press, Avon

Preface

Nearly fourteen years separates the publication of these volumes from the original Plant Molecular Biology—A Practical Approach. The original book, edited by Charlie Shaw, represented a milestone in the Practical Approach series as the first book in the series dedicated to the application of molecular biology to plant biology. In 1988, the field of Plant Molecular Biology was burgeoning. A number of recombinant DNA techniques had been adapted for the analysis of plant genes, and several dozen genes had been isolated and characterised from plants. The techniques for transformation of a few species using Agrobacterium tumefaciens mediated gene transfer had been developed although there were comparatively few laboratories with practical experience of producing transgenic plants. The methods for using endogenous plant transposons for gene tagging had been established and the first steps towards using transposons as tools for heterologous gene tagging had been taken. Several techniques for the analysis of plant organelles were also available, including the post-translational transport of proteins into chloroplasts. Approaches for analysis of the subcellular localization of macromolecules included light, fluorescent and electron microscopy. In addition to the growing use of recombinant DNA approaches to the study of higher plants, these techniques were also being applied to plant viruses as well as other models including Chlamydomonas and Cyanobacteia. All these approaches were featured in Plant Molecular Biology—A Practical Approach, and an appendix reported the use of the 'recently developed reporter gene' β-glucuronidase.

The advances since 1988 are astounding. There have been an unimaginable number of technical developments that have revolutionised the tools available and the scale of experiments that are possible. A large number of Practical Approach Books have been published since 1988 and several of these have focused on aspects of plant science or included approaches applicable to the analysis of plants, but there was never a second edition of *Plant Molecular Biology*—A Practical Approach. In compiling this all-new two-volume book, *Molecular Plant Biology*, we were mindful that this was not a second edition, but a sequel. Many of the approaches and applications described in the initial book are as relevant and important today as they were in 1988. However, as technologies have advanced there are now a myriad of approaches that build on the original techniques, as well as entirely new approaches based on technologies that did not exist at the time of publication of the original book.

The development and wide-spread use of PCR, the advent of automated DNA sequencing, the availability of entire plant genome sequences, the discovery of new reporter genes, the technological advances in bio-robotics and imaging that have resulted in DNA micro-arrays, and the ability to routinely transform a wide range of plant species, including all our major crops, highlight just a few of these developments. Not only have these advances revolutionised the way that fundamental experimental biology can be done, but they have also been applied commercially with the appearance, and in some cases subsequent removal, of products from transgenic plants on our supermarket shelves.

These tremendous advances have occurred in a relatively short time period, and we have both been fortunate enough to have witnessed all these advances during our own research careers. As editors of this new two-volume book, *Molecular Plant Biology* we have drawn on our combined research experiences and contacts to identify and persuade an international array of authors with particular expertise to contribute the chapters that we feel represent the breadth and scope of the field today. We would like to express our extreme gratitude to these authors for contributing to these books. The range of approaches covered in these two volumes is immense. Not only would it be impossible to find an individual with hands-on experience and expertise in all the topics covered in these two books, but there will surely be few laboratories anywhere where all these approaches are in routine use. We would like to thank all the contributors for sharing their expertise and making these books possible.

In editing these books we were keen to incorporate practical approaches as applied to a range of plant species, and not just the common models. Although much can be gained from the focused analysis of one species, there are many aspects of biology that are unique to non-model organisms. Some approaches covered in the original book are not covered in these new volumes. This is not meant to signify that they are no longer relevant or important, it is merely a reflection of the breadth of the field and a lack of space to incorporate and update approaches for specific organisms such as *Chlamydomonas* and *Cyanobacteria*. An exception to this has been the inclusion of a new chapter dealing with the moss *Physcomitrella patens*, which as a relatively new molecular model and represents the only plant in which homologous recombination is routine.

The two volumes are used to divide the topics into two themes, gene identification and isolation (Volume 1) and gene expression and gene product analysis (Volume 2). Each volume is divided into three sections and the rationale for the organisation of topics within these Sections is described below. However, many of the approaches have multiple applications, and others may therefore see alternative logic.

In Volume 1, the chapters in Section 1 (Gene Identification) provide methods and considerations for gene identification by classical mutagenesis. Also included here is a chapter on plant transformation, because it is an essential tool for the following two chapters which cover gene tagging strategies in transgenic plants. The final chapter in Section 1 focuses on genomic subtraction for gene identification. In Section 2 (Gene Organization) we have included a chapter on approaches for gene mapping, and a further chapter on the techniques for construction and screening of YAC, BAC and cosmid libraries. Also covered are approaches for chromosome *in situ* analysis. Section 3 (Library Screening and cDNA Isolation) there are three chapters that describe PCR cloning strategies, western and South-western library screens and complementation cloning.

In Volume 2, the first Section (Gene Expression) covers a range of methodologies for the analysis of gene expression, with chapters on transcript analysis, *in situ* RNA hybridisation and DNA micro-arrays. Approaches for the *in vitro* analysis of DNA:protein interactions are also covered, along with a final chapter on inducible gene expression in plants. In Section 2 (Gene Product Analysis) there are six chapters covering heterologous expression of recombinant proteins, analysis of proteins and protein targeting to chloroplasts, and biochemical approaches for the analysis of plant tissues. The application of the yeast two-hybrid system for analysis of protein:protein interactions is also included with the final two chapters in this Section devoted to antibody techniques and the construction and application of phage display libraries. The final Section of the book is devoted to functional analysis *in vivo*. In this Section we have included three chapters where the approaches primarily permit analyses in living cells. The first of these focuses on calcium imaging, the second describes a range of approaches using various reporter genes and the third is devoted to *Physcomitrella patens* as a model organism in which homologous recombination provides an exciting new *in vivo* approach. As the final chapter in the two volumes, this brings us full circle back to the first chapter in Volume 1 that is dedicated to mutagenesis.

We hope that this organisation illustrates that the approaches described in these volumes are not a linear progression through genetic, biochemical and molecular techniques, but that integration of these approaches and reiteration of analyses are essential to generate a global perspective of plant biology. In compiling the indexes for these two volumes we have provided one combined index for both books to facilitate full cross-referencing between the two volumes and we hope that this proves to be useful.

In 1988 when we first used *Plant Molecular biology—A Practical Approach*, one of us had just started a PhD, the other had just started a postdoc. We hope that the readers of these new books will find them as useful as we found the original book during our own time at the bench.

February 2002

P. M. G.
C. B.

Contents

SECTION 2 GENE ORGANIZATION

Protocol list

Abbreviations

ABRC	Arabidopsis Biological Resource Centre
AFLP	amplified fragment length polymorphism
AIMS	amplification of insertion mutagenized sites
ALS	amino lactone synthetase
BAC	bacterial artificial chromosome
BAR	phosphinothricin acetyl transferase
CaMV	cauliflower mosaic virus
CAPS	co-dominant amplified polymorphic sequence
cDNA	complementary DNA
CHEF	contour-clamped homogeneous electric field electrophoresis
CIAP	calf intestinal alkaline phosphatase
CTAB	hexadecyltrimethylammonium bromide
dH_2O	deionized H_2O
DNA	deoxyribonucleic acid
ECS	embryogenic cell suspension
EDS	empty donor site
EMS	ethyl-methane sulfonate
EPPS	N-(2-hydroxyethyl)piperazine-N′-3-propanesulfonic acid
EST	expressed sequence tag
FN	fast neutrons
GECN	genetically effective cell number
GFP	green fluorescent protein
GM	germination medium
GSP	gene specific primer
GUS	β-glucuronidase
GW	genome walker
HPT	hygromycin phosphotranserase
HYG	hygromycin resistance marker
Hyg^R	hygromycin resistance
iPCR	inverse PCR
IPTG	isopropyl-1-thio-galactopyranoside
LB	Luria broth

luc	firefly luciferase
M1	generation parental genotype
M2	generation plants that grow out of the seeds resulting from the M1 generation
Mb	megabase
MBG	MADS-box gene
MSARI	callus medium
MSARII	shoot medium
MSARIII	rooting medium
NILs	near isogenic lines
NPTII	neomycin phosphotransferase
OD	optical density
oriT	conjugation transfer origin
PACE	programmable automonously-controlled electrophoresis
PAT	phosphinothricin acetyl transferase
PCR	polymerase chain reaction
PEG	polyethylene glycol
PFGE	pulsed field gel electrophoresis
QTL	Quantitative trait loci
RACE	rapid amplification of cDNA ends
RAPD	random amplified polymorphic DNA
RDA	representational difference analysis
RFLP	restriction fragment length polymorphism
SG	seed germination
Sm^R	streptomycin resistant
Sm^S	streptomycin sensitive
SPEC	spectinomycin phosphotransferase
SSLP	single strand length polymorphism
SSR	simple sequence repeats
TAFE	transverse alternating field electrophoresis
T_{ann}	annealing temperature
TD	touch down
TDF	transcript-derived fragment
T-DNA	transfer-DNA
Ti	tumour-inducing
TIR	terminal inverted repeat
T_m	melting temperature
TSD	target site duplication
TSGI	target-selected gene inactivation
WT	wild-type
X-Gal	5-bromo-4-chloro-3-indolyl-galactopyranoside
YAC	yeast artificial chromosome

Chapter 1
Classical mutagenesis in higher plants

Maarten Koornneef
Department of Genetics, Wageningen University, Dreijenlaan 2, 6703 HA Wageningen, The Netherlands

1 Introduction

Mutagenesis aims at the disruption or alteration of genes. Mutants are the basis of genetic variation and therefore have attracted the interest of plant breeders. For a long time, mutagenesis research in plants focused on crop improvement and, especially for crop plants, optimized protocols were developed with barley being one of the favourite species. However, the interest in mutagenesis has shifted to basic plant research in the last 20 years, when the power of mutant approaches in combination with molecular techniques to investigate the molecular nature of the genes became fully appreciated.

Mutations can be induced by chemical, physical, and biological means. The latter technique, where genes are disrupted by insertion of DNA, originating from primary transformation events (see chapter 2 and 3) or from transposable elements (see chapter 4), became very popular because of the subsequent ease to clone the mutated gene. Although the application of the biological mutagens differs from the use of the 'classical' chemical and physical mutagens, which are the topic of this chapter, many of the basic principles are the same.

The present outline on mutagenesis procedures is based on experience with self-fertilizing species, such as the model plants *Arabidopsis thaliana* and tomato. The basic principles of mutagenesis in these species also hold for other self-fertilizing species. Several reviews and detailed protocols for classical mutagenesis of *Arabidopsis* have been published (1–3).

Differences in mutagenesis procedures between self-fertilizing species, cross-fertilizing species, and plants that are vegetatively propagated, are that, for the latter two types of species, the parental plants are often heterozygous at many loci. As a consequence, mutations in dominant alleles at these loci are recognizable as mutants in the mutagen-treated plants. In the case of vegetatively propagated plants, plant parts instead of seeds are treated. Because mutations are induced in single cells, mutations in such multi-cellular structures will appear as mutant sectors (chimeras) deriving from the cell that contains the mutation. Non-chimeric mutants can be obtained by propagating the mutant sectors vegetatively. This can be done with classical techniques or *in vitro* culture. For self-fertilizing species, dominant mutations are visible in the plants derived from mutagenized seed, but recessive mutations can only be identified in their progeny. A handbook describing many details of mutagenesis of such plants and crops has been published by van Harten (4).

Cross-pollinators that can be selfed, such as maize, *Brassica* sp., etc., can be treated in mutagenesis experiments as self-pollinators, following similar procedures as described in this Chapter, although generating selfed progenies will often be more laborious. When selfing is not possible, one will need other inbreeding procedures such as sib-mating to reveal the presence of homozygous recessive mutants in the progeny of mutagen-treated plants.

Tetraploid and allotetraploid plants are not very useful for mutagenesis experiments because they contain four copies of every gene in their genomes. This implies that genotypes that are homozygous for the recessive mutant allele will not appear in the selfed progeny when the parental genotype contained four wild-type alleles.

2 General principles and nomenclature

Most mutations lead to a loss of function of a functional gene. Genetically, such mutations behave as recessive, which implies that the mutant phenotype is not observed in the plant in which the mutations occurred. This generation of plants grown from the mutagen-treated parental genotype is called the M1 generation and is heterozygous for mutations. The embryos within the seeds formed on such plants and the plants that grow out of these seeds represent the M2 generation, in which homozygous recessive mutations will segregate. It is relevant to realize that the outer layer of seeds (testa) is of maternal origin. This means that mutants defective in the seed coat (of which many seed colour mutants are examples) can only be detected among the seeds developing on the M2 plants, which themselves have M3 embryos.

The material to be treated with mutagen is usually seed, which are multicellular structures. Since the genome is damaged randomly in each cell, different cells of the same seed will contain different mutations. M1 plants, derived from the mutagenesis of seeds, will therefore be chimeric. For the detection of mutations only those that occurred in cells that form the germline can be detected in the M2 generation. The number of cells that contribute to the germline was called the genetically effective cell number (GECN) by Li and Rédei (5), and estimated to be, on average, two in *Arabidopsis*. The consequence of this is that half of the progeny of an individual M1 plant derives from a sector in which a specific gene (*A*) is not mutated (*AA*), and the other half from the sector that was heterozygous for this gene (*Aa*). The segregation ratio observed for *aa* in the progeny of this chimeric plant is then 7 *A* (*AA* + *Aa*) : 1 *aa*. This 7 : 1 ratio is obtained from a 4 : 0 ratio from the non-mutated sector and a 3 : 1 ratio from the sector carrying the mutated allele *a*. The sectors that originate from the germline cells present in seeds are not identical in size and their contribution to the main inflorescence on which seeds are harvested changes during development. Parts of the plant that are formed later (the top) are mostly composed of only one sector derived from only one germline cell. The consequence of this is that variation occurs between M1 plants and that the harvesting policy also determines how many sectors contribute to the progeny. Chimerism disappears when the mutation is transmitted through the progeny because every individual of this progeny is derived from a single zygote cell.

In monoecious plants, such as maize, where male and female flowers are separated and almost always derive from different germline cells, the selfed progeny of a mutagen-treated seed will not segregate for the recessive mutation because a recessive allele is only provided by one of the gametes. Another generation of selfing will reveal the recessive mutants (in the M3). To bypass this chimerism problem, maize geneticists generally treat pollen with the mutagens, which they apply to a pistillate parent on which the M1 seeds are thereafter harvested (6). Each M1 plant will then be non-chimeric and recessive mutants will appear in the selfed progeny (M2 generation). Protocols for EMS mutagenesis of maize pollen and seeds have been published by Neuffer *et al.* (6).

Chimerism should also be taken into account when transposable elements are used as mutagens (see chapter 4). In this case, sector size is much more variable and depends on when during development the insertion occurred. In the case of transposable elements, chimerism is reintroduced when the element moves again.

Mutation frequencies (m) should be based on the frequency at which mutations occur per treated cell. Usually, these are expressed per locus or per group of mutants.

In practice, these estimates are based on the estimate of the number of mutants found in the M2 in relation to the number of M2 plants that are screened. It is also possible to base the estimate on the number of M1 plants whose progeny were screened for mutants.

The method of Gaul (see 7, 8, and references therein) is most simple and expresses the number of mutants found per number of M2 plants divided by f, which is the average mutant frequency in the progeny of a heterozygote ($f = 0.25$ for recessive mutants).

Therefore $m = m'/n \times f$ in which m' is the number of mutants found per n M2 plants screened. This estimate is independent of the number of M1 plants from which these are derived and also from the degree of chimerism. By dividing m by 2, the frequency of mutation per haploid genome can be obtained. Li and Rédei (5) derived the same frequency from the number of M1 plants whose progeny were tested. This procedure takes into account the GECN, and also requires that all mutations will be detected and that, therefore, the size of the progeny of individual M1 progenies should be large enough not to miss the mutant by chance. When GECN = 2, the mutant occurs only at a frequency of 1/8 which requires 23 plants not to miss the *aa* plant by chance in a given progeny at $P < 0.05$.

The mutation frequency per cell is calculated according to the method of Li and Rédei (5) with the formula: $m = M/(S \times \text{GECN})$, where M is the number of M1 progeny segregating for a mutation (type) among S progeny that were screened.

Estimates for a number of loci and mutant groups are given for EMS and irradiation by different authors (1, 3, 5, 7) to be between 10^{-5} and 10^{-3} per locus per haploid genome for effective treatments. This implies that mutants with lesions at specific loci can be found with a reasonable chance in the progeny of a few thousand M1 plants (see discussion on numbers hereafter). These frequencies are lower when only specific base pair changes give a certain, often dominant, phenotype.

Although some literature has suggested that the mutation frequency per locus depends on the genome size, data provide no evidence for this (7).

3 Choices in a mutagenesis experiment

Important decisions to be made in a mutagenesis experiment are the choice of the parental genotype, the type of mutagen, the dose of mutagen, the number of M1 plants to be grown, and the number of M2 plants to be screened.

3.1 Parental genotypes

The choice of parental genotype determines which mutations can be found. An obvious, but often neglected explanation for this is that so-called wild type plants can be mutant for specific loci (reviewed in 9). In *Arabidopsis*, the frequently used line Landsberg *erecta* (L*er*) behaves as a mutant of the flowering inhibitor gene *FLC*. The latter gene is epistatic to several other flowering time genes, which implies that mutants in such genes are not expressed in the L*er* background. Another standard line Wassilewskija (Ws1) is mutant for the phytochrome D (*PHYD*) and the cauliflower (*CAL*) gene. The latter mutation is not observed because it is only expressed in an *ap1* mutant background, which indicates that *AP1* and *CAL* have a similar function. This gene redundancy, for which genotypes can differ, provides another reason for differences in the possibility to find specific mutants in different genetic backgrounds. The various wild types often differ in genes that modify the expression of the mutant phenotype. This leads to complications when one compares mutants in different backgrounds and in addition gives problems when one combines the new mutant with existing mutants in a different genetic background. This aspect makes it useful to apply mutagenesis on commonly used standard wild types, when there are not specific reasons to use a different parental genotype. In *Arabidopsis* these are Columbia (Col), L*er* and Ws.

3.2 Choice of mutagens

The available mutagens can be divided into physical, chemical, and biological mutagens. The most commonly used physical mutagens are ionizing radiation, such as γ and X-rays, and fast neutrons. Each type of radiation produces deletions at a high frequency. The deletions are supposed to be larger after fast neutron radiation. The latter type of radiation yields densely ionized tracks compared to the mainly sparse ionizations caused by γ and X-rays. For this reason fast neutrons are sometimes preferred. However, facilities where one can apply this type of radiation are rare in comparison to sources of γ and X-rays. Deletions are attractive for many experiments because the mutants are often true nulls. In addition, a deletion is often recognizable with Southern blot and PCR analysis, which facilitates the recognition of the gene to be cloned using map based cloning (see chapter 6) and genomic subtraction procedures (see chapter 5). Although large deletions, e.g. parts of chromosome arms, can occur they are rarely observed in M2 populations because a large deletion in most cases affects the viability of germinating pollen grains that then will not contribute to fertilization. This effect of reduced transmission of mutated pollen is called certation.

Most chemical mutagens lead to base pair substitutions, especially GC→ AT in the case of ethyl-methane sulfonate (EMS). Physical and chemical mutagens have a different mutant spectrum, which is most obvious when one compares the ratio M1 sterility versus embryo lethal and chlorophyll mutants. For example, in an experiment (10) where EMS and X-rays were compared, an EMS treatment resulting in 44% sterility (percentage reduction of fertilized ovules compared to control) gave 18.2% embryonic lethals and 9.4% chlorophyll mutants, whereas an X-ray treatment that gave almost the same sterility produced only 5.9% embryonic lethals and 1.0% chlorophyll mutants. This relative high sterility is due to the fact that irradiation leads to chromosome breaks and chromosomal aberrations, giving rise to meiotic disturbances and, therefore, sterility. Base pair changes may lead to specific amino acid changes, which may alter the function of proteins, but do not abolish their function as deletions and frame-shift mutations mostly do. Plant species differ in their sensitivity for specific chemicals. In barley, sodium azide is a very potent mutagen (11), but it is hardly effective in *Arabidopsis* (12), where EMS is preferred (see *Protocol 1*). However, several other chemical mutagens have been shown to be effective (1, 2).

Chemical mutagens are extremely toxic and, therefore, require more care in their application, compared with physical mutagens. The latter can also be easily applied to pollen (13, 14). Pollen treatment is attractive for the generation of additional alleles of existing recessive mutants. To achieve this objective wild-type pollen is irradiated and then used to pollinate a mutant. The F1 progeny will be wild type, except when a pollen grain was mutated in that specific gene. In practice, this procedure yields many large deletions, which are not transmitted to the next generations (13, 14, and references therein). The effectiveness of radiation treatments depends heavily on the moisture and oxygen content of the treated material.

Tissue culture can induce so-called somaclonal variation which can be considered in the context of biological mutagens. This method might be useful, when a high enough mutation frequency and preferably, a different spectrum of mutants can be obtained. Although there is no doubt that genuine mutations occur after tissue culture, their frequency is much lower than with classical mutagens. Without question the amount of work to generate large numbers of

M1 plants, or in the case of tissue culture induced mutations R0 (regenerant), R1, or SC1 (somaclonal) plants that subsequently need to be selfed, is much more than a seed treatment with EMS or irradiation. Another negative aspect of this approach is that tissue culture produces many more polyploids in comparison with classical mutagens. A comparison of EMS and tissue culture mutagenesis in tomato has been described by van den Bulk *et al.* (15).

3.3 Mutagen dose

The higher the mutagen dose, the more efficient the experiment, because it provides a higher number of possible mutants. However, there are good reasons not to use the highest mutagen dose possible because this also will lead to an increase in unwanted mutations at other loci, many of which lead to sterility or even lethality. This makes a relatively low dose attractive in those cases where the amount of work and costs involved in growing more M1 plants and screening more M2 plants is not excessive. In those cases where it is expensive to grow many M1 plants and when laborious mutant screens are applied, the costs involved justify a higher dose. This approach then leads to more 'dirty' mutants that need 'cleaning up' by subsequent backcrosses to the wild type.

The determination of the optimal mutagen dose is not easy because of the reasons mentioned above. However, two criteria can be used to access the effectiveness of a treatment. These are the sterility of the M1 plants, which should be significant. Approximate numbers are 30–60% seed set of the control, based on percentage of fertilized ovules, or 20–50% of plants producing no offspring. As mentioned before, one requires considerably higher levels of sterility after irradiation compared to chemical mutagenesis to obtain a reasonable number of mutants. In M1 plants one also should see sectors with pigment defects in up to 1% of the plants. Pigment phenotypes screened for in the M2 population at the seedling level (or within the immature siliques of *Arabidopsis*) have mutation frequencies of 2–15% in populations that have been mutagenized effectively (3, 10). Survival or final germination percentage of the mutagen-treated seeds is not a good criterion because, even in highly mutagenized material, germination is only delayed in *Arabidopsis*. However, when treated seeds are not kept wet enough, germination can be strongly reduced.

Protocol 1

Ethyl-methane sulfonate (EMS) treatment for Arabidopsis

Equipment and Reagents

- Protective clothing and chemical hood
- Buchner vacuum flask
- Cloth bag
- Seeds pre-imbibed in water
- Freshly prepared ethyl-methane sulfonate (EMS) solution of 0.4–0.6% [v/v]
- 1M NaOH
- 0.15–0.2% [w/v] water agar solution

Protocol 1 continued

EMS is highly carcinogenic and volatile. Experiments should, therefore, be performed with all necessary safety precautions such as wearing protective clothing and working in a chemical hood. One should take all required precautions for waste disposal. Decontaminate EMS solutions with 1M NaOH.

Method

1. To allow good uptake of EMS, use seeds pre-imbibed in water. Imbibition at low temperature allows dormancy breaking (when necessary).
2. Prepare from a stock solution a freshly prepared EMS (EMS hydrolyses in water) solution of 0.4–0.6% [v/v]. EMS batches sometimes differ in effectiveness.
3. Stir the solution thoroughly because EMS does not dissolve very well in water.
4. Add a volume of the EMS solution similar to the volume in which the seeds are imbibed so that a 0.2–0.3% [v/v] solution is obtained.
5. Mix the seeds thoroughly with the solution.
6. An alternative procedure (3) is to add the stock solution directly to the seeds imbibed in water to produce a 0.2–0.3% [v/v] solution.
7. Imbibe the seeds for 12–24 hours at room temperature in darkness. The effectiveness of the treatment is the product of EMS concentration $\times$ duration of the treatment (7).
8. Remove the EMS solution by carefully decanting or by passing the seeds over a Buchner vacuum flask. Add a volume of water similar to the volume used for the treatment. Repeat these washes 5–10 times. Collect all EMS and wash waste in a waste container. Add NaOH to the waste to a concentration of 1 M to destroy the EMS.
9. Some researchers (1) pack the seeds in a cloth bag, which facilitates rinsing by transferring the packet of seeds to the wash solutions.
10. Mix the remaining seeds in 0.15–0.2% [w/v] water agar solution, in which the seeds can be pipetted directly to soil. A concentration of 100 seeds per 5 ml is sufficient for 100 cm^2 of soil surface. Keep the seeds moist after planting.
11. An effective EMS treatment will lead to reduced germination speed and reduced seedling growth, the presence of sectors with dominant colour mutations, and growth aberrations and sterility.
12. Procedures for different seed types mainly require adjustment of the treatment and wash volumes for species with large seeds. Planting of individual seeds will mostly be practiced instead of distributing the seeds in liquid agar.

Protocol 2

Radiation mutagenesis of Arabidopsis seeds

Equipment and Reagents

- Non-dormant seeds
- 10 ml 0.15–0.2% [w/v] water-agar in tubes
- Gamma or hard X-rays at a dose between 200 and 300 Gy (10 Gy = 1 Krad)

Method

1. Use non-dormant seeds.
2. Add 1000–5000 seeds to 10 ml 0.15–0.2% [w/v] water-agar in tubes.
3. If dry seeds are the starting material, wait 3–6 hours to allow imbibition of the seeds and then irradiate the tubes with gamma or hard X-rays with a dose between 200 and 300 Gy (10 Gy = 1 Krad).
4. After irradiation, it is better not to store the seeds in the agar solution, dilute the suspension to a density of 100 seeds per 5 ml, which is sufficient for 100 cm^2 of soil surface, and pipette the suspension on moist soil in pots or trays.
5. The seeds will develop into M1 plants, and delayed germination and growth together with sterility will indicate that the treatment was effective.
6. Pollen irradiation can be applied in *Arabidopsis* to inflorescences placed in plastic bags using a dose of 300 and 600 Gy and used immediately after the treatment for pollination (12), preferably of a male sterile genotype. When the pollen carries a dominant marker it allows seeds obtained by crossing from unwanted selfing events to be distinguished.

3.4 Numbers of M1 and M2 plants

Mutation frequencies that are obtained after mutagenesis are often around 5×10^{-4} per locus per genome (7). In cases where the germline contains two diploid cells, with two genomes per cell, there are four genomes per M1 plant. The mutant frequency per M1 plant is then 2×10^{-3}. To obtain a 95% probability of finding a specific mutation in a sample, one requires 1500 M1 plants when the mutant frequency is 2×10^{-3}. Since recessive mutants segregate at a frequency of $^1/_4$ per sector and, therefore, $^1/_8$ per M1 progeny, some mutants will not be found by chance when the M2 population size relative to the number of M1 plants is too small. One can increase the number of mutants by increasing the number of M1 plants and also, to a certain degree, the number of M2 plants. However, it was shown by Rédei (1) that an optimal ratio between the number of M1 plants and the number of M2 plants to be screened exists in order to obtain mutants at the lowest possible costs. This ratio depends on the relative costs of growing the M1 and the M2 plants. When the difference between these is low, screening a small number of M2 plants per M1 progeny is most cost effective.

This is often the case in *Arabidopsis*, where researchers prefer growing large numbers of M1 plants and screen between 2000 and 125 000 (3) M2 plants, depending on the ease of the screen. When one screens 10 000 M2 plants, with the assumption that no M1 plants are represented more than once, which is the case with very high numbers of M1 plants used to obtain this M2, this gives $\frac{1}{8} \times 2 \times 10^{-3} \times 10\,000 = 2.5$ mutants per locus. In cases where growing M1 plants and harvesting M2 seeds is relatively expensive, for example, as in tomato, it is worth increasing the number of M2 plants per M1 plant in order not to miss any mutants by chance. This is especially relevant when M2 screening can be done at the seedling stage.

In maize, 1000 non-chimeric M1 plants obtained after pollen mutagenesis was suggested to be sufficient to obtain 1 recessive mutant per locus (6).

3.5 The handling of M1 plants

For *Arabidopsis*, M1 plants are often grown in large numbers per pot or flat, and harvesting these as bulk or in pools is most efficient. However, pooling M1 plants implies that when a mutant is sterile or lethal it is lost. Furthermore, finding a mutant at the same locus twice may be because the two mutants derive from the same cell and, therefore, the same mutation event. Harvesting individual M1 plants and keeping spare seeds will allow the recovery of heterozygous wild type sister-plants, which are within the M1 progeny with twice the frequency of the recessive mutant. Furthermore, finding the same mutant in different progenies provides unambiguous proof for them originating from independent mutation events. Pooling the progenies of a certain number of plants provides a compromise between harvesting individual M1 plants, also called 'pedigreeing' (3) and harvesting all M1 plants as a bulk. However, for efficient recovery of heterozygous sister-plants the number of M1 plants per pool will quickly become too large and, therefore, 'pedigreeing' is necessary for this purpose. To be sure that a sufficient number of independent mutants are isolated, pools of up to a few hundred M1 plants are efficient. These numbers are often obtained when all M1 plants in one pot or tray are harvested. When two phenotypically identical mutants are found in one pool, they are often considered to be derived from the same mutation event. Only one mutant will then be used for further analysis. For large seeded plants one mostly collects single M1 progenies, for example as single maize cobs.

4 Further analysis of mutants

Since mutant screens depend on what the individual researcher is looking for the M2 screening methods can differ very much. These will not be discussed as part of this chapter.

However, when a certain mutant is isolated, a number of generally applicable procedures are required for their standard genetic analyses, which are described in more detail elsewhere (16, 17).

The main points are summarized below:

- In case the mutant is sterile, which mostly is due to male sterility, pollination with wild type pollen will save the mutant. Male sterility is indicated by the absence or poor quality of pollen. When the pollen of a sterile plant looks fertile, it can be used to pollinate a wild type plant. Female sterility is often observed in megaspore and ovule mutants. Harvesting fertile (heterozygous) sister-plants from the same M2 family also helps to save the mutant.
- Confirm the mutant phenotype in the M3 generation or in the generation derived from selfing the hybrid obtained by crossing the mutant with wild type or from the heterozygous sister-plant.
- Determine the segregation ratio of the mutant by crossing it with the parental wild type and by analysing the F2 generation. A backcross of the F1 to the recessive parent to analyse the genetics of the mutant is also valuable. The analysis of a segregating population, in addition, allows the analysis of putative pleiotropic effects observed in the selected mutant. Absolute linkage of different characters associated with the mutant will indicate true pleiotropism, as well as observing the same association of characters in independent allelic mutants. Segregation of the various traits implies that mutations in different genes are responsible for the syndrome observed in the original mutant. The phenotype of the F1 will already show whether the mutant is dominant, recessive, or co-dominant.
- Cross the mutant with all available mutants that have a similar phenotype. The absence of complementation with an existing mutant indicates allelism in the case of recessive mutants. For dominant mutants, the segregation of wild type plants in the F2 progeny will indicate non-allelism.
- When a new locus is found, its map position should be determined. A specific location on the genetic map may also suggest that other previously isolated mutants, even those with a (slightly) different phenotype, may represent mutants at the same locus, which can be confirmed with allelism tests.

References

1. Rédei, G. P. and Koncz, C. (1992). In *Methods in Arabidopsis research* (ed. C. Koncz, N-H. Chua, and J. Schell), p. 16. World Scientific Publishing, Singapore.
2. Feldmann, K. A., Malmberg, R. L., and Dean, C. (1994). In *Arabidopsis* (ed. E. M. Meyerowitz and C. Somerville), p. 137. Cold Spring Harbor Laboratory Press, Cold Spring Harbor, NY.
3. Lightner, J. and Caspar, T. (1998). In *Methods in molecular biology: Arabidopsis protocols* (ed. J. Martinez-Zapater and J. Salinas), Vol. 82, p. 91. Humana Press Inc, Totowa, NJ.
4. Van Harten, A. M. (1998). *Mutation breeding: theory and practical applications*. Cambridge University Press, Cambridge.
5. Li, S. L. and Rédei, G. P. (1969). *Radiat. Bot.*, **9**, 129.
6. Neuffer, M. G., Coe, E. W., and Wessler S. R. (1997). *Mutants of maize*, pp. 396–401. Cold Spring Harbor Laboratory Press, Cold Spring Harbor, NY.
7. Koornneef, M., Dellaert, L. M. W., and van der Veen, J. H. (1982). *Mutation Research*, **93**, 109.

8. Yonezawa, K. and Yamagata H. (1975). *Radiat. Bot.,* **15**, 241.
9. Alonso-Blanco, C. and Koornneef, M. (2000). *Trends Plant Sci.*, **5**, 22.
10. Van der Veen, J. H. (1966) *Arabidopsis Information Service*, **3**, 26.
11. Kleinhofs, A., Owais, W. M., and Nilan, R. A. (1978). *Mutat. Res.*, **55**, 165.
12. Gichner, T. and Veleminsky, J. (1977). *Biologia. Plantarum (Praha)* **19**, 153.
13. Liharska, T.B., Hontelez, J., van Kammen, A., Zabel, P., and Koornneef, M. (1997). *Theor. Appl. Genet.*, **95**, 969.
14. Vizir, I. Y., Anderson, M. L., Wilson, Z. A., and Mulligan, B. J. (1994). *Genetics*, **137**, 1111.
15. Van den Bulk, R. W., Löffler, H. J. M., Lindhout, W. H., and Koornneef, M. (1990). *Theor. Appl. Genet.*, **80**, 817.
16. Meinke, D. and Koornneef, M. (1997). *Plant J.*, **12**, 247.
17. Koornneef, M., Alonso-Blanco C., and Stam, P. (1998). In *Methods in molecular biology: Arabidopsis protocols* (ed. J. Martinez-Zapater and J. Salinas), Vol. 82, p. 105. Humana Press Inc, Totowa, NJ.

Chapter 2
Plant Transformation

Karabi Datta and Swapan K. Datta
Plant Breeding, Genetics and Biochemistry Division,
International Rice Research Institute, 6776 Ayala Avenue, Suite 1009,
Condominium Centre, Makati City, Philippines

1 Introduction

Genetic transformation involves the incorporation of genes into genomes by means other than fusion of gametes or somatic cells. There are different methods for stable introduction of foreign genes into the genomes of plant species, which are now possible in at least 35 families. The incorporation and expression of foreign genes in plants was described for tobacco in 1984 (1, 2). The introduction and expression of specific genes in plants provides a powerful tool for plant improvement and other molecular biology studies. The plant transformation approach is being used to generate plants possessing traits unachievable by conventional plant breeding, especially in cases where there is no source of the desired trait in the gene pool. Transgenic crops have now been field-tested in many countries, covering 39.9 million ha globally. Seven transgenic crops—maize, cotton, canola, rapeseed, potato, squash, and papaya—are grown commercially in 12 countries (3). Future prospects involve novel genes that can be introduced to generate plant lines useful for producing materials ranging from pharmaceuticals (4) to biodegradable plastics (5).

Different systems are now available for gene transfer and successive regeneration of transgenic plants, the most common being *Agrobacterium*-mediated transformation (6, 7, 8), particle bombardment (9), and direct gene transfer into protoplasts (10). In a gene transfer system, the three basic requirements for the production of transgenic plants are (a) the availability of target tissues competent for plant regeneration, (b) a suitable method to introduce DNA into cells that can regenerate, and (c) a procedure to select and regenerate transformed plants with a reasonable frequency. During the DNA transfer process, only a small portion of the target cells receive the DNA and only a small portion of these cells stably integrate the introduced DNA to survive the selection treatment. Efficient systems for selecting transformed cells among a large number of untransformed cells and for establishing regeneration conditions that allow recovery of plants derived from single transformed cells are therefore essential. In this chapter, we will deal with protocols for three different transformation systems for cereals, using rice as a case study (*Figure 1*) (11). Transformation of Arabidopsis is discussed in Chapter 3.

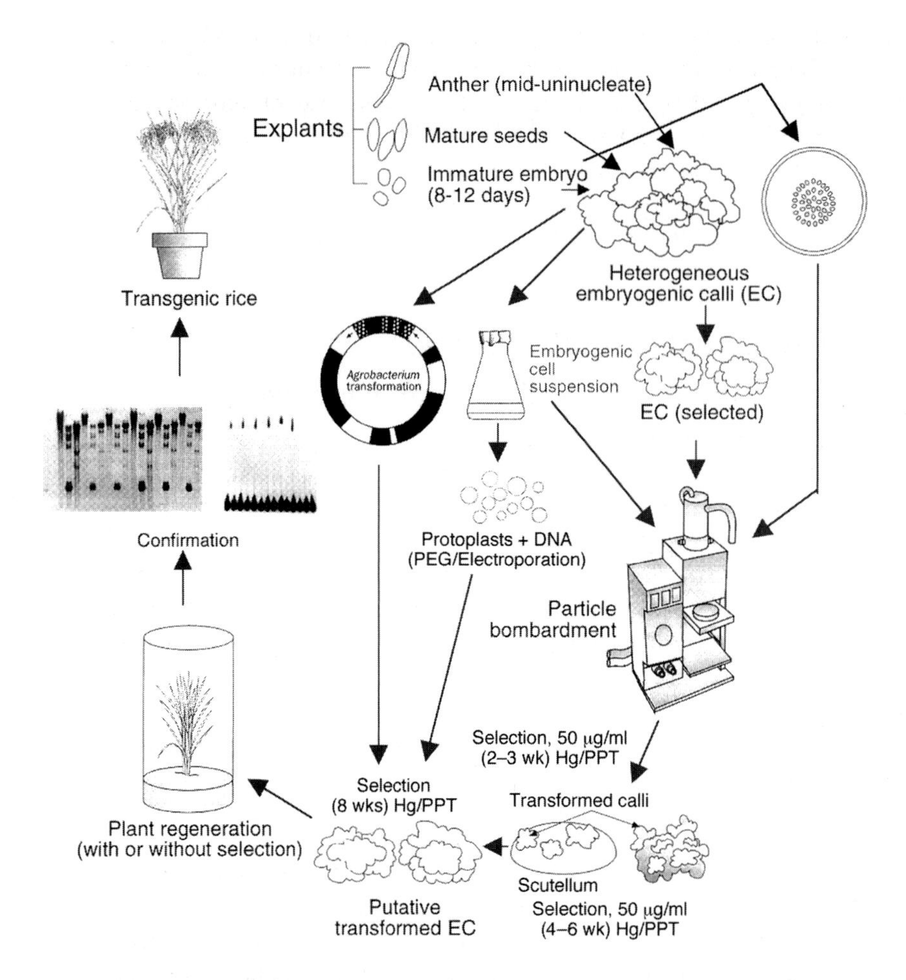

Figure 1 A schematic protocol for production of transgenic rice plants using biolistic, protoplast, and *Agrobacterium* systems.

Stable transfer of high-molecular-weight DNA into dicotyledonous plants using a binary bacterial artificial chromosome (binary BAC) vector by *Agrobacterium*-mediated plant transformation has been reported (12). This transfer of large insert genomic DNA by *Agrobacterium*-mediated transformation could be useful in a functional genomics programme, e.g., for inserting genetic regions containing quantitative trait loci, which are likely to comprise several agronomically important genes known only by their phenotypes (see Chapter 6).

1.1 Embryogenic culture and recalcitrance

Not all genotypes respond in the same way in embryogenic culture and during plant regeneration. Many cells receive foreign DNA irrespective of the DNA delivery system, but only a few cells (variable in different genotypes) are competent enough to produce fertile plants. The success of a transformation event depends on the normal development of the plant from the transformed cell(s). Consequently, the development and judicious use of embryogenic cultures can increase transformation frequencies. The high regeneration ability of *Arabidopsis*, tobacco, and japonica rice enables their use as model plants for gene expression. However, other recalcitrant species, such as pea, barley, and indica rice can be transformed, but with only limited success.

2. *Agrobacterium*-mediated transformation

2.1 *Agrobacterium-tumefaciens* and the Ti plasmid

In nature, the Gram-negative soil bacterium *Agrobacterium tumefaciens* causes crown gall formation in dicots by a multi-step transformation process. This naturally occurring mechanism of gene transfer has been extensively used in plant biotechnology. *A. tumefaciens* has long been used as a vector to transfer foreign genes into dicots, the preferred hosts of *Agrobacterium* (6, 13, 14). Although monocotyledonous plants are generally not considered within the host range of crown gall, it has been shown recently that *A. tumefaciens* can also be used as a vector for the successful transformation of monocotyledonous species, particularly rice (15–21).

Virulent strains of *A. tumefaciens* possess an extra-chromosomal plasmid that is involved in crown gall formation. Because of its role in tumour induction, this plasmid is called the tumour-inducing (Ti) plasmid. During infection, a small segment of the Ti plasmid, the T-DNA, surrounded by a 24-bp border repeat, is transferred to the plant cell and is integrated into a host chromosome in the nucleus. The T-DNA contains *onc* genes that encode phytohormone biosynthetic enzymes whose expression leads to tumour formation. Deletion of the *onc* genes of the Ti plasmid results in a non-oncogenic strain. Wild-type *A. tumefaciens* can be classified as octopine, nopaline, succinamopine or LL-succinamopine types according to the opine synthesizing capabilities encoded by their T-DNAs. Opines are tumour specific products that are synthesized by plants infected with *A. tumefaciens*, and used by the bacteria as carbon and nitrogen sources. Opine biosynthetic genes have eukaryotic regulatory sequences for expression. Besides the *onc* genes, a large number of other genes involved in tumourigenicity are present on the Ti plasmid in a segment called the virulence (*vir*) region. The T-DNA transfer system is determined by the *vir* genes, while the flanking 24-bp direct repeats are essential as recognition signals for the transfer apparatus.

2.2 Vector systems for *Agrobacterium*-mediated transformation

On the basis of the naturally occurring gene transfer mechanism of crown gall formation, various plant transformation mechanisms have been designed for genetic engineering of plants. A key development in this field was the removal of wild type T-DNA from Ti plasmids to create 'disarmed strains' such as LBA4404 (22) and C58C1 (pGV3850) (23). Various genes, including selectable marker genes and genes of interest, can be placed in the vector and introduced into the disarmed strain. The process of transfer of T-DNA will take place even if the virulence genes and the T-DNA are located on separate replicons in an *Agrobacterium* cell (22). The widely used binary vector system consists of a plasmid providing the virulence functions needed for transfer and a small vector carrying an artificial T-DNA (24, 25). The binary vectors replicate in *Escherichia coli*, as well as in *A. tumefaciens* and allow easy cloning of genes of interest between the T-DNA borders. During *Agrobacterium*-mediated gene integration, the DNA between the two defined T-DNA border sequences is transferred precisely and exclusively. Vectors such as pBin19 (24), pGA482 (25), and pBI121 (26) are well-known vectors used in various *Agro*-transformation systems, especially in dicots. Recently, a new binary vector pGA1611 has been constructed (27) that contains the unique *Hind*III, *Sac*I, *Hpa*I, and *Kpn*I sites for cloning of a gene between the maize *Ubi* promoter and *nos* terminator.

2.3 Supervirulent strains and superbinary vectors

In addition to the conventional *Agrobacterium* strains and vectors described in 2.2, a range of improved systems have been described. The supervirulent *Agrobacterium* strain A281 (28) has a wider host range and its transformation efficiency is higher than other strains due to its Ti plasmid, pTiBo542 (29–31). The strain EHA101 (32) and EHA105 (33) also carry disarmed versions of pTiBo542. In one superbinary vector system, the *virB*, *virC*, and *virG* genes from the virulence region of pTiBo542 are inserted into a small T-DNA-containing plasmid in addition to the disarmed Ti plasmid with its full set of virulence genes. *virB* and *virG* are the genes responsible for the supervirulence of strain A281. The supervirulent *Agrobacterium* strain AGLO was used to transform *Dianthus caryophyllus* by leaf infection (34). Induction of hairy and normal roots on *Picea abies*, *Pinus sylvestris*, and *Pinus contorta* by supervirulent *Agrobacterium rhizogenes* R1600 has also been reported (35).

2.4 Transformation using *Agrobacterium* as a vector

This protocol is used for the transformation of several rice cultivars. It involves the inoculation of immature embryo-derived calli with disarmed strains of *A. tumefaciens* carrying the vector of choice. The plant tissue is co-cultivated with *Agrobacterium* for 3 days. After the co-cultivation period, the bacterial population is removed by bacteriostatic antibiotics (cefotaxime and carbenicillin) and transformants are selected and regenerated on media containing selective agents. In total, 2–3-months post-co-cultivation are required to obtain rooted plantlets that can be transferred to soil.

Protocol 1

Transformation using *Agrobacterium* as a vector

Explants, equipment and reagents

A. Callus induction from immature embryos

- *Plant material*: plant material is *Oryza sativa*: indica-type rice cultivars, IR64, IR72, Basmati 122, Tulsi, Vaidehi, and japonica-type rice cultivar Taipei 309; immature embryos (panicles collected 7–10 days after anthesis)[a]
- *Soil and fertilizers*: 2.5 g $(NH_4)_2SO_4$, 1.25 g P_2O_5, and 0.75 g K_2O mixed per 2 kg of soil
- *General equipment*:

  Laminar flow cabinet, stereomicroscope, incubator, autoclave

  Plastic Petri dishes (50 mm; 90 mm diameter), pipettes

  Greenhouse facilities.
- *Sterilizing solutions*:

  Ethanol (70%)

  Sodium hypochlorite solution (e.g. 100 ml 1.8% sodium hypochlorite with two drops Tween 20 or 50% Chlorox containing 1–2 drops of Tween 20).
- *Media*:

  Callus induction and proliferation medium, e.g. modified MS medium (36)[b] (*Table 1*).

B. Agrobacterium culture for transformation

- Agrobacterium *strains*: a number of different strains of *A. tumefaciens* containing different binary vectors need to be considered. *A. tumefaciens* strain LBA4404 (pTOK 233) from Japan Tobacco Inc., contains pTOK 233, a superbinary vector (18, 37), whose T-DNA carries (a) the β-*glucuronidase* (*gus*) reporter gene with an intron in the N-terminal region, (b) a hygromycin resistance gene, and (c) a kanamycin resistance gene. The intron-*gus* gene expresses GUS activity in plant cells, but not in *A. tumefaciens* cells. The binary vector contains *virB*, *virC*, and *virG* genes derived from supervirulent Ti plasmid pTiB0542 (30).
- Laminar flow cabinet
- Rotary shaker at 28 °C
- 20-mm high Petri dishes (Greiner)
- For growth of *Agrobacterium*, use liquid AAM medium (*Table 2*) (38) or Luria broth (LB) medium [1% Bacto-peptone (Difco Laboratories), 0.5% Bacto- yeast extract (Difco Laboratories), 1% NaCl] together with suitable antibiotics for plasmid selection.
- For maintenance of *Agrobacterium*, use solid AB (*Table 3*) or LB medium with agar (15 g/l).

C. Transformation and regeneration

- General equipment:

  Centrifuge (speed-and temperature-controlled)

  (b) Spectrophotometer for measuring optical density of bacterial solution

  (c) Vacuum infiltration apparatus

  (d) Plastic centrifuge tubes of 50 ml capacity

  (e) 10 ml pipettes

  (f) Automatic pipette

  (g) Petri dishes (90 mm)
- Solutions and media:

  (a) 10-mM $MgSO_4$ solution

  (b) N6-AS medium (39) with acetosyringone for infiltration and co-cultivation[c] (*Table 4*). For infiltration, N6-AS liquid medium should be used (without agar)

  (c) Cefotaxime: 250 mg/l (Duchefa, Haarlem, The Netherlands, or Sigma) filter sterilized and stored at −20 °C

Protocol 1 continued

(d) GUS staining solution (26) (see Volume 2, Chapter 13)

(e) MS-CH medium for selection (*Table 4*)

(f) Regeneration medium (*Table 4*)

(g) Rooting medium—MS basal medium without hormones

(h) Culture solution (40) (*Table 5*)

Methods

A. Growth of donor plants

1 Break the dormancy of clean seeds of the desired pure rice line by incubating them at 50 °C for 3–5 days.

2 Sow the seed in seed boxes in damp soil.

3 Twenty-one days after sowing, transplant the seedlings to pots (one seedling per pot) containing soil and fertilizer.

4 Grow the plants in a glasshouse and keep them well watered. The glasshouse should have full sunshine and sufficient ventilation to maintain day-time temperature of 27–29 °C with a 90% humidity level.

B. Establishment of embryogenic calli from immature embryos

1 Sterilize dehulled immature (9–14 days after anthesis) seeds in 70% ethanol for 1 min and then in 1.8% sodium hypochlorite for 30 min. Wash three times with sterile double distilled water.

2 Isolate the embryos from the sterile seeds and place them on Petri dishes containing 20 ml modified MS medium (*Table 1*). The distance between each plated embryo should be at least 5 mm. The scutellar tissue should be upward, since callus growth occurs only with this orientation.

3 Incubate the embryos in the dark at 25 °C.

4 After 3–4 days, cut off the emerging shoots and roots, and subculture the remaining tissues on fresh medium.

5 Make at least two subcultures at 2-week intervals until yellowish-white soft embryogenic calli develop on the surface of the scutella.

6 Select and collect the embryogenic calli from the 4–6-weeks old culture only on the day of transformation.

C. Preparation of *Agrobacterium* culture

1 Maintain the *Agrobacterium* culture in AB or LB solid medium and subculture once a month.

2 Preculture: 4 days prior to transformation, grow a single clone of the bacteria on 10 ml AAM medium or LB medium with the appropriate selective agent[d].

3 Incubate with shaking (250 r.p.m.) at 28 °C overnight.

Protocol 1 continued

4 Transfer 500 μl of the culture to 20 ml AAM or LB medium with appropriate antibiotics and acetosyringone.

5 From the preculture medium, streak the culture again onto AB or LB solid medium.

6 Incubate both solid and liquid cultures at 28 °C for 2 days. Liquid cultures should be shaken at 250 r.p.m..

D. Transformation

1 Centrifuge the bacterial culture in liquid medium at 3800 g for 30 min at 10 °C.

2 Discard the medium and harvest the bacteria in 20 ml 10mM $MgSO_4$ in a 50-ml centrifuge tube[e]. In the case of bacteria growing on solid AB medium, harvest directly in 10 mM $MgSO_4$ solution[f].

3 Prepare a homogeneous bacterial suspension in 10 mM $MgSO_4$ by pipetting up and down several times.

4 Pipette 2.7 ml 10 mM $MgSO_4$ solution into another tube and add 300 μl of the bacterial suspension in 10 mM $MgSO_4$ (total volume 3 ml).

5 Take optical density (OD) 600 nm measurement of the mixture from step no. 4.

6 Centrifuge the original harvested bacterial suspension in 20 ml 10 mM $MgSO_4$ at 3800 g for 30 min.

7 Discard the supernatant.

8 Resuspend the bacterial pellet by pipetting up and down in a small volume (5 ml) of liquid N6-AS infiltration medium.

9 Adjust the final optical density (600 nm) to approximately 2 by adding N6-AS medium[g].

10 Add the embryogenic calli to the bacterial suspension in a Petri dish.

11 For vacuum infiltration, place the Petri dish (without covering) in a vacuum dessicator for 10 min.

12 Let the embryogenic calli stand in the bacterial suspension for 20 min after removal from the vacuum dessicator.

13 Remove the bacterial suspension from the Petri dish with a pipette.

14 Blot the calli on filter paper to remove excess inoculum and transfer the calli to co-cultivation medium (N6-AS solid medium).

15 Keep the cultures in the dark at 25 °C for 3 days.

E. Washing and selection

1 Wash the embryogenic calli twice with sterile water containing 250 mg/l cefotaxime to remove the bacteria[h].

2 After washing, remove the excess water adhering to the calli with sterile blotting paper.

Protocol 1 continued

3 Place some of the calli in GUS staining solution to check GUS expression if the T-DNA carries the β-glucuronidase reporter. Keep overnight at 37 °C.

4 Place the remaining calli into MS-CH selection medium containing cefotaxime (the bacteriostatic antibiotic) and hygromycin (the selective antibiotic). Keep for 15 days in the dark at 27 °C.

5 Check the stained calli from 3 for the presence of blue spots.

F. Regeneration

1 Following the first selection, put the growing calli through three more selection cycles in the same medium, without cefotaxime, under the same conditions.

2 Transfer the selected embryogenic colonies with developing somatic embryos to regeneration medium (*Table 4*). Incubate in darkness at 25 °C for 2 weeks.

3 Transfer the developing embryos onto fresh plates of the same medium and culture in the light (24 $\mu mol/m^2/s$), with a 16-hours photoperiod at 25 °C to obtain plantlets.

4 Transfer the plantlets to the same medium without hormones to obtain a well-developed root system.

5 Transfer the plants to a culture solution (39) in a plastic tray, place in the phytotron or greenhouse (with 29 °C light/21 °C dark period) in a 16-h photoperiod and 70–95% relative humidity. The plants should set seed 3–4 months following transfer to soil.

[a] Mature seeds can also be used for callus induction.

[b] Some rice cultivars respond better in N6 medium. In that case, depending on genotype, the callus induction medium may be either N6 or MS medium.

[c] Acetosyringone is used in the medium to induce *vir* gene function.

[d] For LBA4404 (pTOK233), 50 mg/l kanamycin and 50 mg/l hygromycin are used. In case of LBA4404 (pNO1), 50 mg/l hygromycin and 20 mg/l gentamycin are used.

[e] It is not necessary to make the volume to 20 ml. This is only for convenience of the calculation.

[f] Some *Agrobacterium* strains, when grown in liquid medium, do not give homogeneous suspensions. They give a clumpy appearance. In this case, it is better to use solid culture and scrape off the bacteria with a scalpel.

[g] Desired OD × final volume = OD reading × dilution × initial volume.

[h] Sometimes, 250 mg/l cefotaxime is not enough for removal of bacteria. If this is the case use 250 mg/l carbenicillin in addition to cefotaxime.

3. Transformation by biolistic method

In a biolistic transformation system, metallic particles are coated with DNA and transferred to target plant cells by shooting with a particle bombardment machine. Immature embryos and embryogenic calli from immature embryos or mature seeds, have been widely used as explants for biolistic transformation. Immature embryo explants, especially from monocotyledonous plants, have

Table 1 Callus induction and proliferation medium

Component	Quantity (mg/l)	Component	Quantity (mg/l)
NH_4NO_3	1650	$FeSO_4.7H_2O$	27.8
KNO_3	1900	Na_2 EDTA	37.3
$CaCl_2.2H_2O$	440	Nicotinic acid	0.5
$MgSO_4.7H_2O$	370	Pyridoxine-HCl	0.5
KH_2PO_4	170	Thiamine-HCl	1.0
KI	0.83	Glycine	2.0
H_3BO_3	6.3	Casein hydrolysate	300
$MnSO_4.4H_2O$	22.3	Myo-inositol	100
$ZnSO_4.7H_2O$	8.6	2,4-D	2
$Na_2M_0O_4.2H_2O$	0.25	Sucrose/maltose	30 g/l
$CuSO_4.5H_2O$	0.025	Agar	8 g/l
$CoCl_2.6H_2O$	0.025		

pH should be 5.8, sterilize by autoclaving.

Table 2 AAM medium

Component	Amount (mg/l)	Component	Amount (mg/l)
Macronutrients			
$CaCl_2.2H_2O$	150.0	$NaH_2PO_4.H_2O$	150.0
$MgSO_4.7H_2O$	250.0	KCl	2950.0
Micronutrients			
KI	0.75	$Na_2MoO_4.2H_2O$	0.25
H_3BO_3	3.0	$CuSO_4.5H_2O$	0.025
$MnSO_4.H_2O$	10.0	$CoCl_2.6H_2O$	0.025
$ZnSO_4.7H_2O$	2.0		
Iron composition			
Na_2 EDTA	37.3	$FeSO_4.7H_2O$	27.8
Vitamins			
Nicotinic acid	0.5	Glycine	2.0
Pyridoxine HCl	0.5	Inositol	100.0
Thiamine HCl	1.0		
Others			
L-glutamine	876.0	Sucrose	68.5 g/l
Aspartic cid	266.0	Glucose	36 g/l
Arginine	174.0	Acetosyringone	200 μM
Casamino acid	500.0	pH 5.2	

Mix components before adjusting pH. Filter sterilize.

Add appropriate antibiotic for selection.

Table 3 AB medium

Component		Component	
Stock I	g/500 ml		
K_2HPO_4	30.0	NaH_2PO_4	10.0
Stock II	g/500 ml		
NH_4Cl	10.0	$CaCl_2$	0.1
$MgSO_4.7H_2O$	3.0	$FeSO_4.7H_2O$	025
KCl	1.5		
Stock III	g/900 ml		
Glucose	5.0	Agar	15.0

Autoclave the three stock solutions separately.
Add 25 ml each of stocks I and II to 450 ml of stock III.
Add appropriate antibiotics as required.

Table 4 Composition of media for growth of transformants

Component	N6-AS (mg/l)	MS-CH (mg/l)	Regeneration medium (mg/l)
$(NH_4)_2 SO_4$	463.0	–	–
KH_2PO_4	400.0	170.0	170.0
KNO_3	2830.0	1900.0	1900.0
NH_4NO_3	–	1650.0	1650.0
$CaCl_2.2H_2O$	166.0	440.0	440.0
$MgSO_4.7H_2O$	185.0	370.0	370.0
Na_2EDTA	37.3	37.3	37.3
$FeSO_4.7H_2O$	27.8	27.8	27.8
$MnSO_4.4H_2O$	4.4	22.3	22.3
H_3BO_3	1.6	6.3	6.3
$ZnSO_4.7H_2O$	1.5	8.6	8.6
KI	0.8	0.83	0.83
$CoCl_2.6H_2O$	–	0.025	0.025
$CuSO_4.5H_2O$	–	0.025	0.025
$Na_2MoO_4.2H_2O$	–	0.25	0.25
Thiamine–HCl	1.0	1.0	1.0
Nicotinic acid	0.5	0.5	0.5
Pyridoxine – HCl	0.5	0.5	0.5
Glycine	2.0	2.0	2.0
Myo-inositol	100.0	100.0	100.0
Casamino acids	1000.0	500.0	500.0
Kinetin	–	–	2.0
NAA	–	–	1.0
2,4–D	2.0	2.0	–
Cefotaxime	–	250	–
Hygromycin	–	50	–
Glucose (g/l)	10	–	–

Table 4 (*contd*)

Component	N6-AS (mg/l)	MS-CH (mg/l)	Regeneration medium (mg/l)
Maltose (g/l)	–	30	–
Sucrose (g/l)	30	–	30
Agar (g/l)	8	7	–
Agarose (g/l)	–	–	6
Acetosyringone	200 μM	–	–
	pH 5.2	pH 5.8	pH 5.8

Table 5 Culture solution

Component	Quantity (mg/l)	Component	Quantity (mg/l)
NH_4NO_3	114.25	$(NH_4)_6Mo_7O_{24}4H_2O$	0.09
$NaH_2PO_4.H_2O$	50.37	H_3BO_3	1.16
K_2SO_4	89.25	$ZnSO_4.7H_2O$	0.04
$CaCl_2.2H_2O$	110.75	$CuSO_4.5H_2O$	0.04
$MgSO_4.7H_2O$	405.0	$FeCl_3.6H_2O$	9.62
$MnCl_24H_2O$	1.87	Citric acid (monohydrate)	14.87

pH should be 5.0.

been frequently used for studies of gene delivery by biolistic bombardment. Stable transformants using this method have been reported for papaya (41), barley (42), rice (9, 43, 44), wheat (45), and maize (46), using immature embryos as a recipient for gene delivery. In this protocol, we describe the genetic transformation of rice by particle bombardment using the PDS-1000/He (Bio-Rad) system.

3.1 Biolistic transformation

Protocol 2

Biolistic transformation

Explants, equipment, and reagents

A Immature embryos/immature embryo-derived calli, 4–6 weeks old

B Preparation and delivery of DNA-coated gold particles and bombardment.

- Gold particles (approximately 1 μm diameter)
- $CaCl_2$ (2.5 M, filter-sterilized) in 1 ml aliquots, stored at −20 °C
- Spermidine-free base (0.1 M, filter-sterilized) in 1 ml aliquots, stored at −20 °C
- Ethanol (100%)
- DNA in TE (10 mM Tris-HCl, 1 mM EDTA, pH 8.0)
- Helium gas cylinder
- Particle bombardment machine (Bio-Rad): BIOLISTIC® PDS-1000/He

Protocol 2 continued

Methods

A. Preparation of immature embryos/embryogenic calli

1 Collect spikelets at 9–14 days after anthesis and sterilize as described in *Protocol 1*, step B1.

2 Isolate immature embryos and arrange 100–120 of them, with scutellum facing upwards, in a 2-cm diameter circle at the centre of a Petri dish containing MS medium supplemented with 100 mg/l each of benlate (anti-fungal), cefotaxime, and carbenicillin (antibacterial) to avoid contamination. Embryogenic calli can be generated following the method described in *Protocol 1*, step B. Antibiotics should not be used with embryogenic calli.

3 Incubate the samples overnight in the dark.

B. Sterilization of consumables and preparation of micro-carrier particles

1 Soak macro-carrier holders, macro-carriers, stopping screen, and rupture disks in 70% ethanol for at least 15 min and air-dry in a laminar flow hood.

2 Suspend 30 mg gold particles in 0.5 ml 100% ethanol and vortex for 1–2 min.

3 Centrifuge at 3000 g for 1 min and discard the supernatant.

4 Wash the gold particles three times by vortexing, centrifugation in a microfuge at 10,000 g for 1 min, and finally resuspending in 0.5 ml sterile distilled water.

5 Aliquot 50 μl of the final suspension into 1.5-ml microtubes.

6 Store at 4 °C or room temperature.

C. DNA precipitation

1 Under continuous vortexing, add to a 50-μl aliquot of gold particles the following components in this order: 8 μl DNA (DNA concentration = 1 μg/μl), 50 μl 2.5 M $CaCl_2$, and 20 μl 0.1 M spermidine.

2 Vortex for 3 min, spin at 3000 g in a microfuge for 5–10 sec.

3 After removing the supernatant, add 250 μl cold (−20 °C) 100% ethanol.

4 Vortex for 1–2 min and centrifuge in a microfuge at 4500 g for 1 min.

5 Remove the supernatant and resuspend the micro-carrier in 60 μl 100% ethanol.

6. Install macro-carriers in the macro-carrier holder.

7. Pipette 10 μl of the DNA-coated micro-carrier (gold particles) onto the centre of each macro-carrier. Each preparation is sufficient for six bombardments.

8. Dry for about 1 min in the laminar flow hood.

D. Operation of particle delivery system

1 Set the delivery pressure from the helium cylinder between 1300 and 1500 psi (200 psi above the desired rupture disc value).

Protocol 2 continued

2 Turn on the vacuum pump.

3 Place the rupture disc in the holder and screw the holder tightly.

4 Transfer the coated macro-carrier disk to the holder (particle side facing down), and place the macro-carrier assembly unit in the chamber at level 2 from the top.

5 Position the sample for bombardment at the desired level.

6 Close the chamber door.

7 Follow the Bio-Rad instruction manual for bombardment protocol.

8 Leave the plates overnight in the dark after bombardment.

E. Selection and regeneration

1 Transfer the bombarded explants to a Petri dish containing a selection medium for callus proliferation, as in *Protocol 1*, step F.

2 After 4–5 selections, further steps for regeneration are the same as in *Protocol 1*, step F.

4 Protoplast transformation

Direct gene transfer to protoplasts exploits the efficient uptake of DNA into the protoplast. It can be enhanced by chemical treatment, for example, with polyethylene glycol (PEG) (47) or by electrical pulses (electroporation) (48). This method does not require any kind of biological vector. For stable transformation of protoplasts, the two most important prerequisites are the ability to isolate and culture protoplasts in large numbers and an efficient system for routine plant regeneration from protoplasts. For dicotyledonous species, a large number of protoplasts can be obtained from leaf mesophyll, from which plant regeneration can be routinely achieved. However, there is still no convincing evidence of sustained divisions in protoplasts isolated from the leaves or shoots of any cereal. Nonetheless, protoplasts isolated from embryogenic suspension cultures can be induced to divide in culture (49). Microspore cultures and immature or mature embryos may be used to obtain embryogenic calli that can be used to establish embryogenic cell suspensions.

The basic procedures involved in the protoplast system of transformation are the preparation of protoplasts from plant tissue (mesophyll for dicotyledons and embryogenic cell suspensions for monocotyledons) by enzymatic digestion, the addition of DNA to protoplast suspension, and the uptake of DNA stimulated by various treatments (PEG or electrical pulses). The selection for expression of the inserted gene is usually applied at some stage during the development of the calli from the protoplasts. Finally, calli formation is stimulated in order to regenerate plants.

PEG-mediated gene transfer to rice protoplasts appears to be an efficient, reliable, and inexpensive method. In this system, suspension culture from either microspores or immature/mature embryo-derived embryogenic calli is used to derive a large population of protoplasts (44, 50, 51). Selection pressure is usually applied at an early stage of callus development from the protoplasts and successive regeneration of transgenic plants is performed as in other methods. However, the tissue-culture response may vary, depending on plant genotype, handling, and the condition of the suspension cells.

4.1 Protoplast transformation using PEG

Protocol 3 is usually used for cereal transformation, especially rice, where mesophyll protoplasts cannot be used for generating transgenic plants. Thus, this protocol starts with the establishment of an embryogenic cell suspension (ECS), which can eventually be used for protoplast isolation.

Protocol 3

Protoplast transformation using PEG

Explants, equipment and reagents

A. Mature/immature embryo-derived calli

- Plant material

  Plant material is *Oryza sativa*, 4–6-week-old embryogenic calli developed by the method described in Protocol 1 step B.

B. Establishment of suspension culture

- Media for culture development and maintenance (*Table* 6) AA medium (38) and R2 medium (52)

C. Protoplast isolation, transformation, and regeneration

- *General equipment:*
  (a) Centrifuge
  (b) Temperature controlled shaker
  (c) 12 ml round-bottom screw-cap centrifuge tubes
  (d) Nylon sieves—50 and 25 μm
  (e) Haemocytometer for counting the protoplasts
- Solutions and media:
  (a) Enzyme solution: 4% w/v cellulase Onozuka RS, 1% w/v macerozyme R10 (both Yakult Hansha Co., Japan), 0.02% pectolyase Y23 (Seishin Pharmaceutical Co., Japan), 0.4 M mannitol, 6.8 mM $CaCl_2$, pH 5.6, filter sterilized
  (b) Wash solution: 0.4 M mannitol, 0.16 M $CaCl_2$, autoclaved
  (c) MaMg buffer solution: 0.4 M mannitol, 15 mM $MgCl_2$, 1% (w/v) MES, pH 5.8, autoclaved
  (d) PEG solution: 40% polyethylene glycol solution in 0.1M Ca $(NO_3)_2.4H_2O$ and 0.4 M mannitol
  (e) Protoplast culture medium, e.g., P1 medium (*Table* 7)
  (f) Agarose-protoplast medium: 600 mg sea plaque agarose melted in 30 ml P1 medium
  (g) P2 medium (*Table* 7)
  (h) DNA for transformation (gene of interest, preferably linearized, together with calf thymus DNA)
  (i) Nurse cells—for rice transformation, OC cell line derived from seedlings of *O. sativa* L. C5924[a]

Protocol 3 continued

Methods

A. Establishment of suspension culture

1. Transfer the embryogenic yellowish-white soft calli developed by the method described in *Protocol 1*, step B to 6 ml of liquid AA medium in 50-mm Petri dishes and incubate on a gyratory shaker at low speed (80 r.p.m.) in diffuse light (3 μmol/ m^2/sec) at 25 °C.
2. Subculture the callus every 7 days with continued manual selection of small and densely cytoplasmic cells, which are transferred to R2 medium[b].

B. Protoplast isolation

1. Add 20 ml of enzyme solution to approximately 3–5 g of suspension cells (3–4 days after subculture) and incubate without shaking at 30 °C in the dark for 3–4 h[c].
2. Add an equal volume of wash solution to the protoplast enzyme mixture.
3. Pass the mixture through 50 and 25 μm sieves.
4. Transfer the filtrate to 12 ml round-bottom screw-cap centrifuge tubes and centrifuge for 10 min at 70 g to separate off the enzyme solution.
5. Wash the pellet containing protoplasts two times by centrifugation and determine the density of protoplasts using a haemocytometer.

C. Transformation using PEG

1. After the final washing resuspend the protoplast pellet in MaMg buffer solution. Adjust the protoplast density to $1.5–2.0 \times 10^6$/0.4 ml.
2. Add 6–10 μg of sterile plasmid DNA and 20 μg of calf thymus DNA (transformation can be done without calf thymus DNA, if not available) to each 0.4 ml aliquots of protoplast suspension.
3. Add 0.5 ml of the PEG solution drop wise and mix gently. Incubate at room temperature for 10 min.
4. Slowly add 10 ml of wash solution and centrifuge at 70 g for 10 min. Discard the supernatant.
5. Resuspend the protoplast pellet in 0.4 ml P1 medium.
6. Mix 0.6 ml of agarose protoplast medium (after melting, cool to 40 °C) with 0.4 ml of P1 medium containing protoplasts and transfer to a 3.5 cm Petri dish. Incubate at 20 °C for 1 h.
7. Transfer the solidified agarose gel to a 5 cm Petri dish containing P1 medium (bead type culture) (53) with or without adding nurse culture. Incubate culture in the dark at 28 °C with slow shaking (40 r.p.m.).
8. After 10 days remove all the nurse cells and add 5 ml fresh P1 medium. Keep in dark at 28 °C with slow shaking.

Protocol 3 continued

D. Selection

1 First selection: 14 days after transformation add the selective agent to the medium (concentration of selective agent depends on the selectable marker gene and the cultivar used). Incubate for 2 weeks without shaking.

2 Second selection: for second selection pressure replace 4 ml of medium with fresh P1 medium with the same concentration of selective agent. Incubate for another 2 weeks in the same conditions.

3 Third selection: transfer the visible calli onto P2 soft agarose medium (0.3% agarose) with same concentration of selective agent and incubate in dark at 25 °C for 2 weeks.

E. Regeneration

1 Transfer the visible colonies to P2 medium containing 0.6% agarose without the selective agent. Keep in same condition for 2 weeks.

2 Transfer the embryogenic colonies with developing somatic embryos to regeneration medium (*Table 4*) and follow the same procedure as described in *Protocol 1*, step F to regenerate transgenic plants.

[a] For nurse cells any actively dividing rice cell suspension can be used. In our case, we use an OC cell line, which is actively dividing, but is non-regenerable.

[b] These suspensions can be maintained for a long time in R2 medium by subculturing every 7 days and incubating with shaking (80 r.p.m.) at 25 °C in dark or diffuse light.

[c] Depending on nature of suspension culture for protoplast release, enzymatic digestion time should not exceed 5 h. If sufficient protoplasts are not released, the experiment should not be continued.

5 Conclusions

Plant transformation is a science and an art that must incorporate considerations of gene constructs, delivery systems, plant regeneration, and phenotypic and genotypic characteristics of the species in question. The end result is the development of new plant varieties. Tremendous improvements in basic transgenic technologies have been made to the extent that all major crop plants are now transformable and available for improvement by these technologies. We are now in the fortunate situation of understanding gene expression and being able to use it to modify a particular crop species. However, molecular characterization and selection of a normal phenotype are essential.

6 Acknowledgements

We thank our laboratory researchers, particularly Lina Torrizo, Norman Oliva, Editha Abrigo, and Niranjan Baisakh, for their technical help and Tess Rola for

Table 6 Media for suspension culture development

Component	AA medium quantity (mg/l)	R2 medium quantity (mg/l)
$CaCl_2.2H_2O$	150.0	147.0
$MgSO_4.7H_2O$	250.0	247.0
$NaH_2PO_4.H_2O$	150.0	240.0
KCl	2950.0	–
KNO_3	–	4044.0
$(NH_4)_2SO_4$	–	330.0
KI	0.75	–
H_3BO_3	3.0	0.50
$MnSO_4.4H_2O$	10.0	0.50
$ZnSO_4.7H_2O$	2.0	0.50
$Na_2MoO_4.2H_2O$	0.25	0.05
$CuSO_4.5H_2O$	0.025	0.05
$CoCl_2.6H_2O$	0.025	–
Nicotinic acid	1.0	0.5
Pyridoxine – HCl	1.0	0.5
Thiamine–HCl	10.0	1.0
Glycine	7.5	2.0
Inositol	100.0	100.0
Na_2 EDTA	37.3	37.3
$FeSO_4.7H_2O$	27.8	27.8
L–glutamine	876.0	–
Aspartic acid	266.0	–
Arginine	174.0	–
2,4–D	1.0	1.0
Kinetin	0.2	–
GA_3	0.1	–
Sucrose/maltose (g/l)	20.0	20.0
	pH 5.6	pH 5.6
	Filter-sterilized	Autoclaved

Table 7 Composition of protoplast culture media

Component	P1 (mg/l)	P2 (mg/l)
$(NH_4)_2 SO_4$	330.0	463.0
KH_2PO_4	–	400.0
KNO_3	4044.0	2830.0
$CaCl_2.2H_2O$	147.0	166.0
$MgSO_4.7H_2O$	247.0	185.0
Na_2EDTA	37.3	37.3
$FeSO_4.7H_2O$	27.8	27.8
$NaH_2PO_4.2H_2O$	240.0	–

Table 7 (*contd*)

Component	P1 (mg/l)	P2 (mg/l)
$MnSO_4.4H_2O$	0.5	4.4
H_3BO_3	0.5	1.6
$ZnSO_4.7H_2O$	0.5	1.5
KI	–	0.8
$CuSO_4.5H_2O$	0.05	–
$Na_2MoO_4.2H_2O$	0.05	–
Thiamine HCl	1.0	1.0
Nicotinic acid	0.5	0.5
Pyridoxine HCl	0.5	0.5
Glycine	2.0	2.0
Myo-inositol	100.0	100.0
2,4-D	2.0	2.0
Sucrose (g/l)*	136.92	60.0
Maltose (g/l) *	144.12	60.0
Agarose (g/l)	–	3 or 6
	pH 5.6	pH 5.8
	Filter-sterilized	Autoclaved

*Either sucrose or maltose can be used.

editorial assistance. Financial assistance from the Rockefeller Foundation, USA and the BMZ/GTZ, Germany, is gratefully acknowledged.

References

1. Fraley, R. T., Rogers, S. G., Horsch, R. B., Sanders, P. R., and Flick, J. S. (1983). *Proc. Natl Acad. Sci. USA*, **80**, 4803–4807.
2. Herrera-Estrella, L., Van den Broeck, G., Maenhaunt, R., Van Montagu, M., and Schell, J. (1984). *Nature* **300**, 160–163.
3. James, C. (1997). *ISAAA Briefs* 4. ISAAA Ithaca, NY.
4. Haq, T. A., Mason, H. S., Clements, J. D., and Arntzen, C. J. (1995). *Science*, **268**, 714.
5. Nawrath, C., Poirier, Y., and Somerville, C. (1995). *Mol. Breed.*, **1**, 105
6. Hooykaas, P. J. J. and Schilperoort, R. A. (1992). *Plant Mol. Biol.*, **19**, 15.
7. Tepfer, D. (1990). *Physiol. Plant.*, **79**, 140.
8. Zupan, J. R. and Zambryski, P. (1995). *Plant Physiol.*, **107**, 1041.
9. Christou, P., Ford, T. L., and Kofron, M. (1992). *TIBTECH*, **10**, 239.
10. Datta, S. K., Peterhans, A., Datta, K., and Potrykus, I. (1990). *Bio/Technology*, **8**, 736.
11. Datta, S. K. (1999). In *Molecular Improvement of cereal crops* (ed. I. K. Vasil), Vol. 5, p. 149. Kluwer Academic Publisher, The Netherlands.
12. Hamilton, C. M., Frary, A., Lewis, C., and Tanksley, S. D. (1996). *Proc. Natl. Acad. Sci. USA*, **18**, 9975.
13. Fraley, R. T., Rogers, S. G., and Horsch, R. B. (1986). *Crit. Rev. Plant Sci.*, **4**, 1.
14. Zambryski, P. (1992). Annual Review of Plant Physiology, *Plant Mol. Biol.*, **43**, 465.
15. Raineri, D. M., Bottino, P., Gordon, M. P., and Nester, E. W. (1990). *Bio/Technology*, **8**, 33.

16. Chan, M-T., Lee, T-M., and Chang, H-H. (1992). *Plant Cell Physiol.*, **33**, 577.
17. Chan, M. T., Chang, H. H., Ho, S. L., Tong, W. F., and Yu, S. M. (1993). *Plant Mol. Biol.*, **22**, 491.
18. Hiei, Y., Ohta, S., Komari, T., and Kumashiro, T. (1994) *Plant J.*, **6**, 271.
19. Dong, J. J., Teng, W. M., Buchholz, W. G., and Hall, T. C. (1996). In *Proceedings of the Third International Rice Genetics Symposium* (ed. G. S. Khush), p. 143. International Rice Research Institute, Los Baños, Philippines.
20. Datta, K., Oliva, N., Torrizo, L., Abrigo, E., Khush, G. S., and Datta, S. K. (1996). *Rice Genet. Newsl.*, **13**, 136.
21. Datta, K., Koukolíková-Nicola, Z., Baisakh, N., Oliva, N., and Datta, S. K. (2000). *Theor. Appl. Genet.*, **100**, 832.
22. Hoekema, A., Hirsch, P. R., Hooykas, P. J. J., and Schilperoort, R. A. (1983). *Nature*, **303**, 179.
23. Zambryski, P., Joos, H., Genetello, C., Leemans, J., Montagu Van, M., and Schell, J. (1983). *EMBO*, **2**, 2143.
24. Bevan, M. (1984). *Nucleic Acids Res.*, **12**, 8711.
25. An, G., Evert, P. R., Mitra, A., and Ha, S. B. (1988). In *Plant Molecular Biology Manual A3* (ed. S. B. Gelvin and R. A. Schilperoort), p. 1. Kluwer Academic Press, Dordrecht, The Netherlands.
26. Jefferson, R. A. (1987). *Plant Mol. Biol. Rep.*, **5**, 385.
27. Lee, S., Jeon, J. S., Jung, K. H., and An, G. (1999). J. Plant Biol. **42**, 310.
28. Watson, B., Currier, T. C., Gordon, M. P., Chilton, M. D., and Nester, E. W. (1975). *J. Bacteriol.*, **123**, 255.
29. Hood, E. E., Jen, G., Kayes, L., Kramer, J., Fraley, R. T., and Chilton, M. D. (1984). *Bio/Technology*, **2**, 702.
30. Jin, S., Komari, T., Gordon, M. P., and Nester, E. W. (1987). *J. Bacteriol.*, **169**, 4417.
31. Komari, T., Halperin, W., and Nester, E. W. (1986). *J. Bacteriol.*, **166**, 88.
32. Hood, E. E., Helmer, G. L., Fraley, R. T., and Chilton, M. D. (1986). *J. Bacteriol.*, **168**, 1291.
33. Hood, E. E., Murphy, J. M., and Pendleton, R. C. (1993). *Plant Mol. Biol.*, **4**, 685.
34. Altvorst, A. C., Van Riksen, T., Koehorst, H., and Dons, H. J. M. (1995). *Transgenic Res.*, **2**, 105.
35. Magnussen, D., Clapham, D., Gronroos, R., and Arnold Von, S. (1994). *J. Forest Res.*, **9**, 46.
36. Murashige, T. and Skoog, F. (1962). *Physiol. Plant.*, **15**, 473.
37. Komari, T. (1990). *Plant Cell Rep.*, **9**, 303.
38. Muller, A. J. and Graffe, R. (1978). *Mol. Gen. Genet.*, **161**, 67.
39. Chu, C. C., Wang, C. C., Sun, S. S., Hsu, C., Yin, K. C., Chu, C. Y., and Bi, F. Y. (1975). *Sci. Sin.*, **18**, 659.
40. Yoshida, S., Forno, D. A., Cock, J. H., and Gomez, K. A. (1976). In *Laboratory manual for physiological studies of rice*. p. 61. International Rice Research Institute, Los Baños, Philippines.
41. Fitch, M. M. M., Manshardt, R. M., Gonsalves, D., Slightom, J. L., and Sanford, J. C. (1990). *Plant Cell Rep.*, **9**, 189.
42. Wan, Y. and Lemaux, P. G. (1994). *Plant Physiol.*, **1**, 37.
43. Datta, K., Torrizo, L., Oliva, N., Alam, M. F., Wu, C., Abrigo, E., Vasquez, A., Tu, J., Quimio, C., Alejar, M., Nicola, Z., Khush, G. S., and Datta, S. K. (1997). *Proceedings of the Fifth International Symposium on Rice Molecular Biology*. Yi Hsien Pub. Co., Taipei, Taiwan.
44. Datta, K., Vasquez, A., Tu, J., Torrizo, L., Alam, M. F., Oliva, N., Abrigo, E., Khush, G. S., and Datta, S. K. (1998). *Theor. Appl. Genet.*, **97**, 20.
45. Vasil, V., Castillo, A. M., Fromm, M. E., and Vasil, I. K. (1992). *Bio/Technology*, **6,** 662.
46. Koziel, M. G., Beland, G. L., Bowman, C., Carozzi, N. B., Crenshaw, R., Crossland, L., Dawson, J., Desai, N., Hill, M., Kadwell, S., Lauris, K., Lewis, K., Maddox, D., McPherson,

K., Meghji, M. R., Merlin, E., Rhodes, R., Warren, G. W., Wrights, M., and Evola, S. T. (1993). *Bio/Technology*, **11**, 94.
47. Negrutiu, I., Shillito, R. D., Potrykus, I., Biasini, G., and Sala, F. (1987). *Plant Mol. Biol.*, **8**, 363.
48. Shimamoto, K., Terada, R., Izawa, T., and Fujimoto, H. (1989). *Nature*, **337**, 274.
49. Vasil, I. K. (1987). *J. Plant Physiol.*, **128**, 193.
50. Datta, S. K., Datta, K., Soltanifar, N., Donn, G., and Potrykus, I. (1992). *Plant Mol. Biol.*, **20**, 619.
51. Datta, K., Velazhahan, R., Oliva, N., Mew, T., Khush, G. S., Muthukrishnan, S., and Datta, S. K. (1999). *Theor. Appl. Genet.*, **98**, 1138.
52. Ohira, K., Ojima, K., and Fujiwara, A. (1973). *Plant Cell Physiol.*, **14**, 1113.
53. Shillito, R. D., Paszkowski, J., and Potrykus, I. (1983). *Plant Cell Rep.*, **2**, 244.

Chapter 3
T-DNA tagging

Csaba Koncz and Jeff Schell
Max-Planck Institut für Züchtungsforschung, Carl-von-Linné-Weg 10, D-50829 Köln, Germany

1 Introduction

This Chapter gives a guide to the use of *Agrobacterium* T-DNA as an insertion mutagen in the molecular genetic analysis of plant gene functions. T-DNA represents a segment of Ti and Ri plasmids that is transferred from *Agrobacterium* into the nuclei of infected plant cells where it is randomly integrated into potentially transcribed domains of the chromosomes by non-homologous (i.e. illegitimate) recombination (1, 2). Insertion of T-DNA into plant genes causes gene mutations. Hence, T-DNA is an efficient mutagen (3), see also Chapter 1. T-DNA tagging, which results in the labelling of gene mutations with known sequences and dominant selectable markers, provides a simple means for genetic linkage mapping, functional analysis and molecular characterization of plant gene mutations (4, 5).

Promoterless reporter genes, with or without a translation initiation codon, are inserted into plant genes with the aid of the T-DNA. Thus, gene mutations are identified by selection or screening for the expression of transcriptional or translational reporter gene fusions controlled by T-DNA tagged plant genes (5). Reporter genes bearing only the TATA-region of so-called minimal promoters are used as enhancer traps to identify *cis*-acting regulatory sequences with the help of T-DNA tags in the plant genome (6).

Mutagenesis with T-DNAs, carrying multiple copies of enhancers derived, for example, from the well-known promoter of the Cauliflower Mosaic Virus (CaMV) 35S RNA, are employed to induce *cis*-activation of plant genes located in the vicinity of T-DNA tags (7). This activation T-DNA tagging technology is used for the isolation of dominant mutations that can also be identified in tetraploid and allotetraploid plants. Promoters orientated towards the termini of T-DNA tags provide a means for induction of either constitutive or regulated transcription of neighbouring plant DNA sequences. This may either activate or inactivate genes by transcriptional read-through or interference, as well as lead to alterations in chromatin structure implying long-range *cis*-effects (8).

Applications, aiming at either a removal of certain T-DNA tags from the genome or isolation of a particular type of gene mutation apply conditional 'suicide' genes for negative selection (9–11). A combination of these methods facilitates the isolation of regulatory mutations that affect the activity of single genes, including second site suppressors of known mutations. T-DNA insertions carrying target sites and genes for site-specific recombination also allow chromosome engineering by generation of defined deletions, additions, inversions and translocations (12, 13). Finally, T-DNA serves as a common vehicle to deliver autonomous or defective, but mobilizable, transposons and retrotransposons into diverse plant species as described in Chapter 4.

2 Generation of T-DNA transformants

2.1 The mechanism of T-DNA transformation

The virulence of *Agrobacterium* strains is regulated by the bacterial *vir* loci of Ti and Ri plasmids, and by a few chromosomal genes. The VirA chemosensor histidine protein kinase controls the activation of the VirG transcription factor that binds to *vir*-boxes in the promoters of virulence genes as a positive effector. Activation of the VirA kinase is induced by plant phenolic compounds and sugars, and shows a considerable variation between different *Agrobacterium* isolates. VirG activation by VirA-mediated phosphorylation induces the synthesis of Vir proteins required for T-DNA transfer into plant cells.

The mechanism of T-DNA transfer is analogous to the conjugation of bacterial plasmids, except that it works with two conjugation transfer origins (ori_T) defined by T-DNA 25 bp border repeats (1). During T-DNA transfer, a single-stranded DNA intermediate (T-strand) is produced by strand-replacement DNA synthesis, which is initiated by a single-strand nick at the right T-DNA border and terminated by a similar nick at the left 25 bp border sequence. The VirD2 pilot protein covalently binds to the 5′-end of the T-strand, which is packaged in its entire length by the VirE2 single-stranded DNA-binding protein and transferred into plant cells through pores formed by the VirB proteins. Nuclear localization signals present in VirD2 and VirE2 aid the import of the T-complex into plant cell nuclei (1). Although the plant factors involved in the T-DNA integration are still unknown, the available data show that T-DNA integration requires the function of the VirD2 nicking-closing enzyme and occurs by illegitimate recombination targeting either single-stranded nicks or double-stranded breaks (2). The T-DNA integration process usually generates small target site deletions (extending from one to several hundred base-pairs), but may also cause larger rearrangements such as translocations and inversions. Because the T-DNA integration process is based on DNA replication and repair, it is common that aborted integration events generate footprints causing mutations that are not genetically linked to T-DNA tags.

2.2 *Agrobacterium* hosts

Infection with *Agrobacterium* strains carrying wild type Ti or Ri plasmids leads, respectively, to neoplastic tumour (crown-gall) formation or hairy-root syndrome in many dicotyledonous plants. These disease symptoms result from the expression of T-DNA encoded genes that play no role in the T-DNA transfer process, but cause alterations in the regulation of the hormonal, chiefly auxin and cytokinin, balance of transformed plant cells. Initially, the absence of these typical disease symptoms suggested that monocotyledonous plant species are not suitable hosts for *Agrobacterium*. This view was indeed supported by the identification of several compounds in monocots that inhibit the induction of *Agrobacterium vir* genes. To increase the efficiency of the T-DNA transfer process, if possible independently of the host signals, several wide-host range Ti plasmids were studied. New *Agrobacterium* strains were constructed by engineering of the VirA kinase and over-expression of the VirG regulator that, when produced in excess, can mediate a VirA-independent activation of the virulence genes. To monitor the T-DNA transformation process, intron-containing reporter genes (e.g. *uidA*) silent in *Agrobacterium* were developed (for review see 14), see also Chapter 2.

A break-through resulting from these efforts demonstrated that (i) *Agrobacterium* can transform virtually all plant species, including the most important monocotyledonous crops, and (ii) *Agrobacterium* is capable of systemically infecting most plant tissues, including the reproductive plant organs, protoplasts, cell suspensions, somatic embryos, embryogenic organ cultures, etc. In addition to a huge number of dicots providing food, raw material and pharmaceutical drugs, tissue culture-based T-DNA transformation methods are now available, e.g. for rice (15), maize (16), barley (17), wheat (18), and sugarcane (19) (see Chapter 2). Furthermore, T-DNA transfer into yeast, filamentous fungi and mammalian cell nuclei has been recently achieved thus widely extending the applicability of T-DNA-based gene transfer and insertion mutagenesis technologies (20–22).

3 T-DNA tagging in *Arabidopsis*

To give a guide to the application of T-DNA insertion mutagenesis in plant molecular biology, we describe here the use of T-DNA transformation methods in the model plant *Arabidopsis thaliana*. These methods are adaptable to any plant species for which a suitable *Agrobacterium*-mediated transformation protocol is available yielding fertile transgenic plants.

3.1 Generation of transformants

3.1.1 Transformation of explants

In *Arabidopsis*, as in all other species, the use of tissue culture transformation techniques was first elaborated. These techniques require an initial amplification of *Agrobacterium* transformed cells. This can be conveniently achieved by selection for transformed micro-calli that can then be regenerated into shoots and plants. Therefore, the choice of explants for co-cultivation with *Agrobacterium* largely depends on the response of diverse plant tissues to those hormonal treatments that are employed to induce cell proliferation and subsequent shoot formation. In general, cell proliferation is induced by culturing the explants in media providing a high auxin to low cytokinin ratio. Explants from various *Arabidopsis* ecotypes and mutants respond differently to auxin and cytokinin, and therefore to the induction of cell division. Thus, roots of the ecotype Columbia (Col-0) proliferate much faster than tissues from hypocotyl, cotyledon, petiole, or leaf, whereas acceptable callus formation is achieved from most explants of ecotypes RLD and C24. In comparison, ecotype Landsberg *erecta* (Ler-0) shows a very poor tissue culture response (8). Throughout the last decade, several methods for root transformation were published (14). *Protocol 1* describes a method that has been used by our laboratory to generate a large number of transgenic *Arabidopsis* plants and to search for recessive gene mutations using the T-DNA-aided gene fusion technology.

Protocol 1

Agrobacterium-mediated transformation of *Arabidopsis* root explants

Equipment and Reagents

- Macro-salts: 20 g NH_4NO_3, 40 g KNO_3, 7.4 g $Mg_2SO_4.7H_2O$, 3.4 g KH_2PO_4, and 2 g $Ca(H_2PO_4)_2$ for 1 l
- Micro-salts: 6.2 g H_3BO_3, 16.9 g $MnSO_4.H_2O$, 8.6 g $ZnSO_4.7H_2O$, 0.25 g $Na_2MoO_4.2H_2O$, 0.025 g $CuSO_4.5H_2O$, 0.025 g $CoCl_2.6H_2O$ for 1 l
- Fe-EDTA: dissolve separately 5.56 g $FeSO_4.7H_2O$ and 7.46 g $Na_2EDTA.2H_2O$; mix the two solutions, and adjust the volume to 1 l
- $CaCl_2$: 7.5 g $CaCl_2.2H_2O$ per 1 l.
- KI: 375 mg KI per 1 l.
- Vitamins: 50 g *myo*-inositol, 2.5 g thiamine.HCl, 0.5 g nicotinic acid, and 0.2 g pyridoxine.HCl for 1 l.
- MSAR-medium: to prepare liquid medium, add for 1 l: 50 ml macro-salts, 1 ml micro-salts, 5 ml Fe-EDTA, 5.8 ml $CaCl_2$, 2.2 ml KI, 2 ml vitamins and 30 g sucrose. Adjust the pH to 5.8 with 0.5 M KOH. To prepare solid MSAR media, add for 1 l, either 2.2 g phytagel (essential for high-frequency *Arabidopsis* shoot regeneration; Sigma) or 8 g high quality agar for plant tissue culture. Autoclave the media for 15 min.
- Seed germination (SG) medium: regular MSAR medium, but contains half concentration of macro-salts and 0.5% sucrose.
- Callus medium (MSARI): MSAR medium supplemented with 2.0 mg/l indole-3-acetic acid (IAA), 0.5 mg/l 2,4-dichlorophenoxyacetic acid (2,4-D), 0.2 mg/l kinetin (6-furfurylaminopurine), and 0.2 mg/l N^6-[2-isopentenyl]-adenosine (IPAR).
- Shoot medium (MSARII): MSAR medium supplemented with 2.0 mg/l IPAR and 0.05 mg/l 1-naphtaleneacetic acid (NAA).
- Rooting medium (MSARIII): MSAR medium supplemented with 0.2 mg/l 6-benzylaminopurine (BAP), 0.1 mg/l indole-3-butyric acid (IBA) and 0.05 mg/l NAA.

Protocol 1 continued

- Seed sterilization solution: 5% $Ca(OH)_2$ containing 0.05% Triton X-100. If $Ca(OH)_2$ is not available, use 10% sodium hypochlorite solution containing 0.1% Triton X-100.
- *Agrobacterium* YEB-medium: 5 g beef extract, 1 g yeast extract, 5 g peptone (each from Difco Laboratories) and 5 g sucrose per 1 l water

Methods

A. Preparation of plant material

1. Sterilize 10 mg (about 500) seeds in a microfuge tube after adding 1 ml seed sterilization solution by continuous shaking for 15 min.
2. Pellet the seeds by brief centrifugation (i.e. 1 min in a microfuge at room temperature), remove the sterilization solution, wash the seeds with 4 × 1 ml sterile water, and dry the tubes in a flow-hood overnight.
3. Germinate the seeds on solid SG-medium in Petri dishes for 7–10 days using an 8 h light and 16 h dark period at 25 °C.
4. Place 20 seedlings into a 250 ml Erlenmeyer flask containing 35 ml liquid MSAR-medium with 3.0% sucrose and fix the flasks onto a rotary shaker adjusted to 120 r.p.m. at 25 °C under an 8 h light and 16 h dark period. Within 14–20 days the flasks will be filled with roots. This material can be used either for direct infection with *Agrobacterium* or for generation of cell suspensions as described in *Protocol 2*. Do not collect roots from plants that show senescence.

B. Growth of Agrobacterium

1. Grow *Agrobacterium* to an OD_{550} of 1.5 in YEB-medium supplemented with proper antibiotics to select for the maintenance of binary T-DNA vectors.
2. Pellet the cells by centrifugation (5000 g for 15 min at room temperature).
3. Resuspend cells in MSARI liquid medium at an OD_{550} of 0.5 and distribute 50–80 ml aliquots into large Petri dishes.
4. Collect *Arabidopsis* plants from the Erlenmeyer flasks (see *Protocol 1, A4*), remove their roots, cut the roots in small pieces and place them into the *Agrobacterium* suspension for at least 30 min.
5. Remove the *Agrobacterium* suspension with a pipette from the Petri dishes and plate the roots onto MSARI plates.
6. Alternatively, collect the root segments in an Erlenmeyer flask containing 40 ml MSARI medium and place the flask on a rotary shaker set as described above. Coculture the roots with *Agrobacterium* for 36–48 h.

Protocol 1 continued

C. Selection of transformed plants

1. Transfer the roots from MSARI to MSARI-medium containing 300 mg/l claforan (cefotaxime) and 500 mg/l tricarcillin/clavulanic acid (15:1; Duchefa Biochemie). Alternatively, add claforan alone at 500mg/l concentration.
2. Subculture the liquid root cultures every 3–4 days using the medium above.
3. Root cultures in Petri-dishes do not need to be subcultured. After 14 days, transfer the roots onto MSARII plates containing claforan (300 mg/l), tricarcillin/clavulanic acid (500 mg/l) and, depending on the type of selectable marker encoded by the T-DNA, for example, 15 mg/l hygromycin (Boehringer Mannheim) or 100 mg/l kanamycin (Sigma).
4. Subculture the root explants at 14 days intervals by decreasing the claforan concentration to 300 mg/l after the third subculture.
5. During the first 4 weeks on selective MSARII medium the non-transformed tissues die and green transgenic calli appear on the plates. Within 8 weeks after *Agrobacterium* infection, the transformed calli form embryo-like structures and quickly regenerate to shoots.
6. The regenerating shoots must be continuously transferred onto MSARIII medium in jars and, when they reach a size of 1–2 cm, into large test tubes (e.g. 20 × 2.5 cm diameter) containing 10–20 ml SG-medium to set seed.
7. The test tubes are closed with loose cotton plugs and placed in growth chambers providing a 16 h light and 8 h dark cycle.
8. Typically, root culture from a single Erlenmeyer flask yields explants for 10–12 MSARII plates resulting in an average of 500 to 1000 transformed plants.

3.1.2 Transformation of cell cultures

An advanced version of tissue culture transformation methods is based on the co-cultivation of cell suspensions with *Agrobacterium* that yields an extremely high number of transformants (in the range of 10^6–10^7). Therefore, this method ideally suits the purpose of activation T-DNA tagging, if a proper selection for gene activation is available, for example, using inhibitors of particular enzymes or signalling factors, and plant regeneration is not immediately essential. Although embryogenic cell suspensions can be established with care and transformed similarly by *Agrobacterium* as described in *Protocol 2*, the regeneration of large numbers of transformants is a very laborious process. Moreover, alterations in ploidy (from 2n to 4n or 8n) are unavoidable in cell cultures maintained in the presence of high concentrations of auxin (i.e. 2,4-D). An increase in the ploidy level reduces the recovery of recessive mutations (see also Chapter 1), but permits the identification of transformants that carry dominant mutations induced by activation T-DNA tagging.

Protocol 2

Establishment and transformation of Arabidopsis cell suspensions

Equipment and Reagents

- SC-medium: 4.33 g MS-medium (Sigma), a double concentration of B5 vitamins (Sigma, 2 mg nicotinic acid, 2 mg pyridoxine-HCl, 20 mg thiamine-HCl, and 200 mg *myo*-inositol for 1 l), 30 g sucrose, 0.5 mg/l 2,4-D, 2 mg/l IAA, and 0.5 mg/l IPAR. The vitamins and hormones are prepared as concentrated stocks, filter sterilized, and added to SC after autoclaving.
- Metal sieve: pore size 250 and 850 μm, available from, e.g. Fastnacht Laborbedarf GmbH, Bonn, Germany

Methods

A. Initiation of Arabidopsis cell suspension

1 Continue *Protocol 1A* from step 4. Transfer the roots from an Erlenmeyer flask into a Petri-dish, remove all green tissues.

2 Cut the roots into about 2-mm pieces, and transfer the explants into a 250 ml Erlenmeyer flask that contains 50 ml suspension culture SC-medium. Cover the flasks with aluminium foil and place them on a rotary shaker (120 r.p.m., 24 °C).

3 After 15–20 days, approximately 3–5 g roots (fresh weight) should be available in each flask.

4 Exponential growth rate of roots is important for obtaining fine cell suspensions. Therefore, avoid yellow-brown stationary phase cultures.

B. Subculture of cell suspension

1 After 21 days, harvest the micro-calli released from the roots by passing the root culture medium through a 850 μm sterile metal sieve into a sterile Erlenmeyer flask.

2 Pellet the cells, and replace half volume of the medium with new SC-medium.

3 Add 50 ml fresh SC-medium to the roots and culture them further for 21 days to generate more micro-calli.

4 Subculture the cell suspension every 7 days by sieving and replacing half of the medium with new SC-medium after pelleting the cells.

5 After the fifth subculture, a 50 ml suspension can usually be diluted with 100 ml SC and divided into three equal aliquots into new flasks for amplification.

C. Transformation of cell suspension with Agrobacterium

1 Logarithmically proliferating cell suspensions are usually obtained from *Arabidopsis* ecotype Col-0 after 5–6 weeks.

2 The cell suspensions are then filtered through a 250 μm sieve and subcultured twice.

Protocol 2 continued

3. The resulting fine cell suspension is inoculated with *Agrobacterium*. For this purpose, grow *Agrobacterium* in YEB-medium as described in *Protocol 1B* and pellet the cells from 1.5 ml culture in a sterile microfuge tube by centrifugation for 2 min at room temperature.
4. Suspend the bacterial pellet in 1.5 ml SC-medium and add to 35 ml cell suspension immediately after subculturing.
5. Co-cultivate the cells with *Agrobacterium* for 48 h, then add claforan (300 mg/l) and tricarcillin/clavulanic acid (300 mg/l).
6. Subculture the cells 7 days after the addition of *Agrobacterium* in the same SC-medium.
7. After 1 week transfer the cells into a 50 ml Falcon tube, pellet the cells by centrifugation at 1000 *g* for 3 min and suspend them at a density of 10^3 cells/ml in either 0.2% phytagel dissolved in 3% sucrose or in liquid MSARII medium.
8. Plate the cells on MSARII medium containing 0.2% phytagel, claforan (200 mg/l), tricarcillin/clavulanic acid (200 mg/l), and suitable antibiotics (e.g. hygromycin or kanamycin) as described in *Protocol 1*.
9. If you wish to select for transformed calli without plant regeneration, use MSARI instead of MSARII medium for plating.
10. Transformed calli appear within 2–3 weeks and yield shoots within 35–40 days following the *Agrobacterium* infection.

When using this protocol for selection of mutants resistant to various inhibitors, a proper adjustment of selection conditions should first be performed to ensure 100% killing of non-transformed cells. Some inhibitors are light sensitive, thus they can only be used in combination with MSARI medium permitting the selection of transformed calli in the dark. Further details to this protocol are described by Mathur *et al.* (23).

3.1.3 *In planta* transformation

The first successful approach using a large-scale *in situ* infection of *Arabidopsis* with *Agrobacterium* was the seed transformation method (24). With this method, a larger population of transformants could be raised with less labour than using tissue culture techniques. Furthermore, seed transformation overcame the problems of recognition and removal of polyploids and aneuploids during the genetic analysis (25). The recognition that *Arabidopsis* survives systemic *Agrobacterium* infection led to the development of *in planta* transformation techniques using vacuum infiltration (26). Because the transformation process must target the reproductive organs, this technology was further perfected by dipping only the flower buds into *Agrobacterium* suspension containing 5% sucrose and Silwet L-77 as surfactant (27).

To amplify the number of inflorescence axes produced by a plant for *Agrobacterium* infection, the primary inflorescence stem is often removed. This results in a quick outgrowth of side-inflorescence stems from the rosette providing ample material for transformation. Most binary vectors used currently are devoid of proper partitioning functions, which would provide stable plasmid maintenance. Therefore, to increase the transformation frequency, an application of *in planta* selection for the maintenance of *Agrobacterium* binary T-DNA vectors is advisable. In the case of the pPCV vectors system developed in our laboratory (8), the T-DNA carries an ampicillin/carbenicillin resistance gene. The media for *Agrobacterium* infiltration and floral dip transformation can therefore be supplemented with carbenicillin, which is not deleterious to *Arabidopsis*.

By optimization of the infiltration techniques, on average over 400 transformed seeds are routinely obtained from a single pot containing 10–15 plants. Using 500–1000 pots, one can reproducibly raise a T-DNA mutagenized population of 200 000 transformants within 3–6 months. Based on the expectation that T-DNA integration occurs randomly and the size of *Arabidopsis* genome is 120 Mb, such a population is predicted to carry a T-DNA insert at about every 0.5 to 1 kb in the genome. Therefore, T-DNA tagging may be used as an alternative to chemical mutagens (see Chapter 1) to solve specific genetic problems, such as the isolation of multiple alleles and suppressors for a given mutation.

According to the type of selectable marker gene within the T-DNA, different techniques are used for the selection of primary transformants in the M1 generation. The application of the BASTA resistance marker facilitates the selection of transformed plants in growth chambers by spraying with the herbicide. In contrast, hygromycin or kanamycin selection can be optimally carried out by germination of seeds in large Petri-dishes which, although somewhat more costly, saves considerable time and greenhouse space for a more economic production of the M2 seed generation. Suitable amounts of plant material for subsequent sequence-based PCR screens (see *Protocol 4*) may be collected immediately if the M1 generation is grown initially in short day conditions, which promotes large rosettes. In practice, growing of single plants in separate quadratic pots (e.g. 5 × 5 cm) yields as much as 0.5–1 g of leaf material that can be pooled for DNA preparations, as well as a large amount of seed essential for further genetic analyses.

Protocol 3

In planta transformation of *Arabidopsis* by infiltration

Equipment and Reagents

- Agrobacterium YEB-medium: see *Protocol 1*
- SG-medium: see *Protocol 1*
- Infiltration medium (IM, pH 5.7): half a concentration of Murashige and Skoog basal salt mixture (Sigma), 1 × B5 vitamins (Sigma), 5% sucrose, and is supplemented after autoclaving with 0.05 μM BAP, and 0.005% (volume/volume) Silwet L-77 (Ambersil)

Protocol 3 continued

Methods

A. Growth of plant material

1. Prepare plastic pots (12 cm diameter) with wet soil and sow 10–15 seeds in each.
2. Grow seedlings in short days using an 8 h light (24 °C) and 16 h (18–20 °C) dark period in a growth chamber or greenhouse.
3. When the diameter of rosettes reaches a size of 3–5 cm (usually after 15–18 days), transfer the pots to long day conditions (16 h light and 8 h dark) to induce flowering.
4. Cut off the primary inflorescence stem when it reaches 1–3 cm, to induce the development of secondary inflorescence from the rosette buds.
5. When the size of inflorescence reaches 5–10 cm, carrying closed flower buds and only a few open flowers, the plant material is ready for transformation.

B. Growth of Agrobacterium

1. Grow 100 ml liquid culture of *Agrobacterium* in YEB medium at 28 °C for 2 days applying a proper selection for the maintenance of binary T-DNA vectors.
2. Best transformation results were reported by using the strain GV3101 with the helper Ti-plasmid pMP90 (26–28). A modified version of this *Agrobacterium* host, GV3101 (pMP90RK) is used in combination with the pPCV-type binary T-DNA vectors that carry a carbenicillin resistance gene within their T-DNA (28).
3. When using pPCV-vectors, supplement both *Agrobacterium* and *in planta* infiltration media with 100 mg/l carbenicillin.
4. Inoculate with 10 ml starter culture several 2-l Erlenmeyer flasks containing 500 ml YEB-medium and grow the bacterial cultures to late logarithmic phase for 12–14 h. Harvest the cells by centrifugation (5000 g for 10 min at room temperature) and resuspend the cells at an OD_{550} of 0.8–1.2 in infiltration medium. Typically, 500 ml *Agrobacterium* culture yields 1 l bacterial suspension for infiltration.

C. Vacuum infiltration

1. Assemble a glass or plastic bell jar with a vacuum pump and place trays (for example, 30 × 10 × 5 cm), or beakers (12 cm) containing the *Agrobacterium* culture (*Protocol 3B*) into the jar.
2. Place two pots in inverted position into each tray, or one pot inverted into each beaker such that only the inflorescence shoots are submerged into the *Agrobacterium* suspension. Apply vacuum (e.g. 16 mbar) for 5 min, then release the vacuum quickly by disconnecting the fitting between the pump and the bell jar.
3. The bacterial suspension can be reused 3 or 4 times. Thus, an infiltration of 16 pots needs about 1–1.5 l *Agrobacterium* suspension in IM.
4. Cover the pots for 24 h with plastic quick-seal bags with the corners cut off. If the humidity in your greenhouse is high enough (i.e. 70–80%), this step is not necessary.

Protocol 3 continued

D. Collection of seeds

1. The vacuum infiltrated plants complete flowering within 10–12 days.
2. When the last flowers are fertilized, place the inflorescence stems of five to eight plants into a translucent paper bag (i.e. 13 × 30 cm).
3. Fix the closed bags with tape to wood or plastic sticks fitted into the pots.
4. Water the plants for an additional 10 days and then let the pots dry out.
5. Collect the seed-bags and clean the seeds.
6. Take 5–10 seed aliquots derived from different pots to test the transformation efficiency in a germination assay.

E. Seed sterilization

1. Sterilize seeds as described in *Protocol 1A* and sow them on SG-medium in Petri-dishes (15 cm diameter) containing suitable antibiotics to select for T-DNA transformed plants.
2. Using hygromycin selection (25 mg/l, Boehringer), non-transformed seedlings are arrested and show bleaching after opening their cotyledons, whereas transformed plants develop leaves and long roots within 5–8 days.

F. Selection of transformants

1. If at least 5–10 transformants were obtained from 500 to 1500 seeds, bulk the seed material, sterilize it in aliquots, and plate the seeds at a high density on selective SG-medium with 1000–5000 seeds per large Petri-dish.
2. To save greenhouse space, the transformants picked from Petri-dishes can be grown in jars on SG-medium with hygromycin, and when required with claforan to eliminate possible *Agrobacterium* infection, under short day conditions for another 10 to 14 days.
3. Transfer the seedlings into soil by planting them individually into 5 × 5 cm (or 7 × 7cm) quadratic pots and grow them for 10–14 days under short day conditions to produce large rosettes (average diameter 6–8 cm).
4. Collect rosette leaves (200–500 mg) for DNA purification using a pooling strategy (4), and transfer the plants to long day conditions to collect M2 progeny from each plant separately within 3–4 weeks.

4 Screening approaches and theoretical considerations

4.1 Phenotype-based screening

Mathematical theories of mutagenesis experiments and classical mutant screening strategies are extensively reviewed in recent *Arabidopsis* handbooks (29, 30;

see also Chapter 1). Dominant mutations affecting, for example, the development of organs, epidermal hair shape and distribution, wax deposition, colour, male and female fertility, and seed morphology are rare in T-DNA tagged populations. Nonetheless, the M1 population should be inspected to score for potential mutants. Once such a potential dominant mutant is found, it should immediately be out-crossed with wild type to speed up the genetic analysis. The application of gene fusion and enhancer trap technologies provides the unique advantage that the M1 population can be screened directly for T-DNA tags in genes or in the vicinity of regulatory DNA sequences, respectively. M1 seedlings and various plant organs can thus be used for monitoring the activation of reporter genes encoding β-glucuronidase (GUS), firefly luciferase (luc) or green fluorescent protein (GFP) *in vivo* and *in vitro* (see Volume 2, Chapter 13).

From a genetic point of view, the examination of immature seeds representing the M2 population in fruit capsules (i.e. siliques) is the most meaningful screening approach in the M1 generation. Using classical 'm' tests (31), at least two siliques are opened and examined for Mendelian segregation of the progeny to find defects in embryo development, embryo lethality, albino, or purple (i.e. *fusca*) colour, abnormal shape and size, viviparity, or alteration in starch, protein, fat, metabolite, hormone levels, etc. In addition, pollen may be collected to perform morphological examinations and *in vitro* germination assays.

Most mutant screens are carried out by germinating M2 seeds on diverse media in either dark or light. It must be noted that *Arabidopsis* seeds require a red light stimulus to germinate synchronously and need vernalization, depending on the ecotype. Dark germination assays combined with the exposure of seedlings to different light sources is used to screen for mutations affecting growth responses to red, far-red, blue, and UV light, or a timely combination of these. Hypocotyl and root elongation, opening the cotyledons, growth responses to gravitropic stimuli, hormones, hormone antagonists, stress factors, sugars, nitrate, ammonia, salts, osmolytes, volatile substances, and many other compounds can be assayed simply in both dark and light germination tests at different temperature regimes. Defects of seed germination, alterations in morphology, colour, elongation, differentiation and development of diverse organs, changes in cell size, number and pattern in various tissues, for example, shoot and root meristems, are also scored in simple germination assays. By modification of the growth medium, the seedlings can also be exposed to various combinations and concentrations of inducers and inhibitors of metabolic and signalling functions, suicide enzyme substrates, cytotoxic and genotoxic agents, or labelled compounds in uptake assays providing a wide choice of phenotypic screens. These screening strategies are illustrated by the steadily growing literature of T-DNA tagged mutants (for references see 4).

Recently, many novel cytological, biochemical and physiological screens were developed to isolate T-DNA tagged genes controlling female or male meiosis (32–36), temporal development of embryos, trichomes, and root hairs (37–41), cellular interactions with nematodes (42, 43), responses to genotoxic stress (44), cytoskeleton organization (45), membrane transport (46), and remarkably T-DNA integration in *Arabidopsis* (47). Based on the availability of sequenced cDNAs as expressed sequence tags (ESTs) and recent progress in genome sequencing, smart phenotypic screens are gradually being replaced by systematic sequence-based screening approaches of reverse genetics.

4.2 Reverse genetics using sequence-based PCR screening approaches

Saturation mutagenesis of the *Arabidopsis* genome can be reproducibly performed using the *in planta* transformation methods. A population of 250 000 transgenic plants is probably sufficient to find a T-DNA insertion at any 0.5 kb region of the *Arabidopsis* genome (about 120 Mb), assuming that T-DNA integration is random. If an average of 1–3 T-DNA inserts is present in each transformed plant and given that the size of an average *Arabidopsis* gene is 2–4 kb, a population of 60 000 plants should already contain more than 90% of all gene mutations. Forward genetic screens, proceeding from mutant isolation to gene cloning, are thus probably less efficient than reverse genetics that uses a target-selected mutant isolation approach to establish a correlation between sequence information and mutant phenotypes (see also Chapter 4 on transposon tagging). Epistasis studies with targeted mutations also provide an excellent means for genetic confirmation of plant protein interactions identified in the yeast two hybrid system. RNA-derived probes prepared from any mutant identified by reverse genetics may also be used to probe cDNA-micro-arrays or gene-chips (see Volume 2, Chapter 3), ideally carrying DNA sequences of all *Arabidopsis* genes, in order to generate digitized molecular phenotypes and identify regulatory functions.

For systematic identification of T-DNA tagged gene mutations, PCR-based screening methods use two PCR primers oriented 5′ to 3′ toward the termini of T-DNA inserts. These T-DNA primers are combined with target-specific primers that are oriented toward a specific gene sequence from 5′ and 3′ directions. Thus, when an insertion is localized in a target gene, a PCR product is obtained with either the 5′ or 3′ gene specific primer, or both. This PCR-based mutant screening technique was first developed for the transposon Tc1 in *Caenorhabditis* and the P-element in *Drosophila* (48–51), and then successfully adapted to the identification of transposon and T-DNA tags in *Petunia* and *Arabidopsis* genes (52–57; see also Chapter 4).

The PCR screening methods are helped by various pooling strategies (see also Chapter 4). The samples, representing either single individuals, or pools of mutants derived from 10 to 100 individuals are arranged in an array to form a 10 × 10 matrix. Samples from each horizontal row and column of an array are pooled for one PCR reaction. Piles are built from the arrays to find a positive sample in a three dimensional matrix. This is usually not required for PCR-screening of T-DNA-tagged *Arabidopsis* lines. It is often sufficient to collect 200 mg leaf material from each M1 plant in order to build a 10 × 10 array in which each sample contains pooled DNA from 100 M1 plants. Thus, pooling rows and columns results in 20 PCR reactions for the screening of 10 000 plants. To find a mutant, 100 plants consisting of a positive sample must be newly grown from M2 seed, arranged in a 10 × 10 array, and pooled in rows and columns to run 20 additional DNA preparations and PCR reactions. Thus, when a population of 100 000 plants is examined, a mutant is found by performing 10 × 20 + 20 = 220 PCR reactions. Because each sample contains 20 g of leaf material from 100 M1 plants, when DNA of high quality is prepared, the material is sufficient for the screening of several thousand gene mutations.

4.3 Identification of gene mutations using sequenced T-DNA tags

Direct sequencing of T-DNA insert junctions is a cumulative approach to sequence-based identification of T-DNA tagged genes. Methods using inverse PCR (iPCR) employ two T-DNA-specific primers that allow the amplification and subsequent bi-directional sequencing of plant DNA fragments that are linked to either the left or right T-DNA ends, or both (23). Alternative techniques, such as the asymmetric interlaced PCR or amplification of insertion mutagenized sites (AIMS), use either a degenerated primer or an adaptor in combination with T-DNA end-primers to amplify plant DNA sequences flanking the T-DNA tags (58, 59). *Protocol 4* describes the application of a long-range iPCR technique that allows routine amplification and sequencing of T-DNA-linked plant DNA fragments ranging in size from 300 bp to 12 kb.

Protocol 4

PCR amplification of T-DNA tagged plant DNA fragments for automatic DNA sequencing

Equipment and Reagents

- SG-medium: see *Protocol 1*.
- DNA extraction buffer: 200 mM Tris–HCl (pH 8.5), 50 mM EDTA, 500 mM NaCl, 14 mM 2-mercaptoethanol and 1% SDS
- HTE-buffer: 100 mM Tris–HCl (pH 8.0), 50 mM EDTA
- LTE-buffer: 10 mM Tris–HCl (pH 7.5), 1 mM EDTA
- RNase A: from SERVA, prepare 10 mg/ml stock solution in LTE-buffer and boil it for 10 min. Store at 4 °C
- Proteinase K: Merck, prepare freshly 2 mg/ml stock solution in LTE-buffer and incubate it at 37 °C for 10 min
- T4 DNA ligase buffer: 66 mM Tris–HCl (pH 7.2), 10 mM $MgCl_2$, 5 mM dithiothreitol, 1 mM ATP
- Nested primers: for binary *Agrobacterium* vectors pPCV6NFHyg and pPCV621 (23) the following left (LB) and right (RB) T-DNA border primers are used:

LB1 (5′-CTCAGACCAATCTGAAGATGAAAT GGGTATCTGGG-3′)

LB2 (5′-CTGGGAATGGCGAAATCAAGGCAT CGATCGTGAAG-3′)

RB1 (5′-GTTTCGCTTGGTGGTCGAATGGGC AGGTAGCC-3′)

RB2 (5′-CAGTCATAGCCGAATAGCCTCTCC ACCC-3′)

Protocol 4 continued

- Elongase PCR buffer: 60 mM Tris-SO_4 (pH 9.1), 18 mM $(NH_4)SO_4$, 1 mM $MgSO_4$, and 250 μM of each dATP, dCTP, dTTP, and dGTP
- ABI Prism-Dye Terminator Cycle Sequencing kit: Perkin-Elmer Ltd

Methods

A. DNA preparation

1. Collect two to four leaves (50–200 mg) from M1 plants in microfuge tubes or M2 seedlings germinated on selective SG-medium.
2. Snap-freeze the tubes in liquid N_2, and grind the material to a fine powder with a glass or Teflon rod fitting the shape of the microfuge tube.
3. Add 750 μl freshly made DNA extraction buffer, mix the extracts well, and incubate the tubes with occasional shaking at 65 °C for 10 min.
4. Add 250 μl 5 M potassium acetate (pH 6.0), mix the tubes, and place them on ice for 20 min.
5. Spin the tubes in a microfuge for 3 min and in the meantime prepare 1 ml pipette tips plugged with small pieces of Miracloth (Calbiochem).
6. Filter the cleared lysates through the Miracloth-plugged tips into new tubes and precipitate the DNA with 700 μl isopropanol on ice for 30 min.
7. Collect the DNA by centrifugation in microfuge for 5 min, wash the pellet with 70% ethanol, and dissolve the DNA in 475 μl HTE-buffer containing 200 μg/ml heat-treated RNase A. Incubate the samples for at least 30 min at 37 °C.
8. Add 125 μl proteinase K solution and incubate the samples for 1 h at 37 °C.
9. Extract the samples twice with 600 μl phenol/chloroform/ isoamylalcohol (25:24:1) and with chloroform/isoamylalcohol (24:1) at 0 °C.
10. Add 60 μl 3 M sodium acetate (pH 6.0) and 400 μl cold isopropanol, mix the tubes and precipitate the DNA on ice for 30 min.
11. Pellet the DNA by centrifugation in a microfuge for 5 min and dissolve in 100 μl LTE-buffer.

This modified version of a plant DNA mini-preparation protocol from Dellaporta *et al.* (60) yields a high quality DNA with an average length of >30 kb. Some other protocols (52, 54, 56) yield less clean DNA with comparable length, whereas most nick-column-based mini-preparation methods yield shorter DNA with many single-strand nicks.

B. DNA digestion

1. Digest 5 μg plant DNA in 200 μl appropriate restriction endonuclease buffer (61) for at least 2 h using an enzyme (100 U), which has either no or only a single cleavage site within the T-DNA.

Protocol 4 continued

2 After testing 10 μl aliquots by gel electrophoresis, extract the samples twice with phenol/chloroform/isoamylalcohol, add 20 μl of 3 M sodium-acetate (pH 6.0), and precipitate the DNA with 150 μl isopropanol on ice for 30 min.

3 Pellet the DNA by centrifugation in a microfuge for 5 min, wash with 70% ethanol, dry briefly, and dissolve in 25 μl sterile water.

C. Ligation of DNA

1 Self-ligate 1 μg digested DNA in 200 μl T4 DNA ligase buffer containing 10% polyethylene glycol (PEG 4000) at 15 °C for 8 h.

2 Extract the samples twice with phenol/chloroform/isoamylalcohol and precipitate the DNA as described above (Protocol 4A9). Digest the samples with an enzyme that cleaves once the T-DNA segment used for self-ligation, but has a cleavage site frequency in the *Arabidopsis* genomic DNA less than 1 for every 50 kb (i.e. *Sma* I, *Sal* I etc.).

3 Linearization of the self-ligated DNA is, however, not essential for iPCR amplification, thus this step may be omitted.

D. Design of PCR primers

The primer design depends on whether the enzyme used in step B cleaves within the T-DNA.

If the enzyme does not cleave in the T-DNA:

1 The self-ligated plant DNA sequence can be PCR amplified using two T-DNA primers that are oriented toward the left and right termini of the T-DNA.

2 Because deletions are often generated at the left border during T-DNA integration, design the left border primer such that it is located at least 100 bp upstream of the left 25 bp border repeat.

3 In contrast, deletions at the right T-DNA end are rare, thus the second primer can safely be placed 20–50 bp upstream of the right 25 bp repeat.

4 It is advisable to design nested primers placed at different distances from the T-DNA ends to ensure the success of iPCR and increase the amount of information obtained by subsequent sequencing of the rescued plant DNA segment using the T-DNA end-primers.

If the enzyme cleaves the T-DNA:

1 Design two primers such that they face the enzyme cleavage site from 5′ and 3′ orientation.

2 Use these primers in pairwise combinations with the T-DNA end-primers for PCR amplification and then independently for PCR sequencing.

E. Polymerase chain reactions

1 The iPCR reactions are performed using either Elongase (GibcoBRL) or Takara LA-PCR (Boehringer Ingelheim) enzyme mixes as recommended by the suppliers.

Protocol 4 continued

2 Amplify 0.5 μg template DNA in 50 μl Elongase PCR buffer containing 1–2.5U enzyme and 0.2 μM of each primer.

3 Following denaturation at 95 °C, run 35 cycles (94 °C for 30 s, 65 °C for 30 s, 68 °C for 8 min) followed by elongation at 68 °C for 10 min.

4 Test 2 μl aliquot from each PCR reaction by gel electrophoresis.

5 In rare cases, when no amplification of defined fragments is obtained, perform a new PCR with the nested primers (see *Protocol 4D*) using either fresh template DNA or 1 μl from a 500-fold diluted PCR reaction mix.

6 Purify the PCR amplified DNA fragments from agarose gel using electroelution (60).

7 Clean the DNA by phenol/chloroform extraction and precipitate with isopropanol as described above (Protocol 4A9–4A11)

8 Sequence the DNA fragments using an ABI Prism-Dye Terminator Cycle Sequencing kit according to the protocol of manufacturer.

F. Sequence analysis

1 To find possible gene hits, analyse the obtained plant DNA sequences using the BLASTN and BLASTX programs by performing searches in the non-redundant, EST and *Saccharomyces* databases (http://www.ncbi. nlm.nih.gov).

2 Alternatively run searches using, for example, databases at http:// genome-www. stanford. edu/Arabidopsis/, or http://www.tigr.org/tbd/.

3 After translating the DNA to protein sequences it is possible to search for motifs at http://coot.embl-heidelberg.de/ SMART/ or http://www.motif.genome.ad.jp/. Other useful sequence analysis facilities are listed at http://www.columbia.edu/~ej67/ dtb_lst.htm, http://www.genome.ad.jp/, http://www.infobiogen.fr/services/dbcat/ and http:// www.molbio.net/genomics.html.

References

1. Zambryski, P. C. (1992). *Annu. Rev. Plant Phys. Plant Mol. Biol.*, **43,** 465.
2. Koncz, C., Németh, K., Rédei, G. P. and Schell, J. (1994). In *Homologous Recombination in Plants* (ed. J. Paszkowski), pp. 167–189. Kluwer Academic Press, Dordrecht.
3. Koncz, C., Martini, N., Mayerhofer, N., Koncz-Kálmán, Z., Körber, H., Rédei, G. P. and Schell, J. (1989). *Proc. Natl Acad. Sci. USA*, **86,** 8467.
4. Azpiroz-Leehan, R. and Feldmann, K. A. (1997). Trends Genet. **13,** 152.
5. Koncz, C., Németh, K., Rédei, G. P. and Schell, J. (1992). *Plant Mol. Biol.*, **20,** 963.
6. Topping, J. F. and Lindsey, K. (1995). *Transgenic Res.*, **4,** 291.
7. Kakimoto, T. (1996). *Science*, **274,** 982.
8. Koncz, C., Martini, N., Szabados, L., Hrouda, M., Bachmair, A. and Schell, J. (1994). In *Plant Molecular Biology Manual* (ed. S. Gelvin, and R. Schilperoort B), Vol. B2, pp. 1–22. Kluwer Academic Press, Dordrect.
9. Depicker, A. G., Jacobs, A. M. and Van Montagu, M. C. (1988). *Plant Cell Rep.*, **7,** 63.
10. Czakó, M. and Márton, L. (1994). *Plant Physiol.*, **104,** 1067.
11. Kobayashi, T., Hisajima, S., Stougaard, J. and Ichikawa, H. (1995). *Jap. J. Genet.*, **70,** 409.

12. Osborne, B. I., Wirtz, U. and Baker, B. (1995). *Plant J.*, **7**, 687.
13. Kilby, N. J., Davies, G. J., Snaith, M. R. and Murray, J. A. H. (1995). *Plant J.*, **8**, 637.
14. Koncz, C., Schell, J. and Rédei, G.P. (1992). In *Methods in Arabidopsis Research* (ed. C. Koncz, N-H. Chua, and J. Schell), pp. 224–273. World Scientific, Singapore.
15. Hiei, Y., Ohta, S., Komari, T. and Kumashiro, T. (1994). *Plant J.*, **6**, 271.
16. Ishida, Y., Saito, H., Ohta, S., Hiei, Y., Komari, T. and Kumashiro, T. (1996). *Nature Biotechnol.*, **14**, 745.
17. Tingay, S., McElroy, D., Kalla, R., Fieg, S., Wang, M. B., Thornton, S. and Brettell, R. (1997). *Plant J.*, **11**, 1369.
18. Cheng, M., Fry, J. E., Pang, S., Zhou, H., Hironaka, C. M., Duncan, D. R., Conner, T. W. and Wan, Y. (1997). *Plant Physiol.*, **115**, 971.
19. Enriquezobregon, G. A., Vazquezpadron, R. I., Prietosamsonov, D. L., Delariva, G. A. and Selmanhousein, G. (1998). *Planta*, **206**, 20.
20. Bundock, P., den Dulk-Ras, A., Beijersbergen, A. and Hooykaas, P. J. J. (1995). *EMBO J.*, **14**, 3206.
21. de Groot, M. J., Bundock, P., Hooykaas, P. J. and Beijersbergen, A. G. (1998). *Nature Biotechnology*, **16**, 839.
22. Ziemienowicz, A., Görlich, D., Lanka, E., Hohn, B. and Rossi, L. (1999). *Proc. Natl Acad. Sci. USA*, **96**, 3729.
23. Mathur, J., Szabados, L., Schaefer, S., Grunenberg, B., Lossow, A., Jonas-Straube, E., Schell, J., Koncz, C. and Koncz-Kálmán, Z. (1998). *Plant J.*, **13**, 707.
24. Feldmann, K. A. and Marks, M. D. (1987). *Mol. Gen. Genet.*, **208**, 1.
25. Feldmann, K. A. (1991). *Plant J.*, **1**, 71–82.
26. Bechtold, N., Ellis, J. and Pelletier, G. (1993). *C. Roy. Acad. Sci.*, **316**, 1188.
27. Clough, S. J. and Bent, A. F. (1998). *Plant J.*, **16**, 735.
28. Koncz, C. and Schell, J. (1986). *Mol. Gen. Genet.*, **204**, 383.
29. Koncz, C., Chua, N-H. and Schell, J. (1992). *Methods in Arabidopsis Research*. World Scientific, Singapore
30. Meyerowitz, E. M. and Somerville, C. R. (1994). *Arabidopsis*. Cold Spring Harbor Laboratory Press, Cold Spring Harbor.
31. Li, S. L. and Rédei, G. P. (1969). *Rad. Bot.*, **9**, 125.
32. Peirson, B. N., Owen, H. A., Feldmann, K. A. and Makaroff, C. A. (1996). *Sexual Plant Rep.*, **9**, 1.
33. Bonhomme, S., Horlow, C., Vezon, D., de Laissardiere, S., Guyon, A., Ferault, M., Marchand, M., Bechtold, N. and Pelletier, G. (1998). *Mol. Gen. Genet.*, **260**, 444.
34. Caiping, H. and Mascarenhas, J. P. (1998). *Sexual Plant Rep.* **11**, 199.
35. Howden, R., Park, S. K., Moore, J. M., Orme, J., Grossniklaus, U. and Twell, D. (1998). *Genetics*, **149**, 621.
36. Sanders, P. M., Bui, A. Q., Weterings, K., McIntire, K. N., Hsu, Y.-C., Lee, P. Y., Truong, M. T., Beals, T. P. and Goldberg, R. B. (1999). *Sexual Plant Rep.*, **11**, 297.
37. Zhang, J. Z. and Somerville, C. R. (1997). *Proc. Natl. Acad. Sci. USA*, **94**, 7349.
38. Uwer, U., Willmitzer, L. and Altmann, T. (1998). *Plant Cell*, **10**, 1277.
39. Albert, S., Després, B., Guilleminot, J., Bechtold, N., Pelletier, G., Delseney, M. and Devic, M. (1999). *Plant J.*, **17**, 169.
40. Oppenheimer, D. G., Pollock, M. A., Vacok, J., Szymanski, D. B., Ericson, B., Feldmann, K. and Marks, D. M. (1997). *Proc. Natl. Acad. Sci. USA*, **94**, 6261.
41. Wang, H., Lockwood, S. K., Hoeltzel, M. F. and Schiefelbein, J. W. (1997). *Genes Devel.*, **11**, 799.
42. Barthels, N., van der Lee, F. M., Klap, J., Goddijn, O. J. M., Karimi, M., Puzio, P., Grundler, F. M. W., Ohl, S. A., Lindsey, K., Roberstson, L., Robertson, W. M., Van Montagu, M., Gheysen, G. and Sijmons, P. C. (1997). *Plant Cell*, **9**, 2119.

43. Favery, B., Lecomte, P., Gil, N., Bechtold, N., Bouchez, D., Dalmasso, A. and Abad, P. (1998). *EMBO J.*, **17,** 6799.
44. Revenkova, E., Masson, J., Koncz, C., Afsar, K., Jakovleva, L. and Paszkowski, J. (1999). *EMBO J.*, **18,** 490.
45. Thion, L., Mazars, C., Nacry, P., Bouchez, D., Moreau, M., Ranjeva, R. and Thuleau, P. (1998). *Plant J.*, **13,** 603.
46. Gaymard, F., Pilot, G., Lacombe, B., Bouchez, D., Bruneau, D., Boucherez, J., Michaux-Ferriére, N., Thibaud, J.-B. and Sentenac, H. (1998). *EMBO J.*, **94,** 647.
47. Nam, J., Mysore, K. S., Zheng, C., Knue, M. K., Matthysse, A. G. and Gelvin, S. B. (1999). *Mol. Gen. Genet.*, **261,** 429.
48. Ballinger, D. G. and Benzer, S. (1989). *Proc. Natl. Acad. Sci. USA*, **86,** 9402.
49. Kaiser, K. and Goodwin, S. F. (1990). *Proc. Natl. Acad. Sci. USA*, **87,** 1686.
50. Rushforth, A. M., Saari, B. and Anderson, P. (1993). *Mol. Cell. Biol.*, **13,** 902.
51. Zwaal, R. R., Broeks, A., van Meurs, J., Groenen, J. T. M. and Plasterk, R. H. A. (1993). *Proc. Natl. Acad. Sci. USA*, **90,** 7431.
52. Koes, R., Souer, E., van Houwelingen, A., Mur, L., Spelt, C., Quattrocchio, F., Wing, J., Oppedijk, B., Ahmed, S., Maes, T., Gerats, T., Hoogeveen, P., Meesters, M., Kloos, D. and Mol, J. (1995). *Proc. Natl. Acad. Sci. USA*, **92,** 8149.
53. McKinney, E. C., Ali, N., Traut, A., Feldmann, K. A., Belostotsky, D. A., McDowell, J. M. and Meagher, R. B. (1995). *Plant J.*, **8,** 613.
54. Krysan, P. J., Young, J. C., Tax, F. and Sussman, M. R. (1996). *Proc. Natl. Acad. Sci. USA*, **93,** 8145.
55. Baumann, E., Lewald, J., Saedler, H., Schulz, B. and Wisman, E. (1998). *Theor. Appl. Genet.*, **97,** 729.
56. Souer, E., Quattrocchio, F., de Vetten, N., Mol, J. and Koes, R. (1995). *Plant J.* **7,** 677.
57. Winkler, R. G., Frank, M. R., Galbraith, D. W., Feyereisen, R. and Feldmann, K. A. (1998). *Plant Physiol.*, **118,** 743.
58. Liu, Y-G., Mitsukawa, N., Oosumi, T. and Whittier, R. F. (1995). *Plant J.*, **8,** 457.
59. Frey, M., Stettner, C. and Gierl, A. (1998). *Plant J.*, **13,** 717.
60. Dellaporta, S. L., Wood, J. and Hicks, J. B. (1983). *Plant Mol. Biol. Rep.*, **1,** 19.
61. Sambrook, J., Fritsch, E. F. and Maniatis, T. (1989). *Molecular Cloning: a Laboratory Manual.* Cold Spring Harbor Laboratory Press, Cold Spring Harbor.

Chapter 4
Transposon tagging methods in heterologous plants

Andy Pereira
Plant Research International, PO Box 16, 6700AA, Wageningen, The Netherlands

1 Introduction

Transposon tagging is a method for the isolation and identification of genes that display a mutant phenotype. It can be used both for random mutagenesis, as well as providing a mechanism to generate mutations in specific genes. A transposon tag is a piece of DNA that moves around the genome and on inserting into genes can cause mutations displaying variant phenotypes. Mutant alleles tagged with the transposon can be identified and their sequences isolated using the cloned transposon sequence as a probe. One characteristic property of a transposon mutant is somatic instability, often visualized as variegation of wild type and mutant sectors. This is common for active transposons that transpose (excise and reinsert) via a cut-and-paste mechanism, in contrast to retro-transposons that move without excision via an RNA intermediate. In some circumstances where a stable mutation is required, this instability can be a disadvantage. In other circumstances where an insertion is linked, but not within a specific gene or where revertant alleles are required, this instability can be used to advantage. Transposons contain inverted repeats at their ends, these sequences are recognized by specific transposase proteins that mediate the transposition event. Autonomous transposons contain a functional transposase gene required for their movement, whereas non-autonomous transposons are transposase defective and can transpose only in the presence of a second autonomous element, which encodes the transposase.

Transposons were discovered by McClintock (1) who described the maize *Activator-Dissociation* (*Ac-Ds*) transposable element system; *Ac* is the autonomous element and *Ds* the non-autonomous element of this two component system. Members of the *Ac-Ds* family contain a specific 11 bp terminal inverted repeat (TIR) and create an 8 bp target site duplication (TSD) on insertion. The 4565 bp long *Ac* element produces a 3.5 kb mRNA, which encodes the transposase (TPase) (2). Another maize system different from *Ac-Ds* is the *En(Spm)* system discovered independently as *Enhancer-Inhibitor* (*En-I*) by Peterson (3) and *Suppressor-Mutator* (*Spm*) by McClintock (4). The *En-I(Spm)* family members possess a 13 bp TIR and create a 3 bp TSD on insertion. The *En(Spm)* element is 8287 bp long and encodes two alternatively spliced gene products (5), which are required for transposition (6). Approximately 10 other active transposon systems have been discovered in maize including the commonly used *Mutator* (*Mu*) system. Active indigenous transposons have also been used for tagging in antirrhinum (7), petunia (8), and Arabidopsis (9).

The maize *Ac* transposon was the first to be shown to transpose in a foreign host (10), which indicated that all necessary functions for heterologous transposition are encoded by the autonomous element. This discovery stimulated the use of *Ac-Ds* in a wide range of plant species (11) including Arabidopsis, carrot, potato, tomato, *Triticum-monoccocum*, rice, soybean, lettuce, and flax, with the objective of developing transposon tagging technology for gene isolation.

This chapter deals with the use of the maize *Ac-Ds* and *En-I* (*Spm*) systems for transposon tagging in heterologous plant systems, illustrated by specific examples and supported by general protocols. For transposon tagging with indigenous transposons the reader is referred to the biology and uses of specific transposon systems (reviewed in 12). Though transposons have been classically used for forward genetics strategies of mutant-to-gene identification (see Sections 4 and 5), recent developments in genomics enable reverse genetics approaches (see Section 6). Both strategies can be further subdivided into random or directed mutational approaches that are further dealt with within in the following Sections.

2 Generation of lines for transposon tagging

2.1 Phenotypic excision assays

In heterologous species, transposons can be employed using one or two component systems. In a one-component system, an autonomous transposon can be used as a mutagen because it provides its own transposase in *cis*. In a two-component system, a non-autonomous transposon must be activated in *trans* by expression of the transposase from, for example, a stable derivative of an autonomous element.

To monitor transposon excision, phenotypic assays have been developed using selectable markers, for example, antibiotic resistance or visual marker genes. In a typical construct (13), the transposon is inserted in the 5′ untranslated leader of a marker gene, blocking its expression (*Figure 1*). Excision of the element in the plant restores the activity of the marker gene, which can be visualized as resistance to the particular antibiotic or activation of a visual marker, for example histochemical GUS activity. As excision normally results in a small footprint changing a few nucleotides, this should not affect the structure of the 5′ untranslated leader and yield a wild-type gene product. The neomycin phosphotransferase gene (*NPTII*) gene conferring kanamycin resistance was first used as a phenotypic excision marker, but was soon replaced by better cell-autonomous markers, for example, the streptomycin resistance gene (*SPT*; 14)

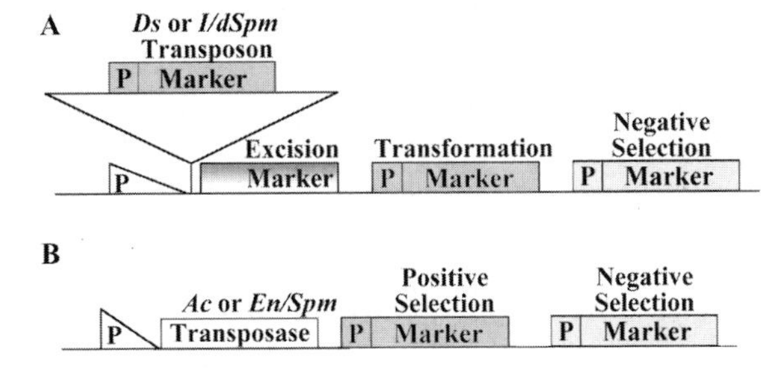

Figure 1 Two component transposon system design for heterologous plants. (A) Mobile (non-autonomous) transposon in the T-DNA, inserted in an excision marker gene, with (excision and selection) markers as described in *Table 1*. (B) Stable TPase source with accompanying positive and negative selection markers that can induce transposition (*in cis* or *in trans*) of the mobile transposon. Promoter sequences are labelled P.

that can distinguish between somatic and germinal excision events. Other engineered resistance genes that serve as selectable markers for excision are hygromycin phosphotranserase (*HPT*; 15), amino lactone synthase (*ALS*; 16) spectinomycin phosphotransferase (*SPEC*; 17) and phosphinothricin acetyl transferase (*PAT*; 18), which confer resistance to hygromycin, chlorsulfuron, spectinomycin, and Basta, respectively. Visual marker genes employed in such assays include β-glucuronidase or GUS (*uidA*; 19), *rolC* (20), waxy/*GBSS* (21) and luciferase (*LUC*; 22). These markers are summarized in *Table 1*.

Table 1 Two component heterologous Ac-Ds tagging systems

Plant	Ac Transposase sources Transposase construct	Positive selection	Negative selection	Ds elements Ds excision marker	Ds selection	Insert type	Ref[a]
Tobacco	35S-Ac cDNA	HPT	None	NPT II	None		98
	35S-Ac cDNA, Nos-Ac, Ocs-Ac	NPT II	GUS	SPT	None		99
	35S-Ac	NPT II	GUS	SPEC/BAR			18
	PR1a-Ac	HPT	None	LUC/GUS	None		22
Arabidopsis	wtAc3′Δ ΔNaeAc	NPT II	GUS	SPT	None		100
	ΔNaeAc	NPT II	IAAH	SPT	None		100
	35S-Ac	HPT	None	NPT II	DHFR	35S readout	101
	sAc, 35S-Ac, nos-Ac, ocs-Ac	NPT II	GUS	SPT	None		31
	wtAc3′Δ	HPT	None	NPT II	BAR		102
	35S-Ac, rbcS-Ac, chs-Ac	DHFR	GUS	NPT II	ALS		103

Table 1 *(contd)*

Plant	Ac Transposase sources: Transposase construct	Positive selection	Negative selection	Ds elements: Ds excision marker	Ds selection	Insert type	Ref[a]
	35S-Ac 35SΔNaeAc	NPT II	IAAH	ALS	HPT	Promoter and enhancer trap	16
	hsp-Ac	NPT II	None	SPT	HPT		34
	35S-Ac	HPT	R-Lc	Cs (ALS)	BAR	Cre-lox recombination	30
	35S-Ac	NPT II	IAAH		NPT II	enhancer/gene trap	77
	35S-Ac	NPT II	GUS	SPT	HPT	35S readout	104
	35S-Ac		GUS	NPT II	HPT	SceI restriction site RS recombination	105
Tomato	Ac			HPT	GUS		56
	sAc	NPT II	GUS	SPEC	BAR		53
				SPT	BAR		
				BAR	SPEC		
	35S-Ac			Lc	GUS		106
	35S-Ac	NPT II	IAAH		NPT II	Enhancer/gene trap	78
	35S-Ac	NPT II	IAAH	ALS	HPT	Enhancer trap	78
Potato	TR2′-Ac	DHFR	None	HPT	NPT II		21
Rice	35S-Ac	None	None	HPT	None		107

[a] References, abbreviations of marker genes as mentioned in references and in the text.

The utility of excision markers is not universal as each plant has its own special differences, for example, antibiotic sensitivity of tissue to be tested. Thus, the *SPT* gene can display excision only in the cotyledons of tobacco and Arabidopsis, but is not suitable for many other plants (18), such as tomato or potato, which are insensitive to streptomycin. The *RolC* gene (20) can be used to monitor excision in adult leaves, but it causes sterility of the plants and, therefore, is not widely applicable.

Excision of a transposon from within a marker/reporter donor site leaves an empty donor site or EDS. Detection of this EDS is achieved by marker gene activation when the transposon is inserted in the 5′ untranslated leader, as in most engineered phenotypic excision assays. The use of excision markers also enables comparison of the germinal excision frequency (14, 23), namely the fraction of seedlings displaying excision events inherited through the germ-line and excluding somatic events, among the total number of seedlings in the progeny of a plant.

The molecular proof of excision and detection of the empty donor site (EDS) can be done by PCR using sets of primers flanking the site of insertion of the transposon in the original construct. DNA is first isolated from single plants (*Protocol 1*) that display excision by a phenotypic assay. The DNA can then be used in a PCR for EDS detection as described in *Protocol 2*, which gives primer combinations suitable for various constructs.

Protocol 1

Miniprep DNA isolation from single plants

Reagents

- DNA extraction buffer: 0.3 M NaCl, 50 mM Tris (pH 7.5), 20 mM EDTA, 2% sarkosyl, 0.5% SDS, 5 M urea, 5% equilibrated phenol (24). The first five components are mixed as a ×2 stock solution, and urea and phenol are added before use.
- Phenol/chloroform, 24:1 (v/v).
- 70% ethanol
- Isopropanol
- TE buffer: 10 mM Tris–HCl (pH 8.0), 1 mM EDTA
- RNase, 10 μg/ml

Method

1. Harvest 100–150 mg young leaf or preferably inflorescence tissue per plant in a micro-centrifuge tube and freeze in liquid nitrogen.
2. Grind tissue to a fine powder in the micro-centrifuge tube.
3. Add 150 μl of DNA extraction buffer and grind once more.
4. Add a further 300 μl of extraction buffer and mix.
5. Leave the samples at room temperature until 18 or 24 samples have been prepared.
6. Extract the samples with 450 μl phenol/chloroform and precipitate the DNA with 0.7 volumes isopropanol.
7. Keep tubes at room temperature for 5 min, then centrifuge in micro-centrifuge at 12 000 *g* for 5 min.
8. Wash the DNA pellets with 70% ethanol and briefly dry them.
9. Dissolve DNA pellets in 100 μl TE containing 10 μg/ml RNase.
10. DNA samples may be stored at 4 °C for a few months or at −20 °C long term.

Protocol 2

Molecular proof of excision by PCR identification of empty donor sites

Reagents

- P Primers: Promoter

CaMV 35S	GGATAGTGGGATTGTGCGTC
TR1′	CTTACGTCACGTCTTGCGCA
Nos	GCGCGTTCAAAAGTCGCCTA

- M Primers: Marker Gene

NPT II	CCAGTCATAGCCGAATAGCC
HPT	GTCAAGCACTTCCGGAATCG
SPT	CCAGCTCGAGTGGGTGGTG AG

- Reagents for PCR amplification

Protocol 2 continued

Method

1. Assemble PCR reactions using 20–200 ng of genomic DNA (see *Protocol 1*), depending on genome size of plant[a]. The combinations of empty donor site P and M primer required depend on the type of construct used. Use Taq polymerase buffer and enzyme as recommended by supplier in a 50 μl reaction with 1.5 mM $MgCl_2$, 0.3 μM each primer and 0.2 mM each nucleotide.
2. Carry out the PCR with 35 cycles of denaturing at 94 °C for 1 min, annealing at 55–60 °C for 1 min and extension at 72 °C for 2 min, followed by a 5 min further extension at 72 °C.
3. Load 10–20 μl of the PCR reaction on an agarose gel and fractionate the products, photograph the gel and cut out the fragments corresponding to the empty donor sites (EDS).
4. Isolate EDS fragment DNA using a gel purification kit such as Qiaex (Qiagen).
5. Clone EDS fragment directly in a PCR product cloning vector, for example, pGEM-T (Promega).
6. Select clones with inserts, and sequence to confirm EDS by identifying the excision footprint.

[a] This range corresponds from Arabidopsis to tobacco.

2.2 Autonomous *Ac* systems

The simplicity in the use of the one element *Ac* system makes it versatile for employment in a wide range of species. *Ac* has been successfully used for transposon tagging in petunia, tobacco, Arabidopsis, and flax. Tagging of the *Ph1* gene in petunia (25) was the first example of heterologous tagging with the *Ac* element. In this random tagging approach, plants that carried transposed *Ac* elements were selected and selfed to yield progeny that were screened for mutant phenotypes.

In other plants, targeted tagging of specific loci has been undertaken. In tobacco, the TMV resistance gene, *N*, was tagged with *Ac* (26) in a screen of 64 000 heterozygous *N*/*n* F_1 plants using a convenient positive selection scheme. A transformed flax line carrying an *Ac* element linked at 29 cM to the target *L6* locus was selected and used in crosses to a recessive genotype to yield about 30 000 progeny, which were screened for mutations in the *L6* resistance locus. Out of the 29 mutants obtained, one was tagged by *Ac* and used to isolate the gene, the rest were *L6* deletions (27).

In contrast to other plant species, *Ac* in Arabidopsis was found to display a very low germinal excision frequency of 0.2–0.5 % (23). This frequency was increased 30-fold by deleting a *Nae* I restriction fragment, which contains a methylation sensitive CpG-rich segment, from the 5′ end of *Ac* (28). A T-DNA construct carrying this modified element inserted into the streptomycin phosphotransferase gene (*SPT*) located 22 cM from the *FAE1* gene was the starting point for a directed tagging experiment in Arabidopsis (29). *Ac* transpositions from the *SPT* gene were selected as 2243 streptomycin resistant (Sm^R) seedlings, containing 500–1000 independent *Ac* insertions at remote loci. These seedlings were directly screened by gas chromatography to identify the *fae1* mutant that is defective in the synthesis of very long chain fatty acids in the seed. This screen yielded one tagged mutation of the *FAE1* gene.

2.3 Two-Component *Ac-Ds* tagging systems

Two-component *Ac-Ds* transposon-tagging systems (*Figure 1*) can be modified before introduction into heterologous hosts, as outlined in *Table 1*. One advantage of separate components is that the *Ac* transposase can be segregated out to give stable *Ds* insertions. Other improvements include *Ds* excision assays, introduction of a selectable marker gene within *Ds* to follow its transposition (30) and control of *Ac* transposase expression with the help of strong or regulated promoters. Early studies with the *Ac-Ds* systems revealed that re-insertion did not always follow excision. A selectable marker gene within *Ds* is therefore useful to select for reintegration of the *Ds* element.

A strategy detailed in *Protocol 3* and summarized in *Table 1* was successfully used to identify mutations in the *ALBINO3* gene in Arabidopsis (31, 32). In this system, the *Ds* element carries a hygromycin resistance marker (*HYG*) that disrupts the streptomycin resistance gene (*SPT*). This construct containing *SPT:Ds(HYG)* confers hygromycin resistance (Hyg^R) and streptomycin sensitivity (Sm^S) before transposition, and both Hyg^R and Sm^R following excision and re-insertion of the *Ds* element. The *Ac* element carries a kanamycin resistance gene, the transposase, and a β-glucuronidase reporter gene that can be used as a negative screenable marker. The CaMV 35S-*Ac* transposase source yields about one independent *Ds(HYG)* transposition per F_2 family due to early transposition in the F_1 plant. Thus, only a few Hyg^R + Sm^R F_2 progeny from each F_1 plant were used to produce the F_3 progeny for mutant screening. From the F_2 families, Hyg^R + Sm^R + GUS^- individuals were selected to produce about 180 transposase free F_3 individuals which contain stable transposed *Ds* elements. These plants were screened for the mutant phenotype in soil and on agar plates. Seven mutants were identified out of which three were shown to be tagged by co-segregation of the mutant phenotype with *Ds*-encoded hygromycin resistance (33). Modifications of this strategy for selection of transposants can be employed with other selectable markers and transposase sources, for example under control of the heat shock promoter (34), as well as in other plant species. As shown in *Table 1*, a number of selectable excision markers are applicable in other species, including phosphinothricin acetyl transferase and amino lactone synthetase (ALS) for selection with the herbicides Basta and chlorosulfuron, which in combination with a screen against the transposase, enable the selection of stable germinal excision events.

Protocol 3

Generation of independent Ds transpositions in Arabidopsis[a]

Reagents

- Germination medium: 1 × MS salts, 1 × Gamborg's B5 vitamins, 10 g/l sucrose, 0.5 g/l MES (2-[N-morpholino]ethane sulfonic acid), 8 g/l agar, pH 5.7
- X-Gluc solution: 5-bromo-4-chloro-3-indolyl B-D-glucronide cyclohexylammonium salt at 0.5 mg/ l, 0.05% Triton X-100 in 50 mM sodium phosphate pH 7.
- 4-MUGluc solution: 4-Methylumbelliferyl-B-D-glucuronide at 2 mM in MS medium.

Method

1. Select kanamycin resistant 35S-*Ac* transposase seedlings on germination medium (GM) containing kanamycin (100 mg/l) and hygromycin resistant *SPT:Ds(HYG)* seedlings on germination medium (GM) containing hygromycin (20 mg/l).
2. Transfer resistant seedlings to greenhouse covering them for about 3–5 days under a plastic cover for high humidity.
3. Make multiple crosses between *Ac* and *Ds* plants by emasculating unopened buds and pollinating with anthers of the male donor. The number of crosses depends on the number of independent transpositions required, each F1 seed will yield a minimum of one independent transposition event.
4. Harvest siliques as they turn yellow, shake out the seed and let them air dry.
5. Plant out F1 seed on GM containing kanamycin (100 mg/l) and hygromycin (20 mg/l) to select for double resistant plants, and transfer the plants to a greenhouse. Each F1 plant will typically give only one independent *Ds* insertion for some 35S *Ac* transposase sources, thus use the same number of seed as the number of independent transpositions required.
6. Harvest seed at maturity, clean, and dry seed.
7. Sow the F2 seed (200–300) on GM containing streptomycin (200 mg/l) and hygromycin (20 mg/l), select fully green Hyg^R + Sm^R seedlings at the two-leaf stage.
8. To employ a non-destructive screen for GUS expression, place the resistant seedlings in closed Petri dishes with their roots immersed in X-Gluc solution at 37 °C for 10 min, or in 4-MUGluc for a few h.
9. Test roots for the appearance of a blue colour (GUS^+) after X-Gluc staining, or visualize under long wave UV light after 4-MUGluc treatment.
10. Transfer GUS^- and some GUS^+ seedlings to GM for 1 week to recover and then to soil in the greenhouse.
11. Harvest seed from at least one GUS^- and GUS^+ plant per F2 family to recover independent transpositions.

Protocol 3 continued

12 Sow out the F3 family from the Hyg^R + Sm^R + GUS^- parent and screen for mutant phenotypes. Co-segregation of mutant and the *Ds* borne Hyg^R marker can be used for initial proof of tagging.

13 The GUS^+ family siblings, which carry both the *Ds* insertion in the gene of interest and the active transposase, can be used to recover revertants once a tagged mutant has been identified.

[a] Adapted from 33.

2.4 *En-I(Spm)* tagging systems

The autonomous *En/Spm* elements were shown to transpose in tobacco (35, 36) and potato (37). However this system was shown to be not as effective as *Ac* in *Solanaceous* species due to the complex alternative splicing that was found to occur (5). The *En* transposon introduced into Arabidopsis demonstrated active transposition (38) with a germinal excision frequency of about 7.5%, which continued over a number of generations.

To avoid the production of separate transformants and crossings, a two-component *En-I* system was assembled in one construct (*Figure 1*). This system contains a stable *En-TPase* under control of the CaMV 35S promoter and an *I* transposon inserted in the coding region of the *NPTII* gene. This construct was introduced into Arabidopsis ecotype Landsberg erecta (39). One transformant displaying high transposition contained two 35S-*En* T-DNA loci. These were termed *TEn5* and *TEn2*. The TEn5 locus contained 5 tandem T-DNA copies and was shown to be about five times as active as the *TEn2* locus that contained two T-DNA copies. The *TPase* loci exhibit constitutive transposition throughout development of the plant, with early as well as late transpositions giving rise to many independent *I* inserts in the progeny. The independent transposition frequency (ITF) calculated as the frequency of new unique inserts in a progeny, was determined by DNA blot analysis using an *I* specific probe (39). Ranges in ITF between 5 and 30% were observed in different families. In general, lines containing more than 10 *I* elements display an average of one new transposed *I* element in every progeny plant.

En-I transposon lines containing independent transposed *I* elements can be continuously generated by selfing *En-TPase* bearing genotypes. This random tagging strategy has produced a number of new tagged genes, for example, *ms2* (40), *cer1* (41), *sap1* (42), *anl2* (43) and tagged alleles of known mutants like *ap1*, *gl2*, *lfy* (M. G. M. Aarts and A. Pereira, unpublished). Lines carrying multiple transposing *I* elements were selected and used to further develop populations of independent transpositions by single seed descent (44). These multiple transposon populations can be cost effectively screened for mutations that are difficult to screen for in large populations. Out-crossing a mutant with wild type for one or two generations will segregate out the TPase source, help to reduce the number of inserts, and yield stable mutants for the isolation of *I* element tagged genes.

The autonomous *En/Spm* transposon has also been successfully used for transposon tagging of a number of genes in Arabidopsis ecotype Columbia (45). This has also been achieved by making advanced generation populations using single seed descent (46), carrying independently transposed *En* elements. After a few generations of selfings, each family contains new independent insertions that are used to screen for mutant phenotypes, the majority of which are tagged, as visualized by the appearance of variegation. This system of producing a large number of independent transpositions beginning from a few active transposons is applicable for all active transposition systems like *Ac* in rice and tomato.

3 Mapping transposon lines

In maize, the *Ac* transposon was first shown to transpose to positions that are closely linked to the original donor site (47). The feature of local transposition has been utilized in maize to increase the tagging efficiency by using transposons linked to a target locus (12). This behaviour of *Ac* has now been observed in tobacco (48), tomato (49), Arabidopsis (50) potato (51), and flax (27). This finding has stimulated the mapping of T-DNA containing transposons in the genome to obtain transposons closely linked to target loci for efficient tagging.

In order to use linked transposons for targeted tagging purposes, a number of research groups working with various plant species have produced large numbers of transformants carrying transposons. These T-DNA containing transposons have been mapped by a number of methods. One method has been the isolation of genomic DNA flanking the T-DNA inserts by inverse PCR (iPCR; 52, 53). The iPCR method is described in two steps, a common step in *Protocol 4* for self-ligation of genomic DNA using appropriate enzymes that cut frequently such as those with 4 bp recognition sites. For this protocol to work, these sites must not occur between the primers and the T-DNA borders. The second step of conducting iPCR is described in *Protocol 5* with primers for two types of commonly used vectors: pPCV derivatives (54) and pBin19 (55). Another method called plasmid rescue (56) uses the incorporated replication origin and antibiotic resistance sequences for cloning in bacteria, that must be present in the T-DNA construct, and preferably near the borders. This plasmid rescue method is described in two steps, *Protocol 4* for preparation of ligated genomic DNA using enzymes that do not cut between the T-DNA replicon and the border, and *Protocol 6* that describes an efficient method for transformation.

Following isolation of sequences flanking the T-DNA insert, the resultant genomic DNA probes are then used in RFLP mapping analysis in a segregating plant population with other standard markers. This is done in two steps. Step 1 uses the probe on blots of polymorphic parents of an RFLP population, restricted with a set of standard enzymes. Step 2 shows that after choosing an enzyme combination that is polymorphic between the two parents, the probe is used on the whole segregating population to score the polymorphisms, the data from which is then uploaded into in a mapping program such as MapMaker (57) or JoinMap (58), appropriate for the population type and dataset. The results give the position in a molecular genetic map of the species and reveals linkage to potential target genes. Standard maps and populations are available for Arabidopsis from the Nottingham Arabidopsis Stock Centre (NASC) (59) where mapping data can be entered online. For other species, such as tomato and potato (60), the Solgenes database will help RFLP mapping to standard maps. Mapped T-DNAs carrying transposons have been produced for a number of plant species, accounting for about a hundred *Ac*/*Ds* elements each in tomato (53, 61) and potato (51). A large number of transposed *Ac-Ds* elements have also been mapped with respect to the donor T-DNA in Arabidopsis (62, 63). In other cases linkage to target genes was sought, as in flax, by genetic means. In the *En-I* system in Arabidopsis about 30 transposed *I* elements have been mapped (39) for use in targeted tagging.

Protocol 4

Self ligation of T-DNA flanking genomic DNA for plasmid rescue and IPCR

Reagents

- Restriction enzymes *Hin* dIII, *Hae* III, *Mse* I or *Hpa* II with appropriate buffers
- Ligation buffer (×10): 500 mM Tris pH 7.5, 100 mM $MgCl_2$, 100 mM DTT, 100 mM ATP
- Equipment and reagents for agarose gel electrophoresis
- Phenol:chloroform (24:1 v/v)
- Sodium acetate: 0.3 M
- Isopropanol
- Sterile distilled H_2O
- 70% (v/v) ethanol

Method

1. Isolate DNA from transposon line plants following *Protocol 1*.
2. Determine appropriate restriction enzyme based on T-DNA construct map/sequence or by Southern analysis. Recommended enzymes are *Hin* dIII for plasmid rescue, and *Hae* III or *Mse* I for IPCR of left borders (LB) of pPCV vectors, and *Hpa* II or *Mse* I for IPCR of pBin19-derived vectors.
3. Digest 200–500 ng genomic DNA with 10–20 units of the appropriate enzyme in 100 μl volume for 2 h at 37 °C. Check 5 μl of the digestion on an agarose gel.
4. Phenol/chloroform extract the DNA and precipitate the DNA by adding 10 μl 0.3 M sodium acetate and 110 μl isopropanol to the extracted aqueous phase.
5. Incubate for 10 min at −20 °C, then centrifuge 12 min at full speed in a microcentrifuge.
6. Resuspend the pellet in 90 μl sterile distilled H_2O, add 10 μl ×10 ligation buffer, add 2.5 units T4 DNA ligase and incubate overnight at 16 °C.

Protocol 4 continued

7 Phenol/chloroform extract and precipitate DNA as described in step 4.

8 Wash the pellet twice with excess 70% ethanol, dry and resuspend in 10 μl H_2O.

9 Transform into *Escherichia coli* by electroporation (*Protocol 6*).

Protocol 5

Amplification of T-DNA flanking DNA by iPCR

Reagents

- pBin19-based LB IPCR primers
- Outward CTGTTGCCCGTCTCGCTGGTG
- Inward GATGGTGGTTCCGAAATCGGC
- pPCV-based vector LB IPCR primers
- G4pA inward CATAACACGCACACTT ACGATAG
- G4pA outward GCTATCATTGCGGCC AAGCTC
- PCR reagents
- Reagents and equipment of agarose gel electrophoresis

Method

1 Use the ligated DNA from *Protocol 4* for a PCR reaction with appropriate primers. Set up 40 μl reactions with 1.5 mM $MgCl_2$, 0.3 μM each primer, 0.2 mM each nucleotide, ×1 Taq polymerase buffer, and 5 μl ligated DNA.

2 Initiate cycles with a hot start by heating reactions for 94 °C for 5 min and then adding 10 μl of ×1 Taq polymerase buffer and 1 unit Taq Polymerase enzyme (Supertaq, HT Biotechnology Ltd, UK) to each tube.

3 Run PCR for 35 cycles of denaturing at 94 °C for 1 min, annealing at 55–60 °C for 1 min, and extension at 72 °C for 2 min, followed by a 5 min extra extension at 72 °C.

4 Load about 15–20 μl of the reaction on an agarose gel, excise specific bands, purify fragments, and clone directly in a PCR fragment cloning vector, e.g. pGEM-T (Promega).

Protocol 6

Preparation and transformation of electrocompetent *E. coli* cells for plasmid rescue[a]

Reagents

- *E. coli* strains MC1061 or DH5α
- Luria Bertani (LB) medium
- Electroporation cuvettes (Biorad)
- Buffer 100 mM HEPES (pH 7.0)
- Buffer 100 mM HEPES (pH 7.0) containing 10% (v/v) glycerol
- SOC medium: 2% bactotryptone, 0.5% yeast extract, 10 mM NaCl, 2.5 mM KCl, 10 mM $MgCl_2$, 10 mM $MgSO_4$, 20 mM glucose

Protocol 6 continued

Method

1. Inoculate 200 ml LB with 2 ml of overnight *E. coli* (MC1061 or DH5α) culture.
2. Grow the culture for 2 h until OD_{550} is 0.5.
3. Collect the cells by centrifugation at 4 °C with 10 000 g for 10 min.
4. Resuspend the cells in 100 ml of 1 mM HEPES (pH 7.0) at 0 °C.
5. Pellet the cells by recentrifugation as described in step 3.
6. Resuspend the cell pellet in 50 ml of 1 mM HEPES (pH 7.0).
7. Pellet the cells by recentrifugation as described in step 3.
8. Resuspend the cells in 5 ml of 1 mM HEPES (pH 7.0) containing 10% glycerol.
9. Transfer the cells to micro-centrifuge tubes and pellet them by centrifugation for 10 min at 2100 g and 4 °C.
10. Resuspend the cells in 400 μl of 1 mM HEPES (pH 7.0) containing 10% glycerol.
11. Mix 40 μl aliquots of the cells with 5 μl ligated DNA in pre-cooled electroporation cuvettes.
12. Electroporate the cells using a Biorad Gene Pulser at 25 μF, 2.5 kV, 200 A, and 4.8 ms.
13. Immediately after electroporation, add 1 ml SOC medium (76) and incubate the cells for 1 h with shaking at 37 °C.
14. Collect the cells by brief centrifugation (20 s) at full speed in a micro-centrifuge.
15. Resuspend the cells in SOC medium and plate aliquots on LB agar plates with ampicillin (100 μg/ml).

[a] Adapted from ref. 54.

4 Transposon tagging strategies

4.1 Random and targeted tagging

The first tagged mutants isolated in the heterologous hosts Arabidopsis (40, 64, 33) and petunia (25) were isolated from a random or non-targeted tagging approach. The strategy used involved the generation of transposants that were then selfed to reveal the mutant phenotypes. Extensive screening of homozygous *Ds* transposants revealed tagged mutation frequencies of 2–4% in Arabidopsis, that are consistent with observations in the *En-I* system and of T-DNA insertions, suggesting that only a fraction of random transposon inserts display major mutant phenotypes. Many mutations might reveal conditional or subtle phenotypes when put through specific screening conditions, as observed in yeast. For saturated mutagenesis in Arabidopsis it is estimated that about 100 000 inserts (65) would be required whereas larger genomes would require a proportionately larger population of inserts. Due to the nature of transposon introductions from only a few initial T-DNA insertion lines into the genome, genome saturation is normally difficult to achieve, unless strategies to efficiently select insertions are employed.

Mutants for a developmental stage, biochemical pathway, physiological process, or microbial interaction can be conveniently isolated by other methods like chemical and physical mutagenesis and their genetic interactions analysed. The genes corresponding to specific mutants of interest may then be isolated. Targeted or directed tagging is a strategy to isolate specific tagged mutant alleles. A simple strategy (*Figure 2*) involves making crosses between a characterised recessive mutant line and active transposon lines containing the corresponding wild type allele that is to be tagged. The use of a transposon line that is linked to the locus of interest will provide tagged mutants at a higher frequency (1 in 10^{-3}) than would be expected at random (1 in 10^{-5}). The efficiency of tagging depends on the linkage and the frequency of independent transpositions (66).

The transposition behaviour of *Ac-Ds* in heterologous hosts has been studied in great detail. Most T-DNA inserts introducing the transposons in the genome produce transposants close to the donor locus (62), while other T-DNA loci produce more unlinked transpositions. An analysis of unlinked transpositions (67) also revealed non-randomness and chromosome specificity. All these results suggest that the genomic position contributes to differences in the distribution pattern of transposed elements (49, 68, 69). The effect of genome position is also evident in the differences in excision frequencies from different *Ds* sites (62).

In tomato, an efficient targeted tagging strategy (*Figure 2*) was used to isolate the *Cf-9* resistance gene (70). A *Ds* element located 3 cM from the *Cf-9* locus, was

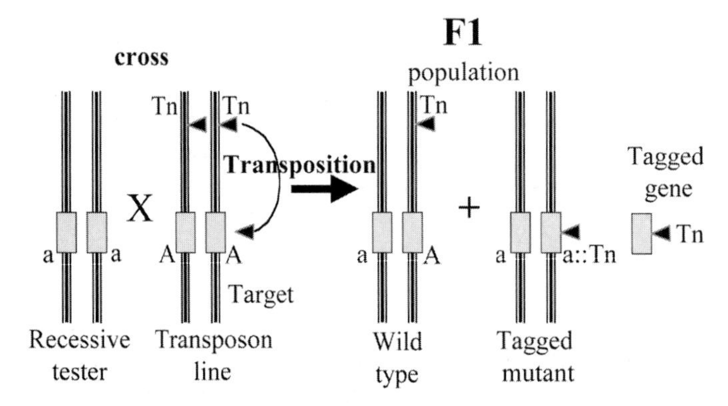

Figure 2 Targeted transposon tagging strategy. A transposon (Tn) is used to isolate mutants at the closely linked target locus '*A*' in a plant genotype containing the TPase to mobilize the Tn. This transposon *A/A* genotype is used to make crosses to a recessive *a/a* genotype to produce an F1 population that is mostly *A/a* heterozygous except for the occurrence of a few tagged *A* alleles (*a::Tn*) that display a mutant phenotype (*a/a::Tn)*. The tagged allele (*a::Tn*) is then used to isolate the transposon flanking DNA and then the tagged gene.

mobilized by a stabilized *Ac* (s*Ac*) locus. The transposon containing lines were crossed to a recessive *cf-9* genotype (expressing the Avr9 protein). In this elegant positive selection scheme for mutants, all the F_1 progeny carrying the functional *Cf-9* and *Avr9* genes become necrotic and died shortly after seed germination. Out of a screen of 160 000 progeny about 63 independent mutations (1/2500) were recovered, out of which at least 37 were tagged with *Ds* (1/4325) and 28 were located within a 3 kb region of the *Cf-9* gene.

Later, in Arabidopsis, an *Ac* element was used to tag the *FAE1* gene located at 22 cM distance (29) and the *L6* resistance gene from flax was isolated (27) using a linked *Ac* (29 cM) element. In tobacco, targeted tagging of the *N* gene (26) using three *Ac* lines at undefined positions, yielded a tagged *N* mutant from 64 000 individuals, probably from a linked donor site.

Not all mutants originating from the *Ac* lines are necessarily tagged, as revealed in flax (27) and tobacco (26). In tomato, somaclonal variation after transformation was attributed to 20% of the mutants observed in progeny of the primary transformants (71). In Arabidopsis, a strategy of sequential propagation of *Ac-Ds* lines for a number of generations, followed by selection for germinal revertants and mutant screening in the progeny yielded many mutants that were not tagged (72). This observation is probably due to the number of consecutive generations of propagation and selection in media, in which transposed *Ds* footprints may also accumulate.

The mapping of transposed *I* elements in Arabidopsis reveals that linked transposition is not as frequent as that of *En* in maize (73) and only about 20% of *I* transpositions are to linked sites (39) around 20 cM from the donor locus. This lower local transposition thus reduces the frequency for targeted tagging with a linked transposon, but is compensated by the ease in producing a large number of new transpositions (74). On average, about 2% of unselected progeny produce transpositions to linked sites, useful for targeted tagging that can be initiated with the 30 mapped *I* elements that have been described (39).

The *En-I* system is more suitable for random tagging. With an independent transposition frequency of 10–30%, it is possible to generate a large population of different inserts in only a few generations. Populations of multiple transposons generated by single seed descent are available using the autonomous *En* transposon (45) and the *En-I* system (44) that are screenable as families allowing recovery of difficult phenotypes.

4.2 Entrapment with transposons

The majority of transposon inserts do not display a mutant phenotype, even if inserts are in or near genes. This finding is probably due to gene redundancy or indicates that some genes are not essential for the organism. While in yeast only a sixth of insertions result in a mutant phenotype, in Arabidopsis less than 4% *Ds* insertions yield mutants (66, 33).

To be able to use insertional mutagenesis in a more efficient way, for example to use transposons to probe the function of a flanking gene, a number of entrapment strategies (75) have been designed. These strategies make use of inserts containing reporter gene constructs, whose expression is dependent on transcriptional regulatory sequences of the adjacent host gene. After insertion nearby or within a gene, the reporter displays the expression pattern of the entrapped gene. The expression pattern of the reporter can give a clue about the function of the host gene in physiology and development. In this way, genes that display a lethal phenotype when mutated can be analysed for their expression pattern, for example, embryo lethals can be analysed to determine in which developmental steps the gene products are involved. The entrapment inserts thus enable study of the function of genes that when mutated do not have major mutant phenotypes. Such systems also allow for the selection of inserts in specific classes of genes, based on their expression pattern and aid subsequent analyses, such as double mutant isolation, which may display phenotypes predicted specifically by the expression pattern.

Enhancer trap elements have been developed to identify enhancers in the genome, which are capable of orientation independent transcriptional activation from a distance. The enhancer trap transposons contain a minimal promoter, upstream of a reporter/marker gene that is placed near one end (left or right end) of the transposon. An *Ac-Ds* system was constructed with a cauliflower mosaic virus (CaMV) minimal promoter GUS placed at the left end of *Ds* (16). The *Ds* element carrying a hygromycin marker is inserted in the chlorsulfuron resistance gene to select excisions, as outlined in *Table 1*. The *Ac TPase* construct contains the *IAAH* gene as a negative selection marker, so that stable transposed *Ds* elements can be selected in the progeny. This selection system led to the isolation of a novel gene, *LRP1*, which is involved in root development (76). Insertion into this gene does not cause a visible mutant phenotype, probably because it belongs to a small gene family and is redundant, thus exemplifying the utility of enhancer entrapment for function identification. Extensive populations of similar stable transposed *Ds* enhancer traps have been generated and used to study the expression pattern conferred by adjacent DNA in Arabidopsis, with about 50% displaying marker gene activity (77). These constructs have also been introduced into Micro-Tom, a model genotype for tomato research (78).

Gene trap inserts are designed to create fusion transcripts with the target gene (75). A splice acceptor site is placed at the 5′ end of the reporter gene, which enables the generation of fusion transcripts (from genomic sequences) starting upstream. Such gene- or exon-trap *Ds* transposons (77) have been produced in Arabidopsis and resulted in the isolation of *PROLIFERA* (79) and *MEDEA* (80), showing the versatility of the system in detecting interesting genes based primarily on their expression pattern. In the system described (77) selection is for transposition only and not for excision, as the use of the marker combination results in selection of transposed *Ds* elements segregated from the donor T-DNA locus. Such stable transposed *Ds* gene traps are convenient to generate and screen for expression patterns of interest, as about 25% show expression in some specific tissue or developmental stage.

4.3 Chromosomal engineering

The initial discovery of transposable elements was based on chromosomal breakage and rearrangements caused by the *Ac-Ds* system. A few of the structures and mechanisms involved in chromosome rearrangements are now understood, mainly due to the use of transgenic plants containing specific constructs to test the possible mechanisms (81). Thus, deletions and inversions have been produced in transgenic plants using specific double-*Ds* like conformations.

The same effect has been produced by the Cre-lox site-specific recombination system, adapted from the *E. coli* bacteriophage P1, using substrates carried by the engineered *Ds* (82). The *Ds*-lox element containing the lox site for recombination, which has transposed to a linked position near the donor T-DNA containing the lox site, is introduced to the Cre recombinase-producing line by crossing. The lox sites on the T-DNA and on the transposed *Ds*, or between two *cis*-linked or unlinked *Ds* elements, are then used to induce lox–lox recombination, which can cause inversions, deletions or translocations depending on the location and orientation of the lox sites to each other. Thus, introduction of a heterologous recombination system within the transposons enables controlled chromosome rearrangement or engineering.

In model plants such chromosomal engineering schemes can help in identifying genes within specific regions, for example, genes where mapping is not easy due to the quantitative nature of the trait. In crop plants this can also provide necessary chromosome structures that can be employed for plant breeding purposes. The prospects of applying these strategies have been discussed (83) and are in progress.

5 Isolation of tagged genes

5.1 Proof that a gene has been tagged

Mutants tagged with autonomous transposons like *Ac* or *En/Spm* and non-autonomous *Ds* and *I/dSpm* in the presence of the TPase, may display a tagged phenotype as characterized by variegation or chimeric wild-type and mutant clonal sectors. This variegation is often the easiest test for a tagged mutant before any genetic tests are carried out, but is mostly expected for genes that display a cell-autonomous mutant phenotype.

The causal relationship of the mutant phenotype and a specific transposon is best tested by co-segregation analysis (12). This is done by DNA blot analysis using as a hybridization probe, first the transposon and then the putative gene after isolation. A transposon-flanking DNA fragment can be obtained by inverse PCR (iPCR) (52) as described in *Protocols* 8 and 9, or plasmid rescue (57) as outlined in *Protocols* 4 and 6 if the transposon contains sequences for maintenance and selection in bacteria (i.e. replication origin and antibiotic resistance). Presence of a marker gene within the transposon can aid analysis when no other transposon copies are present in the genome. PCR-based analysis can also be done, using primers from the transposon directed outwards and primers from the putative gene sequences to distinguish the different alleles in segregation experiments.

The critical proof comes from an analysis of a number of independent revertant alleles whose EDS sequence should reveal a wild-type gene product, for example, identical or similar amino-acid exchange. This establishes the absolute correspondence of genotype and phenotype, and proves that the transposon insert is responsible for the tagged phenotype. Other derivative alleles, which display a mutant phenotype and contain an EDS sequence of a mutant allele, for example, stop-codons, frame-shifts and variant amino-acids, contribute additional evidence that the transposon tagged allele determines the mutant phenotype.

With an efficient targeted tagging system, the isolation of a number of independently tagged mutants with insertions at different positions within the target locus, as with the tagging of the *Cf-9* gene, is a very convincing proof of gene tagging. The isolation of a cDNA corresponding to the tagged locus and analysis of gene structure helps in interpreting the EDS sequence, by identifying the proper reading frame, position of insertion in the gene and prediction of essential portions of the gene. Corroborative evidence of gene function in relation to the mutant phenotype can be provided by RNA blots or PCR based analysis of gene expression. Finally, complementation of a mutant, for example a chemically induced allele if available, can be used as evidence of having tagged the gene. This might be more convenient and faster in cases where germinal reversions are difficult to come by and when generations times are long. Also, the sequencing of a number of independently derived mutant alleles, obtained by chemical mutagenesis or other means, can be taken as substantial evidence in proving the relationship between sequence and phenotype.

Protocol 7

Analysis of a putative transposon tagged mutant

Method

1. Harvest leaf material from a mutant for DNA isolation as described in Protocol 1.
2. Cross the mutant with wild type (Cross 1) and with a TPase line (Cross 2) if the mutant was not known to express TPase.
3. Follow segregation of the phenotype in the progeny and screen for presence of the *TPase* locus, for example, by selecting for antibiotic resistance.
4. If the mutant contains the *TPase* locus, screen 100–1000 progeny for wild-type-looking revertants to test the stability of the mutant phenotype.
5. If the mutant does not contain a *TPase* locus, screen siblings to find a family expressing TPase. When found, screen the progeny as in step 2. Alternatively,

Protocol 7 continued

screen the F_2 from Cross 2 for families with TPase. When found, screen progeny as in step 2. Only proceed when revertants are found.

6 Identify a transposon insertion that co-segregates with the mutation. Preferably, use a population without TPase, for example, an F_2 from Cross 1, that segregates 3:1 for wild type:mutant. Alternatively, use a population with TPase, such as the F_2 from Cross 2, or revertant and mutant progeny of the original mutant when the mutant contained the *TPase* locus.

7 Perform a Southern blot analysis on about 50 plants for multiple transposon sources (*En/I*), or on about 10 plants (half mutant, half wild type revertant) for a *Ds* tagged mutant that carries single or a few inserts. Load equal amounts of DNA per lane to distinguish between homozygous and hemizygous inserts. Make use of other populations if no co-segregating transposon can be identified.

8 Isolate genomic DNA flanking the co-segregating transposable element by iPCR as described in *Protocol 8* and *9* using preferably DNA template from a plant lacking the *TPase* locus and carrying less than five copies of transposon inserts. If no plants with less than five inserts can be found, use a backcross of the mutant with wild type (F_1 Cross 1 × wild type) to reduce the transposon copy number.

9 Confirm the cloning of genomic DNA flanking the transposon by hybridizing the iPCR probe to a Southern blot containing DNA from mutant and revertant plants, to reveal homozygous inserts in mutant and hemizygous or no inserts in revertant plants.

10 To confirm tagging of the gene, first determine the sequence of DNA fragments carrying genomic DNA flanking the transposon insert using the iPCR fragments as template. Then design PCR primers for the amplification of the insertional target site from wild-type DNA.

11 PCR amplify the target site sequences from wild-type, revertant and mutant alleles without inserts.

12 Clone the PCR products and determine their DNA sequence. All revertants should have at least one allele with (near) wild type DNA sequence. All mutants should have only alleles featuring frame shifts, aberrant termination, or amino acid exchanges.

13 Isolate genomic and cDNA clones from appropriate lambda phage libraries using the iCR products as probes, and determine their DNA sequence.

5.2 Cloning a tagged gene

A stepwise procedure outlined in *Protocol 7* can be used as a general strategy for cloning a transposon tagged gene and is applicable for the analysis of recessive mutations, which do not affect fertility. For rare dominant transposon induced mutants, and infertile or unviable mutants (such as embryo lethals), these procedures will have to be adjusted. iPCR (39) is a reliable way to isolate DNA sequences flanking a transposon insert and is described in *Protocol 8* and *9*. Plants with homozygous, as well as hemizygous inserts can be used for iPCR. The iPCR-derived flanking DNA fragments are then used as probes for southern DNA hybridization analysis of wild type, mutant and revertant plants to confirm the successful cloning of fragments from the tagged gene. DNA sequence information of the flanking DNA and gene prediction offers a quick check to test if the flanking DNA corresponds to the mutant. Upon insertion, both *Ds* and *I* transposons generate a target site duplication of 8 bp and 3 bp, respectively, and after excision often delete or add a few base pairs leading to a frameshift, or generate a stop codon. An adequate proof for cloning the correct gene can therefore be obtained by correlating the sequence of excision alleles with the plant phenotype. Revertant plants should have at least one allele encoding a wild-type-like protein, whereas both alleles of a mutant should display an aberrant reading frame.

Protocol 8
iPCR of I transposons

Reagents

- *I* element IPCR primers

1st PCR:	ILJ1:	GAATTTAGGGTAACA TTCATAAGAGTGT
	IRJ1:	TTGTGCTGTTATGGAG GCTTCCCATC
2nd PCR:	ILJ2:	ATTAAAAGCGTCGGT TTCATCGGGA
	IRJ2:	AGGTAGCTTACTGAT GTGCGCGC
3rd PCR:	ITIR:	GACACTCCTTAGATCT TTTCTTGTAGTG

- Taq DNA polymerase and buffer
- *Hin* fI restriction enzyme and buffer
- Spermidine 1 mM
- dNTP mixture containing 2.5. mM each of dATP, dCTP, dGTP and TTP
- DNA polymerase I Klenow fragment I u/μl
- Phenol/chloroform (24:1) (v/v)
- Sodium acetate 0.3 M
- isopropanol
- 70% (v/v) ethanol
- sterile distilled H_2O
- ×10 ligation buffer
- T4 DNA ligase 5u/μl
- NaCl, 0.45 M
- *Sal* 1 restriction endonuclease

Method

1. Digest 300 ng genomic DNA (protocol 1) with 20 units *Hin* fI in 100 μl 1× restriction buffer containing 1 mM spermidine for 3 h at 37 °C.
2. Add 1 μl 2.5 mM dNTPs and 1 unit of DNA polymerase I Klenow fragment. Incubate for 15 min at room temperature.
3. Extract the DNA once with phenol/chloroform.

Protocol 8 continued

4 Precipitate the DNA from the aqueous phase by addition of 1/10 volume 0.3 M sodium acetate and 1 volume isopropanol.

5 Incubate at −20 °C for 20 min (24).

6 Pellet the DNA by centrifugation at full speed in a micro-centrifuge for 15 min.

7 Wash the pellet in 70% ethanol, and air dry.

8 Resuspend the pellet in 90 μl sterile distilled H_2O, add 10 μl 10× ligation buffer (24), 5 units of T4 DNA ligase and self-ligate DNA fragments overnight at 16 °C.

9 Inactivate the DNA ligase by heating the sample at 65 °C for 10 min.

10 Digest half of the DNA sample by adding 25 μl 0.45 M NaCl and 10 units of *Sal* I, and incubate for 3 h at 37 °C.

11 Precipitate both undigested and *Sal* I-digested DNA samples by addition of 1/10 volume 0.3 M sodium acetate and 1 volume of isopropanol.

12 Collect the DNA pellet by centrifugation at full speed in a micro-centrifuge for 15 min.

13 Wash the DNA pellet with 70% ethanol, air dry and resuspend DNA in 30 μl H_2O.

14 Transfer DNA template into a PCR tube and add 4 μl 10× PCR buffer, 25 pmol primer ILJ1 and IRJ1, and 2 μl dNTPs (2.5 mM each).

15 Set up the PCR reaction starting with a 5 min 95 °C hot-start, then add 10 μl of 1× PCR buffer with 1 unit of Taq DNA polymerase (Supertaq). Set 25 cycles of PCR: 1 min 95 °C, 1 min 55 °C, 3 min 72 °C followed by an elongation step for 5 min at 72 °C.

16 Transfer 2 μl aliquots to a new PCR tube. Add 38 μl 1× PCR buffer, containing 25 pmol of primers ILJ2 and IRJ2, 2 μl dNTPs (2.5 mM each) and 1 unit of Taq polymerase. Run the 2nd PCR for 25 cycles using the conditions described in step 7.

17 Take a 2 μl aliquot and conduct a 3rd round of PCR in a 50 μl PCR reaction, containing 20 pmol of ITIR primer with the same conditions as step 7, but with annealing at 50 °C instead of 55 °C.

18 Load 10 μl of the PCR reaction on a 1.2% TBE-agarose gel to identify IPCR fragments and photograph. To clone fragments, there are three alternatives:
   (a) excise from gel, purify and clone the fragments in a PCR fragment cloning vector;
   (b) phenol:chloroform-treat the PCR reaction, sodium acetate/isopropanol precipitate DNA and shotgun clone PCR fragments;
   (c) digest the DNA precipitated from step (b) with *Bgl* II, a restriction site in the primers

Protocol 9

iPCR of Ds elements

Reagents

- *Ds* primer sequences:
  5′ Outward: GATAACGGTCGGTACGGGAT
  5′ Inward: CGGGATGATCCCGTTTCGTT
- Restriction enzymes, *Hae* III, *Sau* 96, *Nla* III, *Bam* HI or *Cla* I and buffer
- Phenol/chloroform (24:1) (v/v)
- Sodium acetate 0.3 M
- isopropanol
- 70% (v/v) ethanol
- sterile distilled H_2O
- 10× ligation buffer
- T4 DNA ligase 5 u/μl
- NaCl, 1 M

Method

1. Digest 200–400 ng genomic DNA with 10 units of *Hae* III, *Sau* 96 or *Nla* III at 37 °C for 2 h in 100 μl final volume.
2. Extract with phenol:chloroform.
3. Precipitate the DNA from the aqueous phase by addition of 1/10 volume 0.3 M sodium acetate and 1 volume isopropanol.
4. Pellet the DNA by centrifugation at full speed in micro-centrifuge for 15 min at room temperature.
5. Wash the DNA pellet with 70% ethanol and air dry.
6. Redissolve the DNA in 90 μl H_2O add 10 μl 10× Ligase buffer and 4 units T4 DNA ligase.
7. Incubate overnight at 16 °C.
8. Heat inactivate the DNA ligase by incubation at 65 °C for 15 min.
9. Add NaCl to 100 mM and digest with 10–20 units *Bam* HI or *Cla* I enzymes for 2 h at 37 °C.
10. Phenol:chloroform extract the DNA sample.
11. Precipitate and wash the DNA as described in steps 3–5.
12. Dissolve the DNA pellet in 10 μl H_2O.
13. Set up PCR with 10 μl DNA in 50 μl reactions with 0.4 μM *Ds* primers and reaction conditions exactly as given in *Protocol 5* for T-DNA iPCR.
14. Clone and sequence the IPCR fragments as in *Protocol 8* step 18.

6 Reverse genetics strategies to identify mutants

6.1 Target selected gene inactivation

Many molecular biology cloning experiments identify genes and produce extensive DNA sequence information, for example, from characterization of specifically expressed cDNAs, from comparison of multigene families or from genome sequencing projects (84). The expression pattern of the genes gives an idea of their involvement in a specific developmental stage or physiological response. In plants, anti-sense or sense co-suppression experiments can be used to reveal a gene inactivation phenotype, but for multi-gene families it often gives inconclusive results. As a general strategy it is too time-consuming for making a large number of transformants that have to be individually tested. Further reverse genetics strategies for functional analysis, like exhibition of a mutant phenotype, can be provided by a knockout mutation of the gene. In bacteria, yeast and animals homologous recombination is a convenient way to produce a mutation in a specific gene of interest, but in plants, this procedure is only routinely possible in the moss *Physcomitrella patens* (see Volume 2, Chapter 14) In the majority of plant species, insertional mutagenesis is an alternative option. A pre-requisite is that a large saturated population of inserts should be available so that a specific insertion in the target gene can be obtained.

The reverse genetics technique to search for insertions in genes whose sequence is known, is termed 'target selected gene inactivation' (TSGI) and has been developed in a number of eukaryotic transposon systems (85, 86). The selection of an insert in the target gene is undertaken by PCR, using a primer from the gene and another primer from the insertion sequence, for example transposon or T-DNA. As the PCR screening cannot be done on numerous individual plants, pools of plants are used for the PCR analysis and an individual is selected that displays a junction fragment of the specific target gene and the transposon. Strategies that systematically pool individuals three-dimensionally in blocks, rows and columns enable a direct identification by PCR of the individual containing the insert. The individuals containing the homozygous insert can be recovered and analysed for mutant phenotype to address the question of the function of the gene.

In maize, snapdragon and petunia, the large number of endogenous transposon copies provides genome saturation. TSGI has been shown to work in these plants to identify knockout mutations in specific genes. In Arabidopsis a population of T-DNA's (5300 transformants) has also been used to identify insertions in two members of the actin multi-gene family (87). The insertion configurations of T-DNA's are often complex, with tandem repeat insertion structures and heterogeneity of the left border junctions. Transposon inserts in contrast have a defined terminal sequence making them all accessible to PCR selection.

Genome saturation with heterologous transposons is not automatically achieved as the T-DNA delivers a few copies, which yield transposed copies in the genome. One method for saturation is accumulation of high copy number transposons by the occasional amplifications or transposon bursts, which have been documented to occur in transgenic plants, for example, *Ac* in tomato (88) and *En-I* in Arabidopsis (39). Subsequent inter-crosses between high element lines increase the element copy additively. The independent transpositions that occur in families made from the high element lines will yield a large population of independent inserts for genome saturation.

Multiple *I/dSpm* element lines with about 20 elements per plant and autonomous *En* lines with about 10 transposons per plant have been produced in Arabidopsis that are similar to the maize *Mutator* (89) and petunia *Tph1* (90) mutagenesis systems. With multiple transposons per plant, populations of between 2500 and 5000 lines are available for screening for specific inserts. The transposon populations are used to make pools of DNA that can be used for screening for inserts, either as a two step process (46) or a one-step process (44). A general method used in screening, such transposon libraries is outlined in *Protocol 10*. The number of primers to use for a gene depends on its size, about 2–2.5 kb can be reproducibly amplified per primer, so for most genes, even with only cDNA/EST information, a primer from the 5′ and the 3′ region should suffice for screening.

The local transposition behaviour of the maize transposons can also be applied to saturate specific genomic regions. Thus, mapped *Ds* or *I/dSpm* transposons in the genome can serve as donor sites for production of saturated populations of inserts in the linked chromosomal regions. This method has been used for local mutagenesis of specific regions in the Arabidopsis genome, for example, on chromosome 3 (91) and chromosome 5 (92).

Protocol 10

Reverse genetics PCR screen for transposon inserts

Reagents

- Transposon primers (*En-I/Spm* system):
  ITIR2: CTTTGACGTTTTCTTGTAGTG
  ITIR3: CTTGCCTTTTTTCTTGTAGTG
- Gene specific primer for annealing at 60 °C, from 5′ or 3′ of gene of interest
- 10× Taq polymerase buffer: 100 mM Tris–HCl, pH 9.0, 15 mM $MgCl_2$, 500 mM KCl, 1% Triton X-100, 0.1% (w/v) gelatin (SuperTaq)
- Sodium acetate 0.3 M
- Isopropanol

Method

1. Grow transposon populations in the greenhouse to make the DNA pools, using a suitable three-dimensional multiplex pooling strategy (46, 90), with each pool containing about 100 plants. A population of 3000 plants would therefore yield three sets of 10 × 10 × 10 (blocks × columns × rows) that make a total of 90 DNA pools.
2. Prepare DNA from pools in a scale-up of the procedure described in *Protocol 1*, followed by a phenol:chloroform extraction and sodium acetate/ isopropanol precipitation (see *Protocol 4*)

Protocol 10 continued

3 Dilute the DNA to a working concentration of about 10 ng/μl.

4 Make a PCR reaction mix for 100 (with 10 extra) reactions of 25 μl final reaction volumes. This would require a total mix of 2 ml comprising: 250 μl of 10× Taq polymerase buffer, 25 μl of dNTPs (25 mM stock), 2 nmol of each transposon primer, 2 nmol of a gene specific primer and 50 units Taq Polymerase (SuperTaq). Set up the PCR reaction in micro-titre plates (Biozyme, No 170651) adding 20 μl reaction mix to each PCR well and finally 5 μl of pool DNA.

5 Run PCR in a thermal cycler at conditions: hot start at 94 °C for 4 min, 30 cycles of 94 °C for 45 s, 60 °C for 45 s, 72 °C for 3 min followed by an extension step of 72 °C for 7 min.

6 Add 5 μl of 6× Loading buffer (24) and load 15 μl of each sample onto a 1.5 % TBE gel. The 90 samples can be loaded in a single gel tray divided into three compartments with each compartment loaded with samples from a set of 3-dimensions (Block, Row, Column). Run gel overnight, alkali transfer (0.4 M NaOH) DNA onto Hybond N+ membrane using a Vacuum blotter (Pharmacia) or by conventional Southern blotting.

7 Hybridize blot with gene specific probe and analyse hybridization patterns. Hybridizing fragments of the same size that are revealed in all three dimensions indicate the 'address' of a putative insertion within the pools.

8 Sow out progeny plants (10–25) from the 'address' specified by the hybridization data. Prepare DNA from individual plants, run PCR reactions (scale down step 3) and continue with steps 4–5.

9 Identify individual plants containing the transposon insert. Examine this progeny population to identify plants homozygous and hemizygous for the insertion. Make crosses to wild-type, if necessary, to cross out the transposase.

10 Examine the plants with various genotypes (e.g. ± transposase, siblings lacking transposon insert and revertant lines) for mutant phenotypes. Confirm tagging by co-segregation and revertant analysis.

6.2 Sequencing of transposon flanking DNA to identify tagged genes

The genomes of a few model plant species are now available, and substantial EST sequence data is available for Arabidopsis, rice, tomato and maize. All these sequencing initiatives are generating a large sequence database resource to which gene functions can be attached.

A number of methods are available to efficiently isolate transposon flanking DNA, for example, iPCR (*Protocol 8* and 9), plasmid rescue (*Protocol 6*), TAIL PCR (93) and linker-ligation-mediated PCR (94). The sequencing of these transposon-flanking fragments followed by searches for homology in sequence databases reveals the identity of the tagged sequence. In compact genomes like Arabidopsis 30–50% of the inserts are in predicted or known genes (95–97, 44). The potential of this approach can now be fully realized with the availability of genome sequences. It is also more cost effective because inserts in genes will be identified only once without researchers trying to repeatedly isolate inserts in the same gene from different insertion populations around the world. Consequently, the availability of such a public resource database for transposon tagged sites will be a tremendous resource for plant biologists in the future.

References

1. McClintock, B. (1948). *Carnegie Inst. Wash. Yearbook*, **47**, 155.
2. Kunze, R., Stochaj, U., Lauf, J., and Starlinger, P. (1987). *EMBO J.*, **6**, 1555.
3. Peterson, P. A. (1953). *Genetics*, **38**, 682.
4. McClintock, B. (1954). *Carnegie Inst. Wash. Yearbook*, **54**, 254.
5. Pereira, A., Cuypers, H., Gierl, A., Schwarz-Sommer, Zs., and Saedler, H. (1986). *EMBO J*, **5**, 835.
6. Masson, P., Rutherford, G., Banks, J., and Fedoroff, N. (1989). *Cell*, **58**, 755.
7. Coen, E. S., Robbins, T. P., Almeida, J., Hudson, A., and Carpenter, R. (1989). In *Mobile DNA* (ed. D. E. Berg and M. M. Howe), p. 413. *Am. Soc. Microbiol.*, Washington DC.
8. Gerats, A. G. M., Huits, H., Vrijlandt, E., Maraña, C., Souer, E., and Beld, M. (1990). *Plant Cell*, **2**, 1121.
9. Tsay, Y-F., Frank, M. J., Page, T., Dean, C., and Crawford, N. M. (1993). *Science*, **260**, 342.
10. Baker, B., Schell, J., Lörz, H., and Fedoroff, N. (1986). *Proc. Natl Acad. Sci. USA*, **83**, 4844.
11. Haring, M. A., Rommens, C. M. T., Nijkamp, H. J. J., and Hille, J. (1991). *Plant Mol. Biol.*, **16**, 449.
12. Walbot, V. (1992). *Annu. Rev. Plant Physiol. Plant Mol. Biol.*, **43**, 49.
13. Baker, B., Coupland, G., Fedoroff, N., Starlinger, P., and Schell, J. (1987). *EMBO J.*, **6**, 1547.
14. Jones, J. D. G., Carland, F. M., Maliga, P., and Dooner, H. K. (1989). *Science*, **244**, 204.
15. Haring, M. A., Gao, J., Volbeda, T., Rommens, C. T. M., Nijkamp, H. J. J., and Hille, J. (1989). *Plant Mol. Biol.*, **13**, 189.
16. Fedoroff, N. V. and Smith, D. L. (1993). *Plant J.*, **3**, 273.
17. Scofield, S. R., Jones, D. A., Harrison, K., and Jones, J. D. G. (1994). *Mol. Gen. Genet.*, **244**, 189.
18. Jones, J. D. G., Jones, D. A., Bishop, G. J., Harrison, K., Carroll, B. J., and Scofield, S. R. (1993). *Transgen. Res.*, **2**, 63.
19. Finnegan, E. J., Taylor, B. H., Craig, S., and Dennis, E. S. (1989). *Plant Cell*, **1**, 757.
20. Spena, A., Aalen, R. B., and Schulze, S. C. (1989). *Plant Cell*, **1**, 1157.
21. Pereira, A., Aarts, M., Van Agtmaal, S., Stiekema, W. J., and Jacobsen, E. (1991). *Maydica*, **36**, 323.
22. Charng, Y-C., Pfitzner, U. M., and Pfitzner, A. J. P. (1995) *Plant Sci.*, **106**, 141
23. Schmidt, R. and Willmitzer, L. (1989). *Mol. Gen. Genet.*, **220**, 17.
24. Sambrook, J., Fritsch, E. F., and Maniatis, T. (ed.) (1989) *Molecular cloning: a laboratory manual*, 2nd edn. Cold Spring Harbor Laboratory Press, New York.
25. Chuck, G., Robbins, T., Nijjar, C., Ralston, E., Courtney-Gutterson, N., and Dooner, H. K. (1993). *Plant Cell*, **5**, 371.
26. Whitham, S., Dinesh-Kumar, S. P., Choi, D., Hehl, R., Corr, C., and Baker, B. (1994). *Cell*, **78**, 1101.
27. Lawrence, G. J., Finnegan, E. J., Ayliff, M. A. and Ellis, J. G. (1995). *Plant Cell*, **7**, 1195.

28. Lawson, E. J. R., Scofield, S. R., Sjodin, C., Jones, J. D. G., and Dean, C. (1994). *Mol. Gen. Genet.*, **245**, 608.
29. James, D. W. Jr, Lim, E., Keller, J., Plooy, I., Ralston, E., and Dooner, H. K. (1995). *Plant Cell*, **7**, 309.
30. Masterson, R. V., Furtek, D. B., Grevelding, C., and Schell, J. (1989). *Mol. Gen. Genet.*, **219**, 461.
31. Swinburne, J., Balcells, L., Scofield, S. R., Jones, J. D. G., and Coupland, G. (1992). *Plant Cell*, **4**, 583.
32. Long, D., Swinburne, J., Martin, M., Wilson, K., Sundberg, E., Lee, K., and Coupland, G. (1993). *Mol. Gen. Genet.*, **241**, 627.
33. Long, D., Martin, M., Sundberg, E., Swinburne, J., Puangsomlee, P., and Coupland, G. (1993). *Proc. Natl Acad. Sci. USA*, **90**, 10370.
34. Balcells, L., Sundberg, E., and Coupland, G. (1994). *Plant J.*, **5**, 755.
35. Pereira, A. and Saedler, H. (1989). *EMBO J.*, **8**, 1315.
36. Masson, P. and Fedoroff, N. (1989). *Proc. Natl Acad. Sci. USA*, **86**, 2219.
37. Frey, M., Tavantzis, S. and Saedler, H. (1989). *Mol. Gen. Genet.*, **217**, 172.
38. Cardon, G. H., Frey, M., Saedler, H., and Gierl, A. (1993). *Plant J.*, **3**, 773.
39. Aarts, M. G. M., Corzaan, P., Stiekema, W. J., and Pereira, A. (1995). *Mol. Gen. Genet.*, **247**, 555.
40. Aarts, M. G. M., Dirkse, W., Stiekema, W. J., and Pereira, A. (1993). *Nature*, **363**, 715.
41. Aarts, M. G. M., Keijzer, C. J., Stiekema, W. J., and Pereira, A. (1995). *Plant Cell*, **7**, 2115.
42. Byzova, M. V., Franken, J., Aarts, M. G. M., de Almeida-Engler, J., Engler, G., Mariani, C., Campagne, M. M. V., and Angenent, G. C. (1999). *Genes Develop.*, **13**, 1002.
43. Kubo, H., Peeters, A. J. M., Aarts, M. G. M., Pereira, A., and Koornneef, M. (1999). *Plant Cell*, **11**, 1217.
44. Speulman, E., Metz, P. L. J., Arkel, G. van, Lintel Hekkert, B. te, Stiekema, W. J., and Pereira, A. (1999). *Plant Cell*, **11**, 1853.
45. Wisman, E., Hartmann, U., Sagasser, M., Baumann, E., Palme, K., Hahlbrock, K., Saedler, H., and Weisshaar, B. (1998). *Proc. Natl Acad. Sci. USA*, **95**, 12432.
46. Baumann, E., Lewald, J., Saedler, H., Schulz, B., and Wisman, E. (1998). *Theor. Appl. Genet.*, **97**, 729.
47. Greenblatt, I. M. (1984). *Genetics*, **108**, 471.
48. Jones, J. D. G., Carland, F. M., Lin, E., Ralston, E., and Dooner, H. K. (1990). *Plant Cell*, **2**, 701.
49. Osborne, B. I., Corr, C. A., Prince, J. P., Hehl, R., Tanskley, S. D., McCormick, S., and Baker, B. (1991). *Genetics*, **129**, 833.
50. Keller, J., Lim, E., James, D. W. Jr, and Dooner, H. K. (1992). *Genetics*, **131**, 449.
51. Jacobs, J. M. E., Van Eck, H. J., Arens, P., Verkerk-Bakker, B., te Lintel-Hekkert, B., Bastiaanssen, H. J. M., *et al.* (1995). *Theor. Appl. Genet.*, **91**, 289.
52. Earp, D. J., Lowe B., and Baker B. (1990). *Nucl. Acids Res.*, **18,** 3271.
53. Thomas, C. M., Jones, D. A., English, J. J., Carroll, B. J., Bennetzen, J. L., Harrison, K., *et al.* (1994). *Mol. Gen. Genet.*, **242**, 573.
54. Koncz, C., Martini, N., Szabados, L., Hrouda, M., Bachmair, A., and Schell, J. (1994). In *Plant molecular biology manual* (ed. S. B. Gelvin and R. A. Schilperoort). p. B2, 1. Kluwer Academic Publishers, Dordrecht.
55. Bevan, M. (1984). *Nucl. Acids Res.*, **12**, 8711.
56. Rommens, C. M. T., Rudenko, G. N., Dijkwel, P. P., van Haaren, M. J. J., Ouwerkerk, P. B. F., Blok, K. M., *et al.* (1992). *Plant Mol. Biol.*, **20**, 61.
57. Lander, E. S., Green, P., Abrahamson, J., Barlow, A., Daly, M. J., Lincoln, S. E., and Newburg, L. (1987). *Genomics*, **1**, 174.
58. Stam, P. (1993). *Plant J.*, **3**, 739.

59. http://nasc. nott. ac. uk/
60. Tanksley, S. D., Ganal, M. W., Prince, J. P., de Vicente, M. C., Bonierbale, M. W., Broun, P., *et al.* (1992). *Genetics*, **132**, 1141.
61. Knapp, S., Larondelle, Y., Rossberg, M., Furtek, D., and Theres, K. (1994). *Mol. Gen. Genet.*, **243**, 666.
62. Bancroft, I. and Dean, C. (1993). *Genetics*, **134**, 1221.
63. Long, L., Goodrich, J., Wilson, K., Sundberg, E., Martin, M., Puangsomlee, P., and Coupland, G. (1997). *Plant J.*, **11**, 145.
64. Bancroft, I., Jones, J. D. G., and Dean, C. (1993). *Plant Cell*, **5**, 631.
65. Feldmann, K. A. (1991). *Plant J.*, **1**, 71.
66. van der Biezen, E. A., van Haaren, M. J. J., Overduin, B., Nijkamp, H. J. J., and Hille, J. (1994). In *Plant molecular biology manual* (ed. R. A. Schilperoort and S. B. Gelvin), p. 1. Kluwer Academic Publishers, Doordrecht.
67. Briza, J., Carroll, B. J., Klimyuk, V. I., Thomas, C. M., Jones, D. A., and Jones, J. D. G. (1995). *Genetics*, **141**, 383.
68. Healy, J., Corr, C., DeYoung, J., and Baker, B. (1993). *Genetics*, **134**, 571.
69. Jacobs, J. M. E., Lintel Hekkert, B. te, El-Kharbotly, A., Jacobsen, E., Stiekema, W. J., and Pereira, A. (1994) In *Molecular and Cellular Biology of the Potato*, 2nd edn (ed. W. R. Belknap, M. E. Vayda, and W. D. Park), p. 21. CAB International, Wallingford.
70. Jones, D. A., Thomas, C. M., Hammond-Kosack, K. E., Balint-Kurti, P. J., and Jones, J. D. G. (1994). *Science*, **266**, 789.
71. Yoder, J. I. (1990). *Theor. Appl. Genet.*, **79**, 657.
72. Altmann, T., Felix, G., Jessop, A., Kauschmann, A., Uwer, U., Peña-Cortés, H. *et al.* (1995). *Mol Gen Genet*, **247**, 646.
73. Nowick, E. M. and Peterson, P. A. (1981). *Mol. Gen. Genet.*, **183**, 440.
74. Falk, A., Feys, B. J., Frost, L. N., Jones, J. D. G., Daniels, M. J., and Parker, J. E. (1999) *Proc. Natl Acad. Sci. USA*, **96**, 3292.
75. Skarnes, W. C. (1990). *Biotechnology*, **8**, 827.
76. Smith, D. L. and Fedoroff (1995). *Plant Cell*, **7**, 735.
77. Sundaresan, V., Springer, P., Volpe, T., Haward, S., Jones, J. D. G., Dean, C., Ma, H., and Martienssen, R. (1995). *Genes Dev.*, **9**, 1797.
78. Meissner, R., Jacobsen, Y., Melamed, S., Levyatuv, S., Shalev, G., Ashri, A., Elkind, Y., and Levy, A. (1997). *Plant J.*, **12**, 1465.
79. Springer, P. S., McCombie, W. R., Sundaresan, V., and Martienssen, R. A. (1995). *Science*, **268**, 877.
80. Grossniklaus, U., Vielle-Calzada, J. P., Hoeppner, M. A., and Gagliano, W. B. (1998). *Science*, **280,** 446.
81. English, J. J., Harrison, K., and Jones, J. D. G. (1995). *Plant Cell*, **7**, 1235.
82. Osborne, B. I., Wirtz, U., and Baker, B. (1995). *Plant J.*, **7**, 687.
83. Haaren, M. J. J. van and Ow, D. W. (1993). *Plant Mol. Biol.*, **23**, 525.
84. Schmidt, R., West, J., Love, K., Lenehan, Z., Lister, C., Thompson, H., *et al.* (1995). *Science*, **270**, 480.
85. Ballinger, D. G. and Benzer, S. (1989). *Proc. Natl Acad. Sci. USA*, **86**, 9402.
86. Zwaal, R. R., Broeks, A., van Meurs, J., Groenin, J. T. M., and Plasterk, R. H. A. (1993). *Proc. Natl Acad. Sci. USA*, **90**, 7431.
87. McKinney, E. C., Ali, N., Traut, A., Feldmann, K. A., Belostotsky, D. A., McDowell, J. M., *et al.* (1995). *Plant J.*, **8**, 613.
88. Yoder, J. I. (1990). *Plant Cell*, **2**, 723.
89. Das, L. and Martienssen, R. (1995). *Plant Cell*, **7**, 287.
90. Koes, R., Souer, E., van Houwelingen, A., Mur, L., Spelt, C., Quattrocchio, F., *et al.* (1995). *Proc. Natl Acad. Sci. USA*, **92**, 8149.

91. Dubois, P., Cutler, S., and Belzile, F. J. (1998). *Plant J.*, **13**, 141.
92. Ito, T., Seki, M., Hayashida, N., Shibata, D., and Shinozaki, K. (1999). *Plant J.*, **17**, 433.
93. Liu, Y. G., Mitsukawa, N., Oosumi, T. and Whittier, R. F. (1995) *Plant J.*, **8**, 457.
94. Frey, M., Stettner, C. and Gierl, A. (1998) *Plant J.*, **13**, 717.
95. Mathur, J., Szabados, L., Schaefer, S., Grunenberg, B., Lossow, A., Jonas-Straube, E., Schell, J., Koncz, C., and Koncz-Kalman, Z. (1998). *Plant J.*, **13**, 707.
96. Martienssen, R. A. (1998) *Proc. Natl Acad. Sci. USA*, **95**, 2021.
97. http://www. jic. bbsrc. ac. uk
98. Coupland, G., Baker, B., Schell, J., and Starlinger, P. (1998). *EMBO J.*, **7**, 3653.
99. Scofield, S. R., Harrison, K., Nurrish, S. J., and Jones, J. D. G. (1992). *Plant Cell*, **4**, 573.
100. Bancroft, I., Bhatt, A. M., Sjodin, C., Scofield, S., Jones, J. D. G., and Dean, C. (1992). *Mol. Gen. Genet.*, **233**, 449.
101. Grevelding, C., Becker, D., Kunze, R., Von Menges, A., Fantes, V., Schell, J., and Masterson, R. (1992). *Proc. Natl. Acad. Sci. USA*, **89**, 6085.
102. Altmann, T., Schmidt, R., and Willmitzer, L. (1992). *Theor. Appl. Genet.*, **84**, 371.
103. Honma, M. A., Baker, B. J., and Waddel, C. S. (1993). *Proc. Natl. Acad. Sci. USA*, **90**, 6242.
104. Wilson, K., Long, D., Swinburne, J., and Coupland, G. (1996). *Plant Cell*, **8**, 659.
105. Machida, C., Onouchi, H., Koizumi, J., Hamada, S., Semiarti, E., Torikai, S., and Machida, Y. (1997). *Proc. Natl. Acad. Sci. USA.*, **94**, 8675.
106. Goldsbrough, A. P., Tong,Y., and Yoder, J. I. (1996). *Plant J.*, **9**, 927.
107. Izawa, T., Ohnishi, T., Nakano, T., Ishida, N., Enoki, H., Hashimoto, H., *et al.* (1997). *Plant Mol. Biol.*, **35**, 219.

Chapter 5

Cloning plant genes by genomic subtraction

Tai-ping Sun
Developmental, Cell and Molecular Biology Group, Department of Biology, Duke University, Durham, NC 27708, USA

1 Introduction

The genomic subtraction method was designed to isolate DNA sequences corresponding to a deleted region in a mutant (1). This method was originally developed using a mutant with a 5 kb deletion in *Saccharomyces cerevisiae*, which has a genome size of 2×10^4 kb (1). To apply this technique to clone genes in Arabidopsis with a genome size of 10^5 kb, we first carried out a reconstruction experiment in which biotinylated Arabidopsis DNA was used to subtract the same non-biotinylated DNA preparation to which a small amount of non-biotinylated adenovirus DNA had been added (2). Approximately, a thousand-fold enrichment of the adenovirus DNA was achieved after three cycles of subtraction. We then employed this technique to clone the *GA1* locus in Arabidopsis (2) using the *ga1-3* mutant, which was presumed to contain a large deletion based on genetic fine-structure mapping (3). After five cycles of subtraction using biotinylated *ga1-3* DNA and non-biotinylated wild-type Landsberg *erecta* (Ler) DNA, the resulting DNA fragments were amplified and cloned in pUC13, and individual clones were radiolabelled and used to probe a DNA blot containing *Hin* dIII-digested Ler and *ga1-3* DNAs. On average, one in twenty clones corresponded to a 5 kb deletion at the *GA1* locus in the *ga1-3* mutant (2). This indicates that the genomic subtraction technique detected a deletion corresponding to only 0.005% of the Arabidopsis genome (5 kb/10^5 kb).

One limitation of the genomic subtraction technique is that the mutant must carry a large deletion (at least 1 kb). Fortunately, ionizing radiation, such as fast neutrons (FN), X-rays and γ-rays, can induce large deletions or rearrangements of the genomic DNA in plants as summarized in *Table 1* (3–6; see also Volume 1 Chapter 1). The 5 kb deletion in the *ga1-3* mutant was generated by FN bombardment (3, 7, 8). Genetic fine-structure mapping as performed by Koornneef *et al.* (3) provided useful information, but it is extremely time-consuming. Recently, we have successfully isolated another Arabidopsis gene (*RGA*) using four arbitrarily chosen FN-induced *rga* mutant alleles without prior genetic evidence that

Table 1 Examples of mutations in Arabidopsis induced by ionizing radiation

Mutant allele	Mutation	Mutagenesis	Reference
abi3–6	0.75 kb deletion	FN (60 Gy[a])	(13)
era1–2 *era1–3*	7.5 kb deletion > 7.5 kb deletion	FN (60 Gy)	(14)
ga1–2 *ga1–3* *ga1–4*	> 3.4 kb insertion or inversion 5 kb deletion 12 bp deletion	FN (67 Gy) FN (67 Gy) FN (47 Gy)	(15, 16)
ga2–1	C to T	FN	(17, 18)
gl1–1	6.5 kb deletion	?	(8, 19)
hy4 (4 alleles)	> 5 kb deletion	FN (60 Gy)	(20)
ndr1	1 kb deletion	FN	(21)
rga-20 *rga-24* *rga-26*	> 33 kb deletion 8.4 kb deletion 5.9 kb deletion	FN (60 Gy)	(9)
tt5	inversion	FN	(22, 23)
chl3 (3 alleles)	> 5 kb deletions	γ-rays (30 kRad)	(24)
gai-d1 *gai-d2* *gai-d5* *gai-d7*	C to T 1 bp deletion 7 bp deletion 1 bp deletion	γ-rays (70 or 90 kRad)	(25, 26)
35S::tms2 (7 or 8 alleles)	= 5 kb deletion	γ-rays (40–60 kRad)	(27)
ctr1–2	17 bp deletion	X-rays (20 kRad)	(28)
gai-1	51 bp deletion	X-rays (44.7 kRad)	(15, 26)
tt3	inversion and deletions (52 bp and 7.4 kb)	X-rays	(22, 23)

[a] 10 Gy = 1 kRad.

any allele contained a deletion (9). Our studies illustrated that one could clone genes by genomic subtraction using several FN-induced mutant alleles without conducting fine-structure genetic mapping.

This chapter provides a detailed protocol for cloning Arabidopsis genes by using the genomic subtraction technique. This method could be applied to isolate genes corresponding to deletion mutations in other organisms, which have similar or smaller genome size than Arabidopsis. In fact, this technique has been applied to isolate strain-specific sequences, such as mating-type genes and avirulence genes, from micro-organisms (10, 11). However, because the degree of enrichment in Arabidopsis only reaches approximately 1000-fold, this method is probably not effective for organisms with relatively large genomes such as maize and peas. To isolate genes from larger genomes based on point mutations, another method called representational difference analysis (RDA) was developed (12). A brief description of RDA, and a comparison between this method and genomic subtraction are also included in this chapter.

2 The genomic subtraction technique

The genomic subtraction method and the subsequent screening procedures are diagrammed in *Figure 1*. An excess of sheared, biotinylated, deletion mutant DNA is denatured in the presence of a small amount of *Sau*3A I digested wild-type DNA (step A in *Figure 1*). The mixture is then allowed to re-associate. Most of the wild-type DNA strands anneal with complementary, biotinylated mutant DNA strands. In contrast, wild-type DNA strands that correspond to the deleted region in the mutant have no biotinylated complementary strand with which to anneal. In the next step, the biotinylated mutant DNA, and any wild-type DNA that has annealed with it, are removed from the sample by incubating the mixture with avidin-coated polystyrene beads (step B in *Figure 1*). The unbound DNA, which is now enriched for sequences that are absent in the deletion mutant, is collected. This fraction is then combined with an excess of fresh, biotinylated deletion mutant DNA and the mixture is again denatured, renatured and depleted of biotinylated sequences (step C in *Figure 1*).

After several cycles of subtraction, the remaining double-stranded DNA fragments are amplified by polymerase chain reaction (PCR) before further analysis. Oligonucleotide adaptors with *Sau*3A I compatible ends are ligated to the unbound *Sau*3A I fragments (step D in *Figure 1*), and one strand of the oligonucleotide adaptor is used as a primer for PCR (step E in *Figure 1*). The amplification products are cloned (step F in *Figure 1*) and individual clones are used as hybridization probes for DNA blot analysis to determine which clones contain sequences that correspond to deletions (step G in *Figure 1*).

When performing subtraction using multiple alleles in parallel, the most time-consuming step in the protocol is step G. In the case of cloning the *RGA* locus, we screened 25–40 clones for each of the four subtractive products (total of 129 clones) before identifying one clone that corresponded to a deletion. Suggestions for improving the subtraction efficiency will be discussed in Section 4.5.

Genomic subtraction allows attainment of much more dramatic enrichment than do previous subtractive hybridization methods (29–34). One feature critical to the success of our method is the use of avidin-coated polystyrene beads that bind a very high percentage of the biotinylated DNA and are easily removed from the sample. A second important feature is the addition of a DNA amplification step that allows multiple cycles of enrichment to be performed using small amounts of DNA while still yielding abundant products for further analysis. A third is that sequences with low melting temperatures that remain single stranded under the re-association conditions and palindromic sequences, which preferentially self-associate will not have the sticky ends needed for adaptor ligation and, consequently, will not be amplified.

3 Deletion mutagenesis

The presence of a deletion one kb or larger is essential for the success of genomic subtraction. In addition to generating the 5 kb deletion in the *ga1-3* mutant, FN

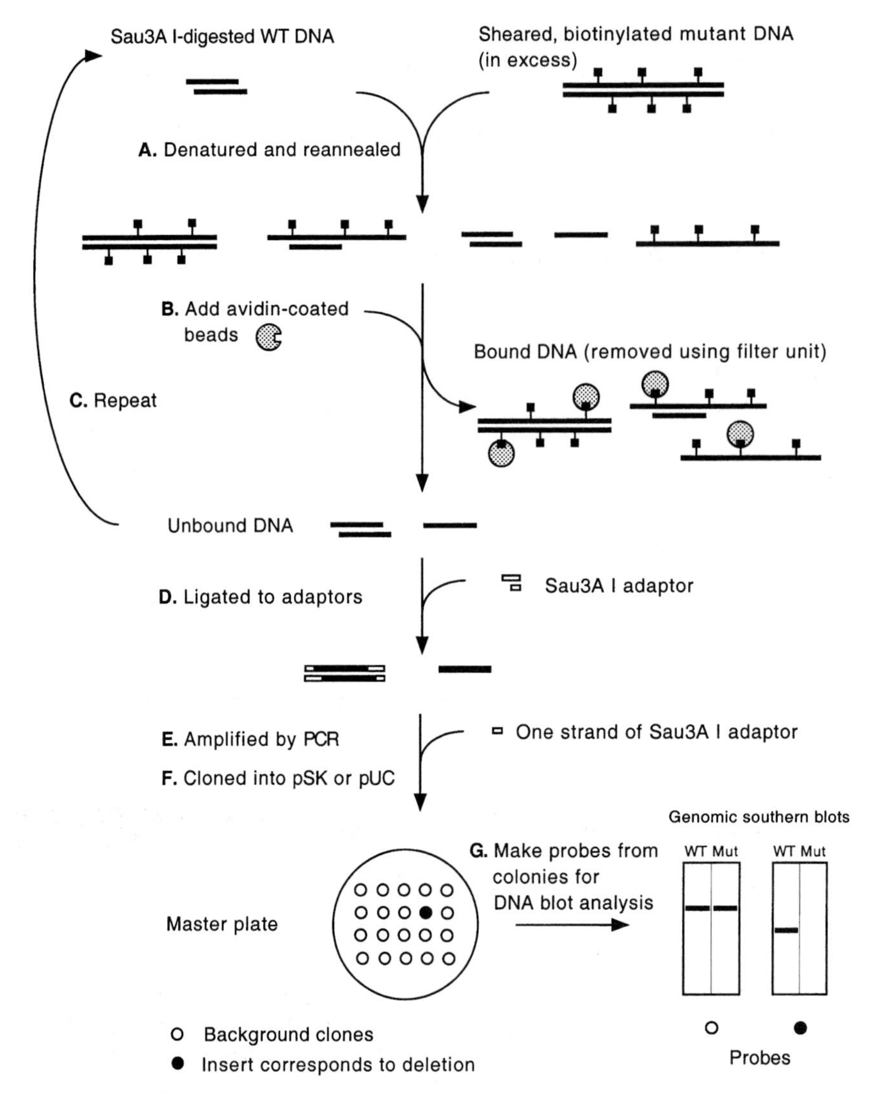

Figure 1. A schematic diagram of cloning genes corresponding to deletions by genomic subtraction. WT, wild-type; Mut, mutant.

bombardment has been shown to induce large deletions at other loci in Arabidopsis (*Table 1*). Bruggemann *et al.* (20) found that four of the nine FN-induced viable *hy4* mutants contain at least 5 kb deletions. In cloning of the *GA1* and *RGA* genes, we showed that one third of the FN alleles at both loci carry large deletions (9, 16). These results suggest that deletions suitable for genomic subtraction are present in 30–50% of the FN mutant alleles.

Other ionizing radiation, such as X-rays and γ-rays have also been illustrated to induce homozygous viable deletions in Arabidopsis at several loci (*Table 1*). Using *35S* promoter::*tms2* transgene as a negative selectable marker, Cecchini *et al.* (27) isolated seven or eight independent γ-ray-induced mutants, and found that all contain at least 5 kb deletions in the transgene. However, γ-ray- induced *gai-d* mutants seem to have very small deletions (1 or 7 bp) or point mutations (26; *Table 1*). These results indicated that the frequency and the extent of deletion mutations are locus dependent.

4 Genomic subtraction protocol for arabidopsis

To date, we have successfully used genomic subtraction to clone two genes from Arabidopsis. Because approximately 30–50% of FN-induced mutants in previous studies contain deletions that are suitable for genomic subtraction (*Table 1*; 9, 20), isolation of at least four FN-induced mutant alleles is desirable before conducting the genomic subtraction experiments. We also strongly recommend working out the experimental conditions using a known Arabidopsis deletion mutant such as the *ga1-3* allele. Alternatively, a reconstruction experiment can be performed in which biotinylated Arabidopsis DNA is used to subtract a mixture of non-biotinylated Arabidopsis DNA spiked with adenovirus DNA.

The following protocols are derived from previously published protocols (1, 2, 9, 35, 36).

4.1 Preparation of genomic DNA from wild-type and putative deletion mutants

Protocol 1

Preparation of DNA from wild-type plants

Reagents

- Qiagen 100/G and 500/G columns
- Hexadecyltrimethylammonium bromide (CTAB, Sigma)
- λ phage DNA (GIBCO, BRL)
- *Sau*3A I (Promega)
- 1 M (Na)EPPS (pH 8.25 at room temperature), adjusted with NaOH. (EPPS = N-(2-hydroxyethyl)piperazine-N′-3-propanesulfonic acid; Sigma E9502)
- 10 x EE: 100 mM (Na)EPPS (pH 8.25), 10 mM EDTA (pH 8.0)

Method

1. Grow wild-type plants on sterile MS plates at 22 °C under 16-h light/8-h dark cycles for 2 weeks. Using sterile wild-type plants to prepare the genomic DNA avoids enriching for foreign DNA sequences due to contaminating micro-organisms.
2. Harvest 3–10 g 2-week-old wild-type seedlings and purify plant DNA by CsCl density gradient centrifugation as described in (37). Alternatively, plant DNA can be purified using Qiagen columns 100/G (for 3 g tissue) or 500/G (for 10 g tissue) by following a procedure including CTAB and chloroform extraction, as recommended by Qiagen,

Protocol 1 continued

except that we preheated the QF buffer to 70 °C before eluting the DNA. Use disposable tubes and pipettes when possible to avoid DNA contamination.

3 Determine the concentration of the purified DNA by measuring OD_{260} and also by comparing to λ phage DNA of known concentration on an 0.8% agarose gel. The plant DNA should appear as a sharp band, which is similar in size as the λ phage DNA.

4 Digest 5 μg wild-type DNA with 10 U of *Sau*3A I at 37 °C for 4 h, phenol/chloroform extract once, ethanol precipitate[a], and resuspend in 40 μl 1 × EE (0.1 μg/μl, assuming 80% recovery). *Sau*3A I can be substituted with other restriction enzymes that have four base recognition sequences as long as suitable adaptors are used in *Protocol 4*.

[a] Standard ethanol precipitation: add sodium acetate (pH 7) to 0.3 M, then add 2 volume of ethanol (−20 °C). Incubate at −20 °C for at least 2 h, spin in a microcentrifuge at 4 °C for 5–10 min, then wash the pellet with 70% ethanol (−20 °C) and dry in the a Speed Vac centrifuge.

Protocol 2

Preparation of DNA from putative deletion mutants

Equipment and reagents

- Sonifier II Cell Disrupter (Model 250, Branson Ultrasonics) or equivalent
- Qiagen 100/G and 500/G columns
- CTAB (Sigma)
- λ phage DNA (GIBCO, BRL)
- 1 M Tris–HCl (pH 9.0)
- 3 M sodium acetate (pH 5.2)
- 10 × EE (see *Protocol 1*)
- Photo-activatable biotin (Pierce #29987G, 0.5 mg), dissolved in sterile deionized water to 2 μg/μl. When freshly made, this solution should be bright orange. The solution is light sensitive and can be kept for several months at −20 °C in the dark
- Reflector Sunlamp Bulb (Clontech #1131-3) or 500 W mercury vapour bulb for photobiotinylation

Method

A. Isolation of mutant plant DNA

1 Grow plants of the four presumptive deletion mutants on soil at 22 °C under 16-h light/8-h dark cycles for 4 weeks (before bolting).

2 Harvest rosette leaves and purify DNA from 10 to15 g of tissue for each mutant using procedures in *Protocol 1*, step 2.

3 Estimate DNA concentration as in *Protocol 1*, step 3.

Protocol 2 continued

B. Sonication of putative deletion mutant DNA

1. Resuspend 100 μg of each mutant DNA in 1 ml of 1 × EE buffer and place into a 15 ml plastic conical tube.
2. Place the tube in a plastic beaker with ice water and insert a clean microtip into the DNA solution without touching the bottom or side of the tube.
3. Shear the DNA by turning on the sonicator for 5–10 s using 60% duty cycle, microtip limit 5 and continuous setting.
4. Check the size of the DNA after sonication by loading 1 μl DNA onto a 0.8% agarose gel. The average size of the DNA should be approximately 2–3 kb. Repeat sonication if the average size is larger than 4 kb. Do not use DNA samples with average size less than 1 kb because the procedure described below in Section C results in the addition of approximately only one biotin moiety per 100–400 bp (38). Small DNA fragments might not be biotinylated and will affect the efficiency of subtraction.

C. Biotinylation of sheared deletion mutant DNA

1. Transfer each sonicated DNA solution to two flat-bottom 2-ml microcentrifuge tubes (0.5 ml each).
2. Denature the DNA by placing the tubes in a boiling water bath for 2 min, and then precipitate the DNA using the standard ethanol/sodium acetate method (see *Protocol 1*, step 4).
3. Resuspend the DNA pellet in each tube in 50 μl dH_2O, add 50 μl (2 μg/μl) photo-activatable biotin to each tube, and place the tubes in a styrofoam float in an ice-water bath. It is important to use the flat-bottomed 2 ml microcentrifuge tubes to ensure even and maximum illumination during the next step.
4. Place the tubes in the ice bath with tops open 10 cm from a sunlamp for 15 min total[a]. Mix the solution briefly every 5 min. The solution should turn dark brown following this treatment.
5. Combine the two aliquots of biotinylated DNA into one of the microcentrifuge tubes and add 10 μl 1 M Tris–HCl (pH 9.0).
6. Extract the biotinylated DNA several times with an equal volume of water-saturated 2-butanol (or *n*-butanol) until the butanol phase is colourless. Discard the butanol phase (upper phase) each time.
7. Add 11 μl 3 M sodium acetate (pH 5.2) to each DNA solution, and precipitate the DNA with ethanol. The DNA pellet should be dark brown.
8. Resuspend each mutant DNA in 36 μl 1 × EE (2.5 μg/μl assuming 90% recovery).

[a] Pierce suggests an alternative treatment by using a photographic flash (5–10 flashes). However, this method is not recommended because it was not as efficient as using the sunlamp in our laboratory.

4.2 Subtractive hybridization

Protocol 3
Subtractive Hybridization

Equipment and reagents

- 10 × EE (see *Protocol 1*)
- EEN: 0.5 M NaCl, 1 × EE
- 4 × SSE: 4 M NaCl, 0.16 M (Na)EPPS pH 8.25, 20 mM EDTA pH 8.0
- A synthetic oligonucleotide of 80–100 bp (that does not hybridize with Arabidopsis DNA), end-labelled using [γ-^{32}P]ATP and T4 polynucleotide kinase, and purified to remove unincorporated ATP
- Yeast tRNA, 20 μg/μl
- Fluoricon™ avidin polystyrene assay particles 0.7–0.9 μm diameter (IDEXX Corp. #310401). To remove sodium azide and free avidin, microcentrifuge 1 ml beads for 1 min, wash two times in 1 × EE, and resuspend in 0.9 ml EEN. This can be prepared in advance and stored at 4 °C.
- Ultrafree-MC filter units (Millipore #UFC3 OGV 00), or Spin-X Centrifuge Tube Filters (Costar #8160), or equivalent

Method

A. The first cycle of subtraction

1. Mix 0.5 μg *Sau*3A I-digested, wild-type DNA (0.1 μg/μl in 1 × EE, *Protocol 1*) and 12.5 μg biotinylated deletion mutant DNA (2.5 μg/μl in 1 × EE, *Protocol 2*) in a 0.5 ml microcentrifuge tube.
2. Add 50 000 cpm (measuring Cerenkov radiation) of an end-labelled synthetic oligonucleotide (80–100-mer) to each mutant DNA and wild-type DNA mix. This labelled single-stranded oligonucleotide serves as a tracer so that one can estimate the percentage of recovery after each cycle of subtraction.
3. Add 10 × EE to bring the final concentration of buffer to 2 × EE (final volume should be 10–15 μl).
4. Heat the DNA in a boiling water bath for 1 min and immediately dry sample in a Speed Vac centrifuge.
5. Resuspend the DNA completely in 3 μl dH_2O.
6. Add 1 μl 4 × SSE, mix well and then overlay with 10 μl mineral oil to prevent evaporation.
7. Incubate at 65 °C for at least 20 h in an air incubator.
8. Remove the mineral oil and add 2 μl yeast tRNA (20 μg/μl) and 100 μl EEN buffer and pipette up and down to thoroughly mix at room temperature.
9. Add 100 μl washed avidin-coated particles resuspended in EEN. Mix well and then let stand for 30 min at room temperature.
10. Transfer each mixture to a filter unit and spin 10 sec in a microcentrifuge at room temperature.

Protocol 3 continued

11 Transfer supernatant to a fresh 1.5-ml microcentrifuge tube. Wash beads on filter with 0.2 ml EEN, spin for 10 s in a microcentrifuge at room temperature, and combine supernatants.

12 Precipitate DNA in supernatants from step 11 with 0.9 ml ethanol (−20 °C), incubate at −80 °C for 15 min, and spin 5–10 min in a microcentrifuge at 4 °C. Wash the pellet with 70% cold ethanol. There is no need to add sodium acetate because of the high concentration of NaCl in EEN.

13 Count both the used filter units and the dry DNA pellets by measuring Cerenkov radiation. Recovery could be lower in the first cycle (60–70%) due to the presence of unincorporated label used to phosphorylate the tracer oligonucleotide.

14 Resuspend each dry DNA pellet completely in 5 μl 1 × EE, and transfer 4.75 μl to a new 0.5 ml tube for the second cycle of subtraction. Save the remaining DNA in the −20 °C freezer until *Protocol* 7, step 4.

B. Cycles of subtraction two through five

1 Add 4 μl (2.5 μg/μl) fresh biotinylated mutant DNA (same mutant DNA as in the first cycle in *Protocol* 3A, step 1) to each tube containing 4.75 μl DNA from the first cycle in *Protocol* 3A, step 14.

2 Repeat steps 3–14 in *Protocol* 3A four more times. On the fourth repeat, skip step 14 and go directly from step 13 to the step 3 in this section. Do not add additional yeast tRNA in the repeats of step 8 from *Protocol* 3A unless a visible pellet is not observed after ethanol precipitation. Recovery should be higher than 70% per cycle.

3 Resuspend the dry DNA pellets from the last cycle of subtraction in 5 μl 1 × EE.

4.3 Amplification and cloning of the DNA fragments after subtraction

Protocol 4

Ligation of adaptors and PCR amplification

Equipment and reagents

- *Sau*3A I adaptor (1)
  Primer 1 (24-mer): 5′-GACACTCTCGAGACATCACCGTCC
  Primer 2 (26-mer): 5′-GATCGGACGGTGATGTCTCGAGAGTG
- NEE-tRNA (1 M NaCl, 1 × EE, 400 μg/ml yeast tRNA)
- T4 polynucleotide kinase (New England Biolabs)
- Agarose 1000 (GIBCO, BRL)
- AmpliTaq DNA polymerase (5 U/μl, Perkin Elmer)
- dNTPs mix (10 mM, GIBCO, BRL), dilute to 2.5 mM in 10 mM Tris-HCl (pH 7.5)
- T4 DNA ligase (400 U/μl, New England Biolabs)
- Thermal cycler for PCR

Protocol 4 continued

Method

1 The adaptors consist of a 24-mer (Primer 1) and a 26-mer (Primer 2). Phosphorylate the 5′ end of Primer 2 by using T4 polynucleotide kinase (39). After heat inactivating the kinase at 70 °C for 15 min, mix in an equal amount of unphosphorylated Primer 1.

2 Heat the mixture of Primer 1 and 2 in boiling water bath for 2 min and then allow to anneal to form the *Sau*3A I adaptors at room temperature for 15 min.

3 Place 50 μl NEE-tRNA in a 1.5 ml microcentrifuge tube and add 0.5–2 μl of the resuspended subtracted DNA from step 3 in *Protocol* 3B. Mix well and incubate at 80 °C for 30 min, place on ice. Then add 200 μl 1 × EE and ethanol precipitate. Resuspend pellet in 5 μl 1 × EE, and place on ice. Because wild-type sequences with low T_ms may fail to hybridize to biotinylated DNA during subtraction, this step ensures that these sequences are completely denatured so that they will not ligate to the adaptors in the next step.

4 Mix 2.5 μl DNA from step 3 with 50 ng annealed *Sau*3A I adaptors at 25 ng/μl from step 2.

5 Add 200 U T4 DNA ligase, 1 μl 10 × T4 ligase buffer, and bring the volume to 10 μl with dH_2O. Incubate at 15 °C for 6 h. As a control, carry out a parallel ligation with the adaptors alone to determine if they form dimers as monitored on a 2% Agarose 1000 gel.

6 Amplify 1–2.5 μl of the ligation mix using 0.2 μg of the phosphorylated Primer 1, 5 μl 10 × Taq polymerase buffer with 15 mM $MgCl_2$, 2.5 U AmpliTaq polymerase, and 5 μl 2.5 mM dNTPs in a total volume of 50 μl. Amplify for 50 cycles (94 °C, 30 s; 55 °C, 30 s; 72 °C, 3 min), and then at 72 °C for 10 min. In this step, phosphorylated Primer 1 is used because the PCR products will be cloned into vector DNA in *Protocol* 5.

7 Analyse 5 μl PCR products by agarose gel electrophoresis. Typically, you will see a smear around 0.3–0.7 kb. Extract the DNA by phenol/chloroform once and then ethanol precipitate. Resuspend in 25 μl dH_2O and store in the freezer.

Protocol 5

Cloning the PCR products

Equipment and reagents

- Klenow enzyme (New England Biolabs)
- dNTPs mix (10 mM, GIBCO, BRL), dilute to 0.25 mM in 10 mM Tris–HCl (pH 7.5)
- Bacterial alkaline phosphatase (BAP, GIBCO BRL)
- pSK (Stratagene)
- *Sma* I (New England Biolabs)
- Hybridization oven for DNA blots
- TE: 10 mM Tris–HCl (pH 8.0), 1 mM EDTA (pH 8.0)
- X-gal (GIBCO BRL): 20 mg/ml in N,N-dimethylformamide (Sigma)
- IPTG (GIBCO BRL): 0.1 M in dH_2O, filter sterilized
- LB agar: 10 g NaCl, 5 g Yeast Extract (Difco), 10 g Tryptone Peptone (Difco), 13 g agar (Difco) per l dH_2O

Protocol 5 continued

Method

1 Mix 3.6 μl of PCR products from step 7 in *Protocol 4* with 0.5 μl 10 × Klenow buffer, 1 unit Klenow enzyme, and 0.7 μl 0.25 mM dNTPs in a total volume of 5 μl. Incubate at 37 °C for 30 min, heat inactivate the enzyme at 70 °C for 10 min, and let stand at room temperature for 10 min.

2 Cut vector pSK with *Sma* I at 25 °C until digestion is completed.

3 Incubate the *Sma* I-digested pSK DNA with BAP (70 U per 1 pmol 5′ termini) in 10 mM Tris–HCl (pH 8.0) at 65 °C for 1 h to remove the 5′ phosphate. Add 1 μl 0.5 M EDTA (pH 8.0), incubate at 50 °C for 10 min to stop the reaction. Then phenol/chloroform extract twice and ethanol precipitate. The DNA pellet is resuspended in TE (25 ng/μl).

4 Mix 2 μl Klenow-treated blunt-end PCR DNA from step 1 with 12.5 ng *Sma* I- and BAP-treated pSK DNA from step 3, 0.5 μl 10 × T4 ligase buffer, 200 U T4 ligase in a total volume of 5 μl. Incubate for 3 h at room temperature.

5 Transform XL1-Blue or DH5 competent cells using the ligation mixture from step 4. Use LB agar plates containing 40 μg/ml X-gal, 0.5 mM IPTG and 100 μg/ml ampicillin to select for the white colonies with insertions (39).

6 For each mutant, pick 200 white colonies onto master and replica plates (100 colonies/plate). Incubate overnight at 37 °C and store the master plates at 4 °C until *Protocol* 6A, step 1. Lift the colonies on the replica plates with nitrocellulose membrane as described previously (39). Store the dried filters at room temperature until *Protocol* 6B, step 3.

4.4 Screening the clones corresponding to deletion mutations

Protocol 6

Identification of the correct clone by DNA blot analysis

Equipment and reagents

- [α-^{32}P]dCTP, 3,000 Ci/mmol, 10 μCi/μl (Du Pont)
- 5 × random primed DNA labelling mix (40, 41)
- Klenow enzyme (New England Biolabs)
- 5M ammonium acetate (Sigma, filter sterilized)
- Sephadex G-50-spin columns (39)
- 6 × SSC, 5 × Denhardt's (39)
- Thermal cycler for PCR
- Hybridization oven for DNA blots
- Phosphorimager (Molecular Dynamics, model 440E) or equivalent

Protocol 6 continued

Methods

A. Preparation of probes using clones on master plates by PCR and random-primed labeling

(For each mutant, screen 10 clones at a time.)

1. Aliquot 10 μl dH_2O into 10 thin-wall 0.5 ml PCR tubes. Use sterile toothpicks to inoculate cells of 10 colonies from a master plate (*Protocol 5*, step 6) into these tubes, separately. Heat the tubes at 100 °C for 5 min in the thermal cycler. Chill on ice for 5 min, quick spin in a microcentrifuge, then keep on ice.
2. To each tube, add 1.5 μl 10 × Taq Polymerase buffer, 0.75 μl 2.5 mM dNTPs, 60 ng unphosphorylated Primer 1 (*Protocol 4*), 0.75 U AmpliTaq polymerase and water to a total volume of 15 μl. Amplify for 35–40 cycles (94 °C, 30 s; 55 °C, 30 s; 72 °C, 1.5 min), and then at 72 °C for 10 min.
3. Load 2 μl PCR DNA on a 1.5% agarose gel to check the sizes of inserts and the amount of DNA amplified.
4. Precipitate the rest of the PCR DNA (13 μl) by adding 62 μl dH_2O, 50 μl 5M ammonium acetate and 375 μl ethanol (room temperature). Mix well, incubate at 4 °C overnight, and spin in a microcentrifuge at room temperature for 10 min. This ammonium acetate precipitation step is important to remove the unincorporated dNTPs before preparing radiolabelled probes.
5. Wash the pellet with 500 μl 70% ethanol at room temperature. Dry and resuspend the DNA pellet in 5 μl dH_2O. Run 0.5–1 μl sample on a 1.5% agarose gel to estimate the DNA concentration.
6. Prepare radiolabelled probes by random-primed labelling using 5–10 ng (0.5–1 μl) ammonium acetate-precipitated PCR DNA from step 5 as templates (denatured by heating in boiling water bath for 2 min). Add 1 μl 5 × random primed DNA labelling mix, 10 μCi ^{32}P-dCTP, and 1 unit of Klenow enzyme in a 5 μl final volume for 1 h at 37 °C.
7. Purify the labelled DNA from step 6 using the G-50-spin columns.

B. Screening by DNA blot analysis and colony hybridization

1. For each putative deletion mutant, prepare 10 mini-DNA blots each containing 0.2 μg *Hin* dIII-digested genomic DNA from wild-type and the mutant.
2. Prehybridize these blots, separately, in individual heat sealed bags containing 3 ml 6 × SSC, 5 × Denhardt's for 1 h at 65 °C. Use the 10 probes prepared in *Protocol 6A* for the hybridization. Mix each probe (2 × 10^5 cpm/ml) with 0.3 mg denatured salmon sperm DNA, and boil for 5 min before adding to the hybridization solution. Incubate overnight at 65 °C with agitation.
3. At the same time, hybridize the two filters from the colony lifts (see *Protocol 5*, step 6) with pooled 10 probes (2 × 10^5 cpm/ml each) from *Protocol 6A*. To reduce the background, add 0.5 mg denatured salmon sperm DNA and 500 ng Primer 1 to the

Protocol 6 continued

10 ml 6 × SSC, 5 × Denhardt's during prehybridization, and also mix the same amount of denatured salmon sperm DNA and Primer 1 with the probes before boiling.

4. Wash the blots from step 2 and step 3 in 2 × SSC, 1% SDS at 65 °C. Use the phosphorimager to analyse the blots. Clones corresponding to the deleted region will only hybridize to the wild-type DNA, but not to the deletion mutant DNA. If none of the 10 clones tested identifies a deleted region in the mutant, record the positive clones on the colony filters to avoid repeated screening these clones in future analyses.
5. Screen 10 clones for each mutant. Repeat the screening by testing additional clones that are not identical to the previously screened ones until the desired clone is identified.
6. Use the desired clone identified in step 5 as a probe to determine whether this DNA sequence is also missing in additional FN-induced mutant alleles. Additional deletions might be present in the mutant background. If more than one mutant alleles contain overlapping deletions, the clone you identified would be most likely at or near the gene of interest.
7. Alternatively, use the 'desired' clone from step 5 as a probe to isolate overlapping large genomic clones from an Arabidopsis genomic library (see Chapter 7). Then use the genomic clone as a hybridization probe to repeat the DNA blot analysis in step 6. This larger clone will be able to detect deletions in other mutant alleles that are near, but are not covered by the original small *Sau*3A I fragment.

4.5 Leak through background clones

In cloning the Arabidopsis *RGA* locus, we found one out of 25 clones, after five cycles of subtractive hybridization between *ga1-3* and *rga-20/ga1-3* DNA, was located in the deleted region in *rga-20/ga1-3*. According to the previously calculated efficiency based on cloning of *GA1*, we would predict to find one third of the clones after five cycles of subtraction to correspond to the deleted region in *rga-20* because this allele contains an over 33 kb deletion around the *RGA* locus. It is not clear why the fold of enrichment is lower during cloning of *RGA* than *GA1*. However, in our screening by DNA blot analysis (*Protocol 6*), we discovered that many clones (~23%) were present at multiple copies in the Arabidopsis genomic DNA samples. When used as hybridization probes, these sequences showed much stronger signals than the single copy sequence on Southern blots. This made it impossible to pool several clones prior to radio-labelling for use in DNA blot analysis.

To determine the nature of these high-copy number clones, we analysed DNA sequences of 10 of these clones (after five cycles of subtraction). Six of them are chloroplast DNA, one is the spacer DNA between *25S* and *18S* rRNA genes, one is the *18S* rRNA gene, one is the mitochondrial *26S* rRNA gene, and one is an unidentified sequence. In the future, one might be able to eliminate most of these sequences by including biotinylated chloroplast/mitochondria DNA and rRNA genes in the subtractive hybridization mixture. In addition, before screening by DNA blot analysis, labelled chloroplast/mitochondria DNA and rDNA can be used as probes to identify these multi-copy sequences among the clones after subtraction on the master plates. For protocols fro the isolation of mitochondria and chloroplasts see Volume 2, Chapter 7 and Chapter 8. Elimination of these high-copy number clones is essential before the more efficient sib-screening experiments can be performed by pooling several clones together for making hybridization probes.

4.6 Pilot experiments

Before embarking on cloning the gene of interest using putative deletion mutants, a pilot experiment is recommended using either Arabidopsis DNA spiked with adenovirus DNA (a reconstruction experiment) or a defined Arabidopsis mutant with several kb deleted (such as *ga1-3* or *rga-24*). In the reconstruction experiment, use the same preparation of Arabidopsis DNA for both the biotinylated and non-biotinylated samples. The amount of enrichment after each cycle of subtraction can be calculated by cloning the amplified *Sau*3A I fragments in pSK and then probing this library with radiolabelled adenovirus DNA or with radiolabelled DNA corresponding to the known deletion. Alternatively, the amplified products from each cycle of subtraction can be fractionated on an agarose gel, blotted, and probed. In the case of the adenovirus reconstruction experiment, we found that individual adenovirus *Sau*3A I fragments in the amplified product after three cycles of subtraction could be directly visualized on an ethidium bromide-stained agarose gel (2).

Protocol 7

Pilot experiment using Arabidopsis DNA spiked with adenovirus DNA

Reagents

- adenovirus DNA (GIBCO, BRL)
- 1 × EE (see *Protocol 1*)
- LB agar plates containing X-gal, IPTG and ampicillin (see *Protocol 5*)

Method

1. Cut adenovirus DNA and Arabidopsis DNA with *Sau*3A I as in *Protocol 1*, step 4.
2. Prepare sheared and biotinylated Arabidopsis DNA as in *Protocol 2*.
3. Mix 12.5 g of biotinylated Arabidopsis DNA with 0.5 μg of *Sau*3A I-digested Arabidopsis DNA and 0.5 ng Sau3A I-digested adenovirus DNA.

Protocol 7 continued

4 Follow *Protocol* 3A and 3B and carry out 4 cycles of subtraction. Save 1/20th the volume (approx. 0.25 μl) of the resuspended pellet (before adding the biotinylated DNA) for testing the degree of enrichment during each cycle of subtraction in the next step. Dilute the 0.25 μl aliquot in 5 μl 1 × EE.

5 Amplify these samples by PCR as in *Protocol 4* and run 5 μl PCR DNA on an agarose gel. A subset of adenovirus *Sau*3A I fragments could be directly visualized on the ethidium bromide-stained agarose gel after three cycles of subtraction. *Sau*3A I-digested adenovirus DNA should be amplified with *Sau*3A I adaptors as a control.

6 To estimate the efficiency of enrichment, clone PCR DNA fragments into pSK as in *Protocol* 5. Spread ~4000 white colonies from each cycle onto a nitrocellulose filter that is placed on the surface of a LB plate containing X-gal, IPTG and 100 μg/ml ampicillin.

7 Hybridize each filter with radiolabelled adenovirus DNA and calculate the amount of enrichment after each cycle of subtraction.

5 Comparison of genomic subtraction to RDA

RDA was designed to isolate restriction endonuclease fragments that are present in one DNA population, but not in another (12, 42). In this method, wild-type DNA is called the tester and the deletion mutant DNA is the driver. After cutting the two genomic DNAs with an infrequent restriction endonuclease (with a 6-base recognition site) and ligating to Adaptor 1, the complexity of the two populations is then reduced by preferential amplification of smaller DNA fragments during PCR. The adaptor sequences are removed by enzyme digestion after PCR and the resulting DNA populations are called the 'amplicons', which are the 'representations' of the original two genomes. In the second step comprised of subtractive and kinetic enrichment of target sequence, a small amount of the tester amplicon is ligated to new adaptors (Adaptor 2), and then mixed with an excess of driver amplicon. After melting and reannealing, the double-stranded tester DNA is exponentially amplified by PCR using one strand of the adaptor as the primer. Single-stranded DNA is digested by mung-bean nuclease, the remaining DNA is again amplified by PCR, and then digested with the restriction enzyme to remove Adaptor 2 and ligated to Adaptor 3. The resulting DNAs can then go through steps 1 and 2 several times to enrich for fragments that are unique in the tester amplicon. Because each enzyme-generated amplicon only represents a fraction of the original genome, several restriction enzymes need to be used in separate reactions.

Computer simulations of various subtraction techniques, including genomic subtraction and RDA, were reported by Cho and Park (43). The differences between the procedures and efficiency of genomic subtraction and RDA are summarized in *Table 2*. The genomic subtraction procedure has the advantages of being simpler with fewer manipulations and only requires one set of sub-

Table 2 Comparison between genomic subtraction and RDA

Genomic subtraction	RDA
Fewer DNA manipulations	More complicated manipulations
Prepare only one set of DNA from wt and mutant	Multiple steps of restriction enzyme digestion and adaptor ligation
Single restriction enzyme treatment	Requires several sets of restriction enzymes reactions for each mutant
PCR adaptors only ligate to unbound double-stranded DNA after subtraction	PCR adaptors are on all tester DNA
Detects large deletions (> 1 kb)	Detect point mutations, deletions or rearrangments
Subtracts whole genomes	Subtracts among 'representations' of two genomes (less complex populations)
Enriches by removing undesired biotinylated DNA using avidin beads	Enriches by PCR (does not include physical separation step)
Linear enrichment in each cycle	Kinetic enrichment in each cycle

tractions for each mutant. This is especially important when carrying out parallel subtraction reactions using multiple mutant alleles to isolate a gene of interest. However, genomic subtraction described in this chapter can only isolate genes corresponding to a large deletion. A modified version of this technique (RFLP Subtraction) was developed to isolate RFLP markers in mice and in the green alga Volvox (44, 45). RDA could detect not only large deletions, but also other alterations that cause RFLP, including point mutations, small deletions and rearrangements (46, 47). RDA also has the advantage of being able to deal with more complex genomes by selecting representational amplicons (12). This makes RDA more applicable for organisms with larger genome size (48–51). In theory, RDA has the potential of reaching a higher level of enrichment because of the 'kinetic enrichment' feature by PCR in each cycle (12, 43). However, this could be a problem if the bias amplification of the same subset of DNAs by PCR is unreproducible in each cycle.

Acknowledgments

I thank Aron Silverstone for cririal reading of the manuscript. This work was supported by National Science Foundation Grant No. IBN-9723171.

References

1. Straus, D. and Ausubel, F. M. (1990). *Proc. Natl. Acad. Sci. USA*, **87**, 1889.
2. Sun, T-P., Goodman, H. M., and Ausubel, F. M. (1992). *Plant Cell*, **4**, 119.
3. Koornneef, M., van Eden, J., Hanhart, C. J. and de Jongh, A. M. M. (1983). *Genet. Res. Camb.*, **41**, 57.
4. Smith, H. H. and Rossi, H. H. (1966). *Radiat. Res.*, **28**, 302.

5. Wessler, S. R. and Varagona, M. J. (1985). *Proc. Natl. Acad. Sci. USA*, **82**, 4177.
6. Liharska, T. B., Hontelez, J., vanKammen, A., Zabel, P., and Koornneef, M. (1997). *Theor. Appl. Genet.*, **95**, 969.
7. Dellaert, L. W. M. (1980). *X-ray- and fast neutron-induced mutations in Arabidopsis thaliana, and the effect of dithiothreitol upon the mutant spectrum.* PhD Thesis. Agricultural University, Wageningen, The Netherlands.
8. Koornneef, M., Dellaert, L. W. M., and van der Veen, J. H. (1982). *Mutat. Res.*, **93**, 109.
9. Silverstone, A. L., Ciampaglio, C. N., and Sun, T.-p. (1998). *Plant Cell*, **10**, 155.
10. Kang, S. C., Chumley, F. G., and Valent, B. (1994). *Genetics*, **138**, 289.
11. Mao, Y. X. and Tyler, B. M. (1996). *Fungal Genet. Biol.*, **20**, 43.
12. Lisitsyn, N., Lisitsyn, N., and Wigler, M. (1993). *Science*, **259**, 946.
13. Nambara, E., Keith, K., McCourt, P., and Naito, S. (1994). *Plant Cell Physiol.*, **35**, 509.
14. Cutler, S., Ghassemian, M., Bonetta, D., Cooney, S., and McCourt, P. (1996). *Science*, **273**, 1239.
15. Koornneef, M. (1979). *Arab. Inf. Serv.*, **16**, 41.
16. Sun, T-P. and Kamiya, Y. (1994). *Plant Cell*, **6**, 1509.
17. Koornneef, M. and van der Veen, J. H. (1980). *Theor. Appl. Genet.*, **58**, 257.
18. Yamaguchi, S., Sun, T.-p., Kawaide, H., and Kamiya, Y. (1998). *Plant Physiol.*, **116**, 1271.
19. Oppenheimer, D. G., Herman, P. L., Sivakumaran, S., Esch, J., and Marks, M. D. (1991). *Cell*, **67**, 483.
20. Bruggemann, E., Handwerger, K., Essex, C., and Storz, G. (1996). *Plant J.*, **10**, 755.
21. Century, K. S., Shapiro, A. D., Repetti, P. P., Dahlbeck, D., Holub, E., and Staskawicz, B. J. (1997). *Science*, **278**, 1963.
22. Koornneef, M. (1990). *Arab. Inf. Serv.*, **27**, 1.
23. Shirley, B. W., Hanley, S., and Goodman, H. M. (1992). *Plant Cell*, **4**, 333.
24. Wilkinson, J. Q. and Crawford, N. M. (1991). *Plant Cell*, **3**, 461.
25. Peng, J. and Harberd, N. P. (1993). *Plant Cell*, **5**, 351.
26. Peng, J., Carol, P., Richards, D. E., King, K. E., Cowling, R. J., Murphy, G. P., and Harberd, N. P. (1997). *Genes Dev.*, **11**, 3194.
27. Cecchini, E., Mulligan, B. J., Covey, S. N., and Milner, J. J. (1998). *Mutat. Res.*, **401**, 199.
28. Kieber, J. J., Rothenberg, M., Roman, G., Feldmann, K., and Ecker, J. R. (1993). *Cell*, **72**, 427.
29. Bjourson, E. K. F. and Reilly, E. (1988). *Appl. Environ. Microbiol.*, **54**, 2852.
30. Kunkel, L. M., Monaco, A. P., Middlesworth, W., Ochs, H. D., and Latt, S. A. (1985). *Proc. Natl. Acad. Sci. USA*, **82**, 4778.
31. Lamar, E. E. and Palmer, E. (1984). *Cell*, **37**, 171.
32. Nussbaum, R. L., Lesko, J. G., Lewis, R. A., Ledbetter, S. A., and Lednetter, D. H. (1987). *Proc. Natl. Acad. Sci. USA*, **84**, 6521.
33. Welcher, A. A., Torres, A. R., and Ward, D. C. (1986). *Nucl. Acids Res.*, **14**, 10027.
34. Wieland, I., Bolger, G., Asouline, G. and Wigler, M. (1990). *Proc. Natl. Acad. Sci. USA*, **87**, 2720.
35. Straus, D. (1995). In *PCR Strategies* (eds M. A. Innis, D. H. Gelfand, and J. J. Sninsky), p. 220. Academic Press, Inc., San Diego.
36. Sun, T-P., Straus, D., and Ausubel, F. M. (1992). In *Methods in Arabidopsis research* (eds C. Koncz, N-H. Chua, and J. Schell), p. 331. World Publishing Co., Singapore.
37. Ausubel, F. M., Brent, R., Kingston, R. E., Moore, D. D., Seidman, J. G., Smith, J. A., and Struhl, K. (eds) (1990). *Current protocols in molecular biology*. Green Publishing Associates/Wiley-Interscience, New York.
38. Forster, A. C., McInnes, J. L., Skingle, D. C., and Symons, R. H. (1985). *Nucl. Acids Res.*, **13**, 745.
39. Sambrook, J., Fritsch, E. F., and Maniatis, T. (eds) (1989). *Molecular cloning: a laboratory manual.* Cold Spring Harbor Laboratory Press, Cold Spring Harbor.

40. Feinberg, A. P. and Vogelstein, B. (1983). *Anal. Biochem.*, **132**, 6.
41. Feinberg, A. P. and Vogelstein, B. (1984). *Anal. Biochem.*, **137**, 266.
42. Lisitsyn, N. A. (1995). *Trends Genet.*, **11**, 303.
43. Cho, T.-J. and Park, S.-S. (1998). *Nucleic Acids Res.*, **26**, 1440.
44. Rosenberg, M., Przybylska, M., and Straus, D. (1994). *Proc. Natl. Acad. Sci. USA*, **91**, 6113.
45. Corrette-Bennett, J., Rosenberg, M., Przybylska, M., E. A., and Straus, D. (1998). *Nucl. Acids Res.*, **26**, 1812.
46. Lisitsyn, N. A., Segre, J. A., Kusumi, K., Lisitsyn, N. M., Nadeau, J. H. and Frankel, W. N. (1994). *Nature Genet.*, **6**, 57.
47. Kajiya, H., Shimano, T., Monna, L., Yano, M., and Sasaki, T. (1996). *Breeding Sci.*, **46**, 387.
48. Delaney, D. E., Friebe, B. R., Hatchett, J. H., Gill, B. S., and Hulbert, S. H. (1995). *Genome*, **38**, 458.
49. Chen, Z. J., Phillips, R. L., and Rines, H. W. (1998). *Theor. Appl. Genet.*, **97**, 337.
50. Schutte, M., DaCosta, L. T., Hahn, S. A., Moskaluk, C., Hoque, A. T. M. S. and Rozenblum, E. (1995). *Proc. Natl. Acad. Sci. USA*, **92**, 5950.
51. Watson, J. E. V., Gabra, H., Taylor, K.J., Rabiasz, G. J., Morrison, H. and Perry, P. (1999). *Genome Res.*, **9**, 226.

Chapter 6
Approaches to gene mapping

S. Grandillo
Research Institute for Vegetable and Ornamental Plant Breeding, Via Universita', 133, 80055 Portici, Naples, Italy.

Theresa M. Fulton
Institute for Genomic Diversity, 130 Biotechnology Building, Cornell University, Ithaca, NY 14853, USA.

1 Introduction

The basic principles of linkage mapping have been in use since the beginning of this century (1, 2). However, for many years, progress in genetic mapping was slowed by the limited number of available markers. With the advent of recombinant DNA technology and restriction fragment length polymorphism (RFLP) markers (3) genetic maps have been constructed for numerous species (4). Their production and use have been further stimulated by the recent development of polymerase chain reaction technology (PCR; 5) and the derived high-volume marker technologies such as random amplified polymorphic DNA (RAPD; 6, 7), and amplified fragment length polymorphism (AFLP; 8, 9; see Chapter 9), which permit the rapid generation of large numbers of genetic markers.

High-density (>500 markers) molecular linkage maps are important tools in plant research because they facilitate the genomic localization of genes responsible for monogenic and polygenic traits, which in turn allows more efficient marker-assisted breeding for species of interest. These maps are also useful for studying evolutionary relationships in related species, as well as for genome organization and germplasm conservation; moreover, they provide starting points for the positional cloning of genes (see Chapter 7).

In order to utilize high-density molecular maps for map-based cloning and for studies of genome organization, it is necessary to convert the genetic (linkage) map to a physical (base pairs) map. This has become feasible due to the development of techniques for manipulation of large (>100 kb) DNA fragments and related cloning technologies, such as pulsed-field gel electrophoresis (PFGE; 10–12) and yeast artificial chromosome (YAC; 13) and bacterial artificial chromosome (BAC; 14) cloning (see Chapter 7). Physical maps can help confirm the order of DNA markers in the target region and, by determining the relationship between genetic and physical distances, they can influence the choice of the strategy to be pursued in order to clone the genes of interest. Molecular physical maps are already available or are in progress for several species (15–17). It is clear that these resources will soon receive a further boost from the rapidly accumulating information in sequence databases, together forming the basis for future genetic studies. Thus, the importance of these basic genetic techniques cannot be undermined.

In this chapter we will first give an overview of the basic principles underlying molecular mapping in plants. We will then describe strategies and protocols for genetic linkage analysis by means of RFLP and RAPD technologies together with a review of the AFLP approach. Finally, we will cover the main aspects and related protocols for the application of PFGE in physical mapping of plant genomes.

2 Molecular mapping: an overview

2.1 Principles of linkage analysis

Genetic mapping relies on meiotic recombination between homologous chromosomes carrying alternative alleles, which result in genetic exchange, called crossing over. Genetic linkage maps describe both the relative order of genetic markers along each chromosome of an organism and their relative distances from one another. Linkage order is based on maximum likelihood; that is to say the order of markers that results in the shortest distance and requires the fewest multiple crossing-overs between adjacent markers. On the other hand, as Sturtevant (2) suggested, the estimation of the fraction of recombinant gametes (with respect to the total number of gametes) can be used as a quantitative index of the linear distance between two gene pairs on the map. One genetic map unit (m.u.) is the distance between gene pairs for which one product of meiosis out of 100 is recombinant, and is generally referred to as either a recombination frequency (RF) of 0.01 (or 1%) or as one centimorgan (cM). General formulae for calculating recombination frequencies are given by Allard (18). These estimates and confidence intervals assume linkage as the only cause of non-independence among loci and Mendelian segregation at each locus. Although RF increases with map distance, the relationship is not linear, and RF is a good measure of distance only if the distance is small and if no multiple exchanges are likely. Several mapping functions have been devised to estimate most accurately the real relationship between map distance and RF, the form of which depends on the degree of interference in crossing over. The most common of these are the Haldane's (assuming absence of interference; 19) and Kosambi's (assuming interference; 20) mapping functions.

Genetic distances are also dependent on statistical sampling, and hence on the size and type of mapping population. The map obtained can be considered only the most statistically likely, given the available data. The standard error for a recombination frequency of true value p can be calculated as $\sqrt{[p(1 - p)/N]}$, where N is the number of gametes in the population. Several other factors can influence genetic distances, including the accuracy of the data (mis-scores will appear as double recombination, exaggerating map lengths), sex-related differences in recombination frequencies and environmental factors.

The same concepts of classical genetic mapping can be applied to mapping with DNA markers; however, with DNA-based genetic maps, many more markers can be mapped in a single population and, therefore, it becomes necessary to perform the analyses using adequate computer programs. The growing complexity of genetic mapping data sets has required corresponding improvements in computer technology and numerous software packages have been developed. In plants commonly used mapping programs are, for example, MAPMAKER (21) and JOINMAP (22). Other descriptions of linkage and mapping programs can be found at the following addresses:

Europe: ftp://ftp.ebi.ac.uk/pub/software/linkage_and_mapping/

US: http://linkage.rockefeller.edu/

2.2 Molecular markers and genetic maps

There are many types of DNA markers that are produced by a wide variety of techniques (23), yet only a subset has been widely used for whole genome mapping projects. RFLPs, RAPDs, simple sequence repeats (SSR) or microsatellites, and, more recently, AFLPs are the most popular molecular markers used in genome mapping. In several cases, for example, in rice, bean and tomato (24–26), high-resolution molecular linkage maps are being developed by integrating PCR-based markers into previously constructed RFLP maps. RFLPs, the first DNA markers to be developed and used for genome mapping (3), still represent the most reliable polymorphisms. Their locus-specificity and co-dominant nature allows the accurate scoring of all genotypes. The locus-specificity of RFLPs has also promoted their wide application to comparative mapping studies conducted across distantly related species. Nevertheless, RFLP analysis has the disadvantage of being an expensive and labour-intensive procedure requiring large quantities of DNA; in addition, RFLP maps often do not provide detailed coverage throughout the genome. These limitations have been, in part, overcome by the introduction of PCR-based markers, such as RAPDs, microsatellites and AFLPs, though none of these alternative technologies lacks disadvantages.

For example, RAPD markers, being relatively inexpensive, fast and easy to perform, have been widely used to construct linkage maps in plant species (27, 28). However, their general dominant nature combined with their population specificity and, often, poor reproducibility has limited, in many cases, the use of this technology for more extensive mapping purposes. On the other hand, microsatellites (29) are co-dominant DNA markers which detect higher levels of allelic variation than do RFLP or RAPD markers. Microsatellites have been used to construct dense maps of mammalian genomes such as human and mouse (30, 31), and can also detect high levels of allelic variation in plants (32, 33). However, they are expensive to develop since they require allele specific primers and, therefore, they are likely to prove less useful for constructing high-density maps of plant genomes. In addition, microsatellite markers seem to be transferable only between very closely related species (34), which makes them of limited use in comparative mapping. The more recently developed technique of AFLP analysis has the capacity to examine a much higher number of loci for polymorphisms than RAPD and microsatellite markers (8, 9), even in intra-specific comparisons among species where the level of polymorphism is much lower; in addition, AFLPs have the advantage of not requiring *a priori* sequence information. The AFLP approach has recently been used to rapidly create linkage maps in a variety of plant species (26, 35, 36). Nevertheless, these studies indicate that *Eco* RI + *Mse* I AFLP markers are not randomly distributed, but cluster around the centromeres; and that the use of locus-specific AFLP markers is more limited to populations within species or to very closely related species (35).

A newer class of genetic markers is represented by single-nucleotide polymorphisms (SNPs), which are considered the most common form of DNA polymorphisms that can be found in any genome. On average, an SNP is believed to be present in every 300–1000 bases (http://www.ncbi.nlm.nih.gov/ SNP/). A synonymous expression is 'biallelic marker' corresponding to the two alleles that may differ at a given nucleotide position in a diploid cell. Because SNPs are expected to facilitate large-scale association genetics studies, there has been a great interest in SNP discovery and detection. SNPs are viewed as the next generation of molecular markers that would either complement or replace the existing markers that are routinely used in many laboratories. One attractive feature of SNPs is the possibility of detection using non-gel based systems. High-throughput, automized systems have already been developed that allow the parallel assay of SNPs at very high rates. Although, in general, very limited data are available on SNPs in plants, recently in Arabidopsis Cho *et al.* (37) have reported the construction of a biallelic genetic map with a resolution of 3.5 cM.

Here, we will discuss RFLPs together with the two PCR-based marker technologies, RAPD and AFLP.

2.3 Choice of mapping populations

The choice of the mapping population depends primarily on the objectives of the work. If the gene(s) to be mapped show a particular phenotype, e.g. identified as a mutant, then a specific population needs to be developed. In contrast, if only polymorphisms at the DNA level, e.g. cloned sequences, have to be located, one could more efficiently use, when available, mapping populations that have already been scored for many markers. Whenever a new mapping population needs to be created, of foremost importance is the selection of the parents of the cross, which need to be genetically divergent enough to exhibit sufficient DNA sequence polymorphisms, but not so distant as to cause sterility of the progeny or too great a reduction in recombination. The genetic variation present in most naturally out-crossing species is generally sufficient for mapping purposes, so that in maize, for example, it is sufficient to make crosses between unrelated inbreds (38). In contrast, most naturally inbreeding species, such as the cultivated tomato, rice and lettuce, are characterized by lower levels of polymorphisms and wider crosses (inter-racial or interspecific) become necessary (39).

Recombination frequencies between pairs of markers can be scored in many different segregating populations. Those derived from a true F1 are the most straightforward because only two alleles are segregating; they include F2s, backcrosses (BCs), recombinant inbred (RI) lines or doubled haploid (DH) lines, with F2s being the most informative. F2 populations permit detection of more distant linkage and better resolution of tighter linkage because recombination in both mega- and micro-gametophytes can be detected; whereas in a BC, recombination in only one gametophyte can be detected. The major drawback to F2 and backcross populations is that they are ephemeral. This limitation can be overcome by the use of RI populations, constructed by selfing or sibling out of an F2, for example, from which the F8 and onwards result in a number of practically homozygous lines, which provide a permanent mapping resource with fixed segregation (40). Since inbred lines must undergo several rounds of meiosis prior to reaching homozygosity, closely linked markers have more chance of being separated by a recombination event, and the linkage map is expanded for short distances. However, one disadvantage of RI populations is the longer time required for their development. As an alternative, in some cases, production of recombinant lines by the double haploid (DH) procedure can be considered, as the time required to reach homozygosity is shorter (41). However, this material is generally weaker than inbreds and the one-step procedure loses the important advantages associated with expansion of the map. A more complex scenario characterizes outbreeding species, where true F1s are not available. Under these circumstances, a single individual of the population can be considered an F1 and it's selfed progeny represents the F2. If the F1 individual cannot be selfed and the segregating population is obtained by crossing to another individual, then the number of alleles segregating at each locus could range from two to four and, with respect to any one locus, the cross could represent an F2, BC1 or BC2.

Maximizing the amount of recombination is an important consideration in choosing a mapping population, and is dependent on both the cross and the type of markers used. The relative two-point mapping efficiency of F2 and backcross populations has been calculated by Allard (18). F2 populations efficiently map co-dominant markers, such as RFLPs, but map dominant markers less efficiently (42), whereas backcross and RI populations map dominant and co-dominant markers with equal efficiency.

2.4 Population size

The size of the mapping population will greatly affect the resolution of a map and the ability to determine marker order. In theory, the larger the population the better. However, in practice the population size is generally a compromise between the need for reaching the desired resolution, and the technical and economical limitations of processing large numbers of DNA samples. Generally, populations of 50–100 progeny are used for mapping in plants, humans and mice. In a segregating population of 100 meiotic events a target gene can be placed within an interval of 5–10 cM with a high degree of certainty. However, this level of mapping resolution becomes insufficient whenever the aim of the work is fine mapping in specific regions of the genome or mapping quantitative trait loci (QTL) of minor effect. In these cases, much larger populations need to be analysed. For example, in a genome of 1 billion base pairs and 1000 cM, in order to have a 95% chance of recovering a single crossover 0.1 cM or less from a target gene, analysis of more than 3000 meiotic products will be required. Similarly, several hundreds of F2 individuals can be necessary in order to detect QTL explaining as little as 1% of the total phenotypic variation of a quantitative trait (43).

2.5 Tagging genes with tightly linked markers

The construction of genetic linkage maps uniformly covering all chromosomes is an essential step for QTL mapping projects, chromosome characterization and for effective marker-assisted breeding. However, there are special cases that do not require whole genome mapping efforts, since specific regions of the genome represent the real focus. These include, for example, high resolution mapping of a gene targeted for map-based cloning or tagging major genes, that is to say genes affecting the phenotype in a clearly recognizable fashion and which appear to segregate as a single gene. Several strategies have been developed that allow one to rapidly screen a large number of random, unmapped molecular markers in a relatively short time and to select just those few markers that reside near the target locus. Basically, these strategies combine high-volume marker technologies, such as RAPDs and AFLPs, with regional targeting strategies, such as near isogenic lines (NILs) and bulked segregant analysis (BSA), allowing one to efficiently target markers linked to specific loci (44–46). Bulked segregant analysis can be useful for a wide variety of populations and traits. It is the most useful for tagging genes with major phenotypic effect and it has now been used to tag major genes in many plant species (47, 48). In fact, while the generation of NILs is very time consuming, as it requires several cycles of backcrossing, the only prerequisite for BSA is the existence of a population (F2, BC, RI, etc.) segregating for the target locus or region, and originating from a single cross.

The BSA strategy involves the grouping of individuals from the segregating population into two contrasting bulks or pools selected on the basis of screening at a single target locus or region. Consequently, the two resultant bulks or pools will be genetically different only at loci contained in the chromosomal region close to the target region, and will be heterozygous and monomorphic for all other regions (*Figure 1*). The selection of the contrasting individuals can be based either on the phenotype (e.g. resistant versus susceptible to a particular disease) or the genotype (44, 45). Therefore, bulks can be made to identify markers in regions of the genome with poor marker coverage, such as gaps in genetic map or ends of linkage groups. In this case, pooled DNA samples will be selected on the basis of DNA markers flanking the genomic region of interest (45). DNA is extracted from each bulk, and the two samples assayed for the marker system of choice. Any marker that is unlinked to the gene to be located will produce similar banding patterns in both samples because it will be segregating independently of the target gene. On the contrary, for those markers closely linked to the target gene the two samples will show different banding patterns (*Figure 1*).

The optimum bulk size depends on several factors, including genome size of the species, the type of the marker being screened (dominant or co-dominant) and the type of population used to generate the bulks. Plant samples from groups of five to ten individuals should be pooled and extracted in bulks to ensure that any non-targeted genomic region would be represented by both parents.

2.6 Gene mapping in two model plant species

For model plant species, such as Arabidopsis and tomato, several mapping resources are available, which can facilitate the location on the genetic map of a new gene (e.g. a mutant with a specific phenotype) or a DNA clone. Information concerning most genomic resources and data about Arabidopsis, including genetic markers and mapping populations, can be found in the Arabidopsis Information Resource (TAIR) database (URL = http://arabidopsis.org/).

In Arabidopsis the location of a new gene on the genetic map can be achieved by crossing the 'clean' mutant line with an appropriate genotype that differs from the original mutant for several previously mapped loci. For morphological markers, several tester lines, predominantly in ecotypes Landsberg *erecta* (Ler) background, are available, which allow the detection and measurement of linkage. If the scope of the project is to locate the new gene in relation to molecular markers (e.g. RFLPs, RAPDs, SSRs, AFLPs) then, preferentially, a cross can be made with one of the available lab ecotypes Ler, Columbia (Col), Wassilewkija (Ws), and so on, for which information on the various polymorphisms is known already (49).

For an initial rough mapping of a new locus, a robust and cheap method has been developed, based on an *A. thaliana* RFLP mapping set (ARMS) which allows to accelerate and simplify the RFLP analysis (50, 51). The ARMS markers are subcloned DNA fragments detecting *Eco* RI polymorphisms and revealing a single polymorphic band per ecotype. They are, therefore, scorable on a single *Eco* RI blot, and it is also possible to choose and analyse in a single southern experiment those ARMS markers whose signals do not overlap. A total of 32 ARMS markers evenly distributed on the five chromosomes have been selected, which reveal 23 and 27 polymorphisms for the ecotype crosses Ler/Enkhein (En) and Ler/Col, respectively. It is thus sufficient to test about 20 individuals homozygous for the mutant allele, to get a rough mapping of any mutant generated in either Ler, Col or En. All clones for ARMS markers are available through the DNA stock centres at ABRC in Columbus, Ohio and at Köln, Germany.

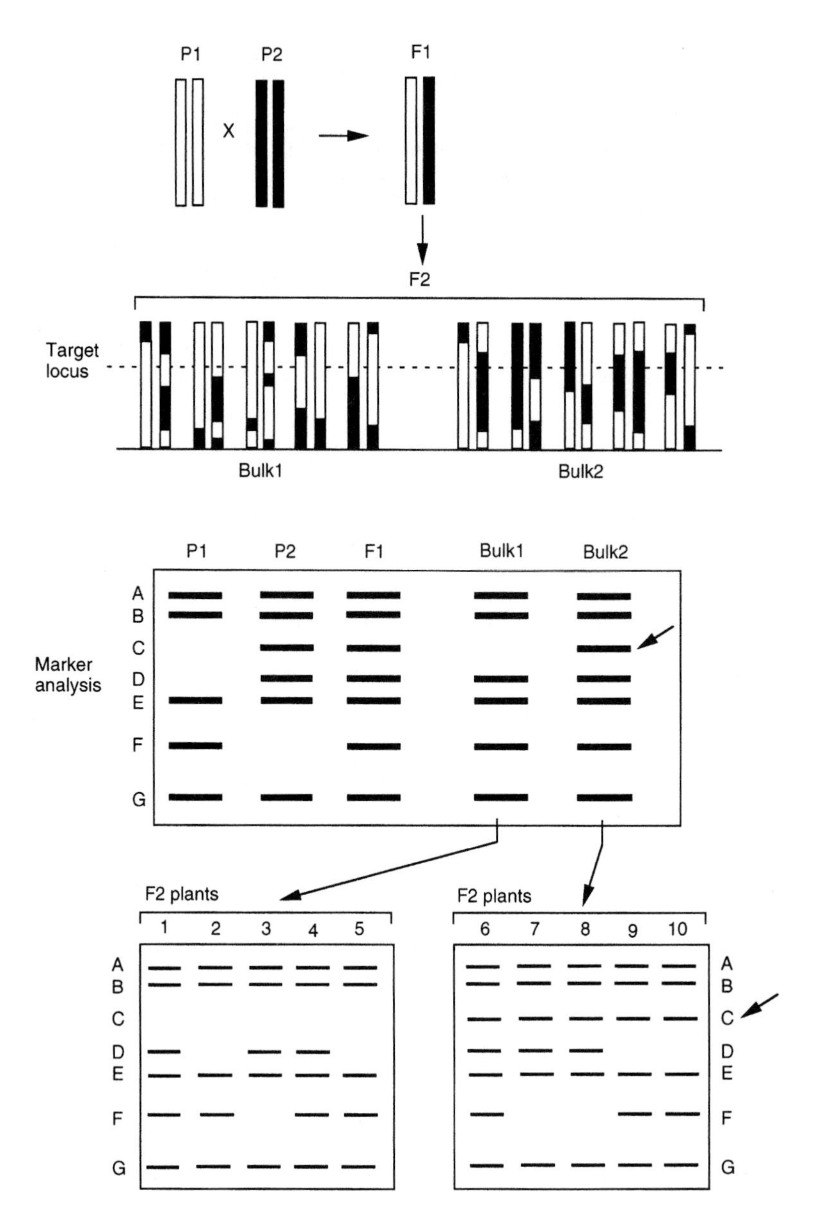

Figure 1. Schematic description of bulked segregant analysis for an F2 population generated from the cross P1 (recessive phenotype) × P2 (dominant phenotype). The schematic shows genotypes at seven loci (A–G) detected in the two parents (P1 and P2), their F1 and F2 progenies, and bulks derived from F2 individuals selected for the recessive phenotype (Bulk 1) or the dominant phenotype (Bulk 2). Bulk 1 plants will contain only P1 alleles near the target locus, while Bulk 2 individuals will contain alleles from both P1 and P2. The dominant allele at locus C is linked in cis to the dominant allele of the target locus and, therefore, is polymorphic between the two bulks. The other two loci that are polymorphic between the parents (D and F) are unlinked to the target locus and, therefore, will both carry alleles from both P1 and P2, and appear monomorphic between the bulks.

Recently PCR-generated markers such as RAPDs (6, 42), AFLPs (8), cleaved amplified polymorphic sequences (CAPS; 52) and SSLPs (53) have been developed for Arabidopsis. Using a set of 18 CAPS markers, Konieczny and Ausubel (52) showed that an Arabidopsis gene could be unambiguously mapped to one of the 10 Arabidopsis chromosome arms using only 28 F2 progeny plants derived from a single cross. New Arabidopsis CAPS markers are continuously being developed, and have been catalogued by Eliana Brenkard and Fred Ausubel in a tabular form including primer sequence, restriction enzyme and number of cuts for each of the examined ecotypes. The list can be accessed via the World Wide Web browser under *http://www.arabidopsis.org/*. SSLPs are used for rapid mapping of mutations and for construction of a physical map by sequence tagged sites (STS) content mapping (53).

The whole-genome sequence information that is being generated for this model plant species is further stimulating the production of large numbers of molecular markers, thus increasing the potential for fast gene mapping approaches. Cereon genomics has produced a collection of approximately 39 000 predicted Arabidopsis single-nucleotide polymorphisms (SNPs) and small insertion/deletions (INDELs) between Col and Ler ecotypes, which can be accessed at *http://www.arabidopsis.org/*. Cho *et al.* (37) have demonstrated that the application of high-density oligonucleotide arrays to the genotyping of hundreds of biallelic SNPs in a single experiment can accelerate the mapping of traits.

If the scope of the project is to genetically locate a cloned DNA sequence on the Arabidopsis molecular map, for Arabidopsis one can use existing recombinant inbred lines (RILs), taking benefit of the available data. Two RIL sets are available through both Arabidopsis stock centres; one set of 300 RILs was derived from the cross Ler × Col (54) and another set of 153 RILs was derived from the cross W100 (mainly Ler background) × Ws (42) (*http://www.arabidopsis. org/*). Files with the segregation data from the W100/Ws and from the Ler/Col RILs are available through AtDB. RIL data can be analysed with the software packages MAPMAKER (21) or JOINMAP (22).

In tomato, cytogenetic stocks have been invaluable tools in the assignment of mutants to particular chromosomes. For morphological markers, linkage tester stocks have also been developed for all 12 chromosomes. In addition, advanced marker stocks have been devised which contain markers which can be evaluated at the seedling stage (Tomato Genetics Resource Centre, Davis, CA; http://tgrc. ucdavis.edu).

If the new gene allele needs to be mapped in relation to molecular markers and no sequence information of the mutant allele is available, a segregating population can be obtained by crossing the mutant line (likely in *L. esculentum* background) with an appropriate wild species, which should allow an easy scoring of the mutant phenotype as well as sufficient DNA polymorphism. Once a population segregating for the gene of interest has been created, bulked-segregant analysis (BSA) could be performed to rapidly identify DNA markers linked to the target locus. The identified RAPD or AFLP markers could then be cloned and used as RFLP probes to be localized on the high density RFLP map of tomato (55, 56). Data on polymorphism for most of the RFLP markers mapped on the high-density tomato map (55, 56) can be found in the Solgenes database (http://genome.cornell.edu/solgenes/). The surveys include the two parental species of the mapping population, *L. esculentum* (LA0490) and *L. pennellii* (LA0716) together with *L. cheesmanii* (LA0483).

In tomato a rapid method for assigning a map position to DNA sequences is based on the use of an introgression line (IL) population that is composed of 75 *L. esculentum* (cv. M82) lines each containing a single RFLP defined introgression from the green fruited species *L. pennellii* (LA716; 57, 58). Each of the ILs is nearly isogenic to the cultivated tomato and together the lines provide complete coverage of the tomato genome. The orientation of the IL population is relative to the F2 RFLP map that includes more than 1500 markers that span 1274 cM (55, 56). The 75 ILs divide the tomato genome into 107 mapping bins, each defined by a unique composition of genome coverage and partitioning the map into segments with an average size of 12 cM. Through probing of the IL membranes with DNA probes it is possible to associate sequences to the specific bins. The high level of polymorphism at the DNA level between the two syntenic species, *L. esculentum* and *L. pennellli*, ensures high mapping efficiency and the perpetual nature of the population allows the accumulation of mapping information from different research group into a single database. The ILs provide an efficient tool for low resolution mapping of DNA clones to the tomato genome. High resolution mapping can be achieved through analysis of F2 generations resulting from crosses of the targeted IL with the cultivar M82. The syntenic relationships of tomato with potato and pepper make the ILs an efficient mapping resource for Solanaceae genetics. The high RFLP variation between the parental species of the ILs allows also for very efficient mapping of heterologous DNA sequences. Together, the set of 75 ILs allow virtually all copies of a heterologous gene family to be mapped in tomato.

2.7 Relationship between genetic and physical maps

Genetic maps, although useful for many purposes, do not tell us how far apart markers within a linkage group really are in actual physical (base pair) distances. Even if an average kb/cM value can be deduced from the genome size and the size of the genetic map, this value can vary substantially (up to two orders of magnitude from the mean value) depending on the regions of the chromosome (59). Very often regions near centromeres, as well as heterochromatic and introgressed regions, show suppressed recombination, whereas other chromosomal regions show a higher rate of recombination. As a consequence, genetic maps appear to have many markers clustered in regions very likely corresponding to centromeres or heterochromatin, while in other regions of the genome markers are separated by larger gaps. Under these circumstances, a marker tightly linked to a gene of interest may in reality be many kilobases away. This may be acceptable for marker-assisted selection, but not for map-based cloning. For cloning purposes it is usually necessary to saturate the map with as many markers as possible and then to integrate genetic and physical maps.

Molecularly-based physical maps are constructed using large fragments of genomic DNA cut with restriction enzymes that can be sized and ordered, usually by pulsed-field gel electrophoresis (PFGE; 60) or related techniques. Another molecular strategy uses cloned DNA such as yeast artificial chromosomes (YACs), bacterial artificial chromosomes (BACs), cosmids, or shorter phage vectors to establish overlapping assemblies of clones, known as contigs (60) (see Chapter 7).

3 Isolation of genomic DNA from plant tissue

Several methods exist for the isolation of high molecular weight genomic DNA suitable for molecular marker procedures such as RFLP analysis and PCR amplification. All DNA preparation methods involve the removal of the envelopes (cell wall and nuclear membrane) around the DNA, the separation of the DNA from all other cell components such as cell wall debris, proteins, lipids or RNA, and the maintenance of the integrity of the DNA during the procedure, i.e. protection from nucleases and mechanical shearing. In the most common method, which is applicable to a wide range of plant material, cells are opened by grinding the plant tissue in liquid nitrogen. The low temperature also prevents nucleases from degrading the DNA.

We describe two DNA extraction protocols, which are modifications of the procedure originally described in Murray and Thompson (61). The large prep procedure (*Protocol 1*) as modified in Bernatzky and Tanksley (62), is useful for extracting large amounts (250–1000 μg) of very pure, high molecular weight DNA, as might be required for large-scale mapping purposes. This method is known to work on tomato, potato, and other succulent plant tissue. However, it is time-consuming and limited by the large volumes necessary to centrifuge. When DNA from large numbers of plant samples is needed, the microprep protocol (63) is more efficient (*Protocol 2*). The general procedure is similar, but it has been adapted to be done entirely in 1.5 ml microfuge tubes. This allows many more samples to be done in a day (up to several hundred per person) with a minimal reduction in quality. The yield is much less, 10–20 μg of DNA per sample, but is sufficient for two to four Southern blots or 50–100 PCR reactions; ample for marker-assisted selection and PCR-based marker technologies. The number of samples can be maximized by using two drills with drill stands concurrently and an on/off switch operated by a foot pedal. This protocol is known to work on tomato, pepper, potato, apple, tobacco, strawberry, *Arabidopsis*, artichoke, rice, and maize.

It is important, especially for the microprep procedure, that the tissue harvested be as young as possible; as plants age, starches and phenolics build up, and the tissue becomes tougher, making it more difficult to lyse the cells. Also, the cells in young meristematic leaves are smaller; therefore there are more nuclei per gram of tissue, thus maximizing yield. For micropreps, four to eight new leaflets, up to 1.5 cm long, are harvested from 1- to 3-week old seedlings. For the large prep procedure, good results can be obtained with somewhat older tissue. However, certain monocots, such as rice and oats, have more lignified tissue, and may require hand grinding in the presence of liquid nitrogen and somewhat altered protocols (64, 65). Other protocols for DNA isolation from plant tissues are described in Chapters 3, 5, and 7.

Protocol 1

Large prep protocol

Equipment and reagents

- Kitchen blender
- Water-bath at 65 °C
- Ultracentrifuge (e.g. Beckman J2-21M, JA10 rotor)
- 250 ml centrifuge bottles
- Paint brushes (with bristles approximately 1–3 cm long)
- Cheesecloth
- Bent Pasteur pipettes
- Centrifuge with swinging buckets for 50 ml tubes (e.g. Beckman TJ-6)
- 50 ml polypropylene conical tubes and microfuge tubes
- Microcentrifuge with a fixed-angle rotor, capable of 10 000 g
- DNA Extraction Buffer (DEB): 0.35 M sorbitol, 0.1 M Tris-base, 5 mM EDTA, pH 7.5 (store at 4 °C)
- Nuclei lysis buffer (NLB): 0.2 M Tris, 0.05 M EDTA, 2M NaCl, 2% (w/v) CTAB, pH 8.25
- N-lauroylsarcosine 5% (w/v)
- TE buffer:10 mM Tris-HCl pH 8.0, 1 mM EDTA
- Chloroform-isoamyl alcohol (24:1) (CI)
- Isopropanol and 70% (v/v) ethanol, prechilled to −20 °C

Method

1. Add 0.3–0.5 g sodium bisulfite/100 ml DEB immediately before use. Blend leaf tissue (20–35 g) in blender using 150 ml of DEB, for approximately 1–1.5 min.
2. Filter mixture through two layers of cheesecloth into 250 ml centrifuge bottles kept on ice.
3. Spin for 15 min, at 700 g, with ultracentrifuge pre-chilled to 4 °C .
4. Gently discard the supernatant and resuspend the pellet in 5 ml DEB using the paint brush.
5. Transfer to a 50 ml polypropylene conical tube containing 2 ml of 5% N-lauroylsarcosine.
6. Add 5 ml of NLB to the centrifuge bottle to rinse and use the paintbrush to transfer the buffer into the 50 ml polypropylene conical tube.
7. Mix gently, cap tightly, and incubate in water-bath at 65 °C for 15–20 min.
8. Add 15 ml of CI, working in the fume-hood. Cap very firmly and mix by inversion 30–40 times to get an emulsion.
9. Spin in centrifuge with swing-out rotor for 15 min at 700 g.

Protocol 1 continued

10 Pipette off the aqueous phase into a new 50 ml polypropylene conical tube using wide bore tips.

11 Add the same volume of chilled isopropanol and gently invert until DNA precipitates.

12 Hook out DNA using a bent Pasteur pipette, rinse the pellet in chilled 70% ethanol, gently dry it on a kimwipe, and transfer it to a new microfuge tube. Resuspend in TE buffer (usually 500 μl), by warming up in 65 °C water-bath for 30–60 min, and vortexing.[a]

13 Once resuspended, centrifuge at 10 000 g for 15 min to pellet out the starch, and (optional) transfer to a clean tube.

[a] If the DNA cannot be hooked, it is also possible to spin it down to a pellet (approximately 15 min at 700 g), gently pour off the supernatant, wash with 70% ethanol, pour off the ethanol, let the pellet dry, and resuspend it in TE directly in the 50-ml tube.

We generally assess concentration and quality (digestibility and eventual shearing or degradation) of the genomic DNA in 0.9% agarose concentration gels by comparing undigested and digested aliquots of DNA with DNA standards.

Protocol 2

Microprep protocol for extraction of DNA from tomato and other herbaceous plants[a]

Equipment and reagents

- Drill: heavy duty household drill with keyless chuck (so pestles can be easily replaced) and plastic disposable pestles (e.g. VWR Scientific, catalogue no. KT95050–99)
- Microcentrifuge: with a fixed-angle rotor, capable of 10,000 g and holding as many samples as possible
- Water-bath at 65 °C
- 1.5 ml microfuge tubes
- DNA extraction buffer (DEB) (see *Protocol 1*)
- Nuclei lysis buffer (NLB) (see *Protocol 1*)
- Chloroform-isoamyl alcohol (24:1) (CI)
- TE buffer (see *Protocol 1*)
- 5% (w/v) N-lauroylsarcosine
- Microprep buffer (MB): 2.5 parts DEB, 2.5 parts NLB, 1.0 part 5% (w/v) N-lauroylsarcosine. Add 0.3–0.5 g sodium bisulfite/100 ml buffer immediately before use
- Isopropanol and 70% (v/v) ethanol, prechilled to −20 °C

Method

1 Collect 50–100 mg of leaf tissue (approximately four to eight new leaflets, up to 1.5 cm long) from a 1- to 3-week-old tomato seedling and nestle loosely in the bottom of a 1.5 ml microfuge tube.[b]

Protocol 2 continued

2 Prepare fresh MB, keep at room temperature.

3 Add 200 μl of MB and grind tissue with power drill and plastic bit (rinsing pestle with water between samples); add another 550 μl of MB and either vortex lightly or shake entire rack by hand.

4 Incubate in 65 °C water-bath for 30–120 min.

5 Fill the tube with CI and mix well.[c]

6 Centrifuge the tubes at 10 000 g for 5 min.

7 Pipette off aqueous phase into new microfuge tube. Add 2/3–1 times the volume of cold isopropanol to each tube. Invert the tubes until DNA precipitates.

8 Immediately spin at 10 000 g for 5 min (no more), pour off the isopropanol and wash the pellet with 70% ethanol. (At this point the DNA pellet can be stored in 70% ethanol at −20 °C indefinitely).

9 Dry the pellet by leaving the tubes upside down on paper towels for approximately 1 h or using a vacuum-centrifuge.

10 Resuspend the DNA in 50 μl of TE at 65 °C for 15 min.

11 Centrifuge at 10 000 g for 10 min and store at 4 °C for up to 1 week, or at −20 °C for longer storage.

12 For RFLP use, digest 15–25 μl for one Southern blot and for PCR use 1 μl.

[a] See Fulton *et al.* (1995) for possible solutions to specific problems.

[b] If only PCR is needed, as little as one cotyledon can be used; resuspend in 50 μl of TE in step 10, but use 5 μl for one PCR reaction. Tissue can be harvested and kept at room temperature for up to 3 h or stored at 4 °C for up to 3 days.

[c] This can be done by vortexing each tube or by sandwiching the tubes between two racks and vigorously inverting or shaking up and down 50–100 times.

4 Restriction fragment length polymorphism (RFLP) analysis

The detection of an RFLP requires extraction and purification of the DNA from an individual (as described above), and digestion of the DNA with a restriction endonuclease to form a mixture of restriction fragments, differing in length according to the specific distribution of cleavage sites in the DNA for that particular enzyme (3). Variation in the distribution of restriction sites can occur through the loss or gain of sites in the genomic DNA by base substitution, insertions, or deletions. This digested DNA is then separated by size in electrophoresis gels and fixed onto a membrane via Southern blotting. Clones representing unique sequences of DNA from the nuclear genome are used as radioactive (or non-radioactive) probes on the Southern blots to detect homologous sequences. Allelic variations are therefore identified as differences in the size of the restriction fragments to which the probe hybridizes.

4.1 Restriction enzyme digestion and gel electrophoresis

The enzymes that are commonly used for RFLP analysis require 4–6 base pair (bp) recognition sequences. Generally, the level of polymorphism for a given species needs to be determined empirically. The chances of finding a useful polymorphism can be increased by using enzymes that cut more frequently or simply by using a greater variety of enzymes.

For the interpretation of RFLP bands, it is absolutely imperative that complete digestion occurs. Generally 2–5 units of enzyme per 1 μg of plant nuclear DNA should guarantee complete digestion. The amount of DNA needed per digestion depends on the genome size of the organism and the purpose of the project. For example, the detection of single copy nuclear genes in plants with smaller genomes, e.g. *Arabidopsis* and rice, can be analysed with approximately 1 and 3 μg of DNA per gel lane, respectively, whereas species with larger genomes will need proportionately more. Larger amounts become necessary if repeat probing of the same membrane is expected. In tomato we generally use 10–15 μg of DNA per lane on both surveying and mapping filters (charged nylon membranes), which allows (radioactively) reprobing of the same membrane up to an average of 15 times.

The total volume of the digestion reaction is flexible (up to the maximum that can be loaded into each well of the electrophoresis gel), depending on the concentration of the DNA needed. In order to avoid non-specific or 'star' activity of some enzymes, the final glycerol concentration should not exceed 5% of the total volume of the reaction and, therefore, the enzyme used should not exceed 10%. Good mixing of the reaction is also very important. Spermidine can be added to help digest relatively crude and difficult to cut DNA preparations. As spermidine can precipitate the DNA, it is advisable to add the reagents in the following order: water, enzyme reaction buffer, spermidine, DNA sample, and restriction enzyme.

Lambda DNA digested with *Hind* III is a convenient general-purpose size standard covering a range of sizes from 500 bp to 23 Kb, which can be run on the same gel as the digested DNA for comparison. For UV visualization use 0.25 μg of lambda; however, only 50 ng is necessary if the standard is to be blotted, radiolabelled, and visualized by autoradiography.

After digestion, the DNA must be run on an electrophoresis gel to be visually checked for complete digestion and concentration, and prepared for Southern blotting (see *Protocol 3*). There are several types of gel systems available; the one we prefer is the submerged, horizontal agarose gel system. The time required for running a gel can be varied. However, running a gel at too high a voltage can generate excessive heat, causing distortion of the DNA bands. The running time can also be reduced by diluting the buffer concentration in the gel and running tray, but this produces some loss in resolution. We recommend that gels to be used for RFLP analysis be run for at least 12 to 14 h, using 1 × NEB as the buffer for the gel and the running tray. We find it convenient to use 20 × 25 cm gel trays, 300-ml gels, and 30–40 well combs.

Protocol 3

Digestion and separation of restriction fragments of genomic DNA by agarose gel electrophoresis

Equipment and reagents

- Apparatus for submerged horizontal gel electrophoresis including 20 × 25-cm gel trays (e.g. the BRL H4 system)
- Erlenmeyer flask
- UV transilluminator
- Rocking platform shaker
- Appropriate restriction endonucleases and their corresponding buffers
- DNA size marker (e.g. *Hind* III digested lambda)
- 10 × neutral electrophoresis buffer (NEB): 1 M Tris, 10 mM EDTA, 125 mM sodium acetate (trihydrate), pH to 8.1 with acetic acid
- Stop buffer: 70% (v/v) glycerol, 0.5 × NEB, 20 mM EDTA, 0.2% (w/v) SDS, 0.6% (w/v) bromophenol blue (Sigma B-6131)
- Agarose (e.g. Boehringer Mannheim, Agarose LE)
- Ethidium bromide stock solution (10 mg/ml)[a]

Method

1. Digest an appropriate amount of genomic DNA with the desired enzymes using a suitable restriction buffer, and the incubation temperature suggested by the manufacturer. Incubate for between 3 h and overnight.
2. After digestion stop the reaction with 1/5 volume of stop buffer. The DNA is ready for electrophoresis (for long-term storage, the samples should be refrigerated or frozen).
3. In a 1 l Erlenmeyer flask combine 300 ml 1 × NEB and 2.7 g for a 0.9% gel.
4. Cover the mouth of the flask with plastic film and heat on a hot plate with continuous stirring or in a microwave with occasional swirling until no agarose particles can be seen. Allow to boil for at least 30 s.
5. Cool the agarose to 65 °C in an ice bath on top of a stir plate with continuous stirring.
6. Prepare the gel form. For example, if using the flat open-end type, tape up the ends securely to prevent leaks.
7. Pour the agarose into the mould and insert the comb very carefully. Allow the gel to harden for at least 30–60 min.
8. Prepare enough 1 × NEB from 10 × stock to fill the gel running tray. When the gel has completely cooled and solidified, remove the tape and place it in the gel running tray. The final level of buffer should be approximately 5 mm above the gel.

Protocol 3 continued

9 Carefully remove the comb. Apply the samples to the wells by using a microlitre pipettor. Be sure to include a lane of 50 ng lambda DNA digested with *Hind* III.

10 Run the DNA samples into the gel at approximately 25 V/50 mA overnight. The blue dye should migrate approximately 10 cm from the origin.

11 Place the gel, bottom side up, in a gel-staining tray (plastic photographic trays work well). A thin sheet of Plexiglass under the gel makes it easier to handle, transport, etc. Stain the gel in 0.5 μg/ml ethidium bromide (25 μl of a 10 mg/ml ethidium bromide stock solution per 500 ml) on a shaker for 15 min.

12 Pour off the stain solution into ethidium bromide liquid waste container[a]. Rinse the gel with distilled water and cover with water. Destain for 15 min. Pour off the water and rinse the gel.

13 Photograph the gel using short-wave UV light. Be sure to ware UV-blocking goggles.

[a] Caution: ethidium bromide is a powerful mutagen and carcinogen. Avoid contact with skin or other surfaces. Handle with extreme care; wear plastic disposable gloves at all times. Dispose of solutions and gels containing ethidium bromide in accordance with local safety arrangements.

4.2 Southern blotting

There are a variety of commercially available membranes for immobilization of nucleic acids, each of which may have different protocols supplied by the manufacturer (66). Nitrocellulose and nylon (charged or uncharged) membranes are the most popular matrices and are suitable for most applications. Nylon filters are preferred if it is likely that the blot will be repeatedly hybridized with different probes, as they are more robust than nitrocellulose ones, which tend to fall apart during multiple probing. Positively-charged nylon has the further advantage that, under alkaline conditions, DNA becomes covalently bound to the matrix. Therefore, with the use of 0.4 M NaOH as a transfer solution, denatured DNA becomes irreversibly coupled to the filter during transfer making additional baking or irradiation steps unnecessary, and reducing the loss of DNA during sequential rounds of probing, washing, and reprobing. In order to improve DNA transfer during blotting, the gel is subjected to an initial acid treatment, which results in partial depurination of DNA fragments. The following alkali treatment denatures the DNA to single strands and causes phosphodiester strand scission at depurinated nucleotides, which in turn results in the production of 1–2-kb fragments. It is important not to over treat the gel to prevent the production of very small fragments of DNA, which will not hybridize well to the probe. The procedure for alkali blotting to a positively-charged nylon membrane is outlined in *Protocol 4*.

Protocol 4

Alkali blotting to a positively-charged nylon membrane

Equipment and reagents

- Rocking platform shaker
- Plastic or glass box (e.g. plastic photographic trays)
- Sheet of Plexiglass
- Nitrocellulose slab sponges
- Whatman 3MM filter paper
- SaranWrap or precut sheets of acetate
- Charged nylon membrane: (e.g., Hybond-N^+, Amersham); cut to the desired filter size, usually 11×20 cm[a]
- 0.25 M HCl
- Transfer solution: 0.4 M NaOH
- $20 \times$ SSC: 3M NaCl, 0.5 M trisodium citrate pH 7.0

Method

1. Acid treatment (depurination): After photographing the gel (remember also to place a ruler by the side for subsequent molecular weight estimations of hybridizing fragments), rinse it with distilled water and soak in 1 l of 0.25 M HCl solution. Leave the gel for 10 min at room temperature on a shaking platform set at low speed.
2. Alkali treatment (denaturation): rinse the gel with water and denature the DNA by soaking the gel in 1 l of 0.4 M NaOH for 10–15 min.
3. Preparing the capillary transfer apparatus (see *Figure 2*): Set up two pieces of nitrocellulose slab sponges in a gel tray and semi-submerge them with 0.4 M NaOH; push out all air bubbles from the sponges first, and add more buffer as needed until the tray is approximately half full. Wet two pieces of Whatman 3MM filter paper in 0.4 M NaOH and use to cover the sponges. Smooth the Whatman 3MM paper over the sponges with a gloved hand by rolling a clean pipette over it. This removes any air bubbles that may be trapped.[b]
4. Assembling the blot: place the denatured gel (DNA side up) on the saturated Whatman 3MM paper on the slab sponges or on the level platform. Remove any air bubbles that may be trapped between the paper and the gel by rolling a clean pipette over the surface of the gel or by gently pressing the gel with a gloved hand. Take the positively charged nylon membrane and place it on the top of the gel. Remove any bubbles that may be trapped between the filter and the gel. Place SaranWrap or pre-cut sheets of acetate around the edges of the gel so that no parts of the nylon filter can directly contact the Whatman 3MM paper. Place two sheets of moist Whatman 3MM paper on top of the filter and stack paper towels to a thickness of 4 cm on top of the Whatman 3MM paper. Put a Plexiglass plate on top of the paper towels and add a small weight (approx. 1 kg) to the top. Leave the blot from a minimum of 3 h up to overnight.

Protocol 4 continued

5 Disassembling the blot: remove the weight, paper towels, and 3MM paper. The gel will have compressed leaving a thin sandwich. Carefully peel off the filter and let it soak in a tray containing 200 ml of 2 × SSC for several min to neutralize the transfer solution and to remove any agarose particles stuck to the filter. Lay the filter on a clean piece of Whatman 3MM paper. The filter may be pre-hybridized or stored (moist) between sheets of Whatman 3MM paper in a cool place.

[a] Always wear disposable plastic gloves while handling the membranes.

[b] Alternatively, the Whatmann 3MM paper can be set up on a flat platform (e.g. sheet of Plexiglass) and the ends of the Whatman 3MM paper should dip into the 0.4 M NaOH reservoir and serve as wick.

4.3 Preparation of nucleic acid probes

There are several methods that can be used to either radioactively or non-radioactively label DNA (67). Although the sensitivity of detection of non-radioactive methods begins to approach that achieved with ^{32}P, we still find isotopic methods more convenient for routine probing of numerous filters. More frequently, ^{32}P-labelled deoxynucleotide triphosphate ([^{32}P]dNTP) is incorporated into a newly synthesized DNA either by a random-priming reaction (68) or by nick translation (69). We use the 'random hexamer' procedure (68), which leads to generally high specific activity probes (10^8 to 3×10^9 cpm/μg) acceptable for most applications (*Protocol 5*). Note that commercial kits are available, reasonably cost-effective, and reliable.

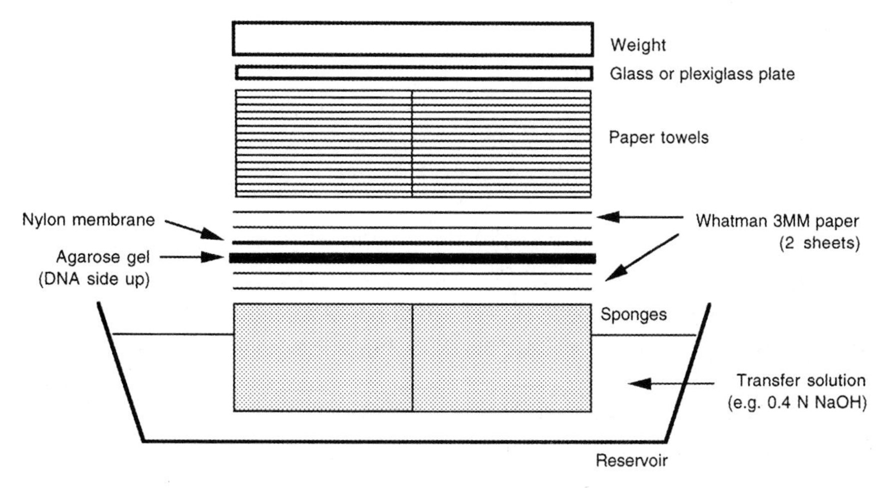

Figure 2 Diagram of Southern blot transfer apparatus.

Protocol 5

Probe labelling: random-priming of DNA

Equipment and reagents

- Heating blocks at 100 °C and 37 °C
- Single-stranded DNA or denatured double-stranded DNA sample
- 5 × oligolabelling buffer (see A below)
- *E. coli* DNA polymerase I, Klenow fragment, 2 units/μl, in 50% glycerol (store at −20 °C)
- Bovine serum albumin (BSA) (fraction V, Sigma) 10 mg/ml
- Sterile double distilled H_20
- [α-^{32}P]dCTP (> 3000 Ci/mmol; 10 mCi/ml)
- NaOH 0.4 N

Method

A. Preparation of oligolabelling buffer (OLB)

1 5 × OLB is made by adding the following three solutions in the ratio of 100:250:150 (A:B:C).

2 Solution A (DTM):

- 1.25 M Tris–HCl, 0.125 M $MgCl_2$, pH 8.0 — 1.0 ml
- 2-mercaptoethanol — 18.0 μl
- 100 mM dATP — 5.0 μl
- 100 mM dGTP — 5.0 μl
- 100 mM dTTP — 5.0 μl

3 Store at −20 °C.

4 Solution B (Hepes):

- 2 M Hepes titrated to pH 6.6 with 5 M NaOH. Store at -20 °C.

5 Solution C (OL):

- Hexadeoxynucleotides [pd$(N)_6$, Pharmacia] resuspended in TE to a concentration of 90 OD units/ml. Store at −20 °C.

B. Oligolabelling

1 Allow the reagents to thaw on ice. Keep the *E. coli* DNA polymerase I (Klenow fragment) at −20 °C and only take it out when necessary, returning it to the freezer immediately after use.

2 In a microfuge tube, add probe DNA and sterile double-distilled water to a final volume of 31 μl (25–100 ng probe).

3 Denature the DNA by heating at 95 °C for 5–10 min and subsequently cooling on ice for at least 1 min. Spin briefly in a microfuge to collect condensate.

4 Add 10 μl of 5 × OLB and 2 μl BSA.

5 Behind appropriate shielding, add 5 μl (50 μCi) of [α-^{32}P]dCTP.

6 Add 2 μl Klenow (final volume of reaction is 50 μl).

Protocol 5 continued

7 Incubate the tube in a 37 °C heating block for 1–2 h (behind appropriate shielding).

8 Remove unincorporated nucleotides using spin column technique (see *Protocol 6*)[a].

9 Denature the purified probe by adding an equal volume of NaOH 0.4 N. Allow to rest for 10 or more min. (Add labelled standard if desired). The labelled probe is now ready for hybridization.

[a] This step can be omitted but then there is no way of knowing if the labelling step has worked.

Protocol 6

Purification of DNA probes on Sephadex spin columns

Equipment and reagents

- Labelled DNA probe mixture (*Protocol 5*)
- 0.5 × 5 cm (1 ml) disposable plastic syringe cut to fit into microcentrifuge.
- Pasteur pipettes (plastic disposable or glass, 1 ml)
- Glass wool
- SDS/EDTA buffer: 1% SDS, 25 mM EDTA
- Sephadex slurry: a 5% (w/v) suspension of Sephadex G50 (Pharmacia) in SDS/EDTA

Method

1 Compact a small portion of glass wool at the bottom of a pre-cut 1 ml disposable syringe. The amount of glass wool should not pass the 0.1 ml mark in the syringe.

2 Using a Pasteur pipette, add Sephadex slurry to the syringe.

3 Place the syringe in a 1.5 ml microfuge tube and spin at 1600 g (for Sephadex G-50) for 1 min.

4 Take the Sephadex column out of the tube and remove excess buffer.

5 Add more Sephadex slurry and repeat the spin. Continue until the column fills the syringe leaving a small gap at the top for addition of the radiolabelled reaction mixture. It is important to consistently spin the columns at the same speed and for the same time each time it is done.

6 Take the Sephadex column out of the tube and place it in a new and labelled 1.5 ml microfuge tube.

7 Gently pipette the entire radiolabelled reaction mixture onto the column.

8 Add 50 μl of SDS/EDTA directly on top. Wait about 30 s, then add another 50 μl.

9 Spin at 1600 g for 20–30 s in microcentrifuge. If necessary add another 50 μl EDTA/SDS and spin again.

10 Remove and discard the columns and add 100 μl SDS/EDTA to the tube, and mix.

11 Count 2 μl of sample in H_2O in scintillation counter and calculate specific activity. You should have a specific activity between 10^8 and 10^9 cpm per μg of DNA template.

4.4 Hybridization analysis of Southern blots

The hybridization between membrane-bound single stranded DNA and labelled DNA probes is regulated by many factors including temperature, ionic strength, GC content, probe length, probe concentration, and time (70). There are a number of protocols for Southern blot hybridization using a variety of reagents and formulations. We have been using the one described by Church and Gilbert (71), which has proven to be simple, versatile and robust.

Due to the large number of filters that often need to be probed at one time (up to 10–15 filters), we find it easier to handle Southern hybridizations in plastic boxes with tight-fitting lids. For example, the 15 × 22-cm hard plastic 'drawer organizer' boxes made by Rubbermaid can be stacked and a box on top can serve as a lid; thus, allowing many boxes to fit in a single hot-air incubator. A minimum of 50 ml of hybridization buffer is used for a single 20 × 11 cm filter and as much as 100–150 ml is used for 10 filters in the same box. Drying of the top filter can be prevented either by laying a sheet of acetate on top of the buffer or by sealing the boxes in plastic bags. Shaking during prehybridization or hybridization is optional, but it is important during washing. Nylon window screening cut to the dimensions of the washing box and placed between the filters helps to reduce background by keeping the filters separate during the wash.

Protocol 7

Southern hybridization

Equipment and reagents

- Southern blot (from *Protocol 4*)
- Plastic hybridization boxes, such as Rubbermaid, and one larger one for washing (approximate capacity of 3–4 l)
- X-ray film (e.g. Kodak)
- X-ray cassette with intensifying screen
- Cling film (or any similar UV-transparent film)
- Whatman 3MM paper
- −80 °C freezer
- Radioactive DNA probe (from *Protocol 5*)
- Geiger counter
- Hybridization buffer (HB): 7% (w/v) SDS, 0.5 M Na_2HPO_4 pH 7.2, 1% bovine serum albumin (BSA)[a]
- Rotating hot-air incubator (e.g. Lab-line orbit environ-shaker)
- Denatured salmon testes DNA (ST). Dissolve 5 g salmon testes in 1 l water. Autoclave. Store aliquots at −20 °C. Immediately prior to use, denature the DNA in a boiling water-bath for 10 min, then snap-cool on ice.
- 20 × SSC (see *Protocol 4*)
- Nylon window screenings cut to the dimension of the washing box (optional)
- Wash solution (2×): 2 × SSC, 0.1% (w/v) SDS
- Wash solution (1×): 1 × SSC, 0.1% (w/v) SDS
- Wash solution (0.5×): 0.5 × SSC, 0.1% (w/v) SDS

Method

1. Heat HB to 65 °C in water bath.
2. Boil ST DNA for 10 min, put on ice.

Protocol 7 continued

3 To a hybridization box, add HB (approximately 50 ml for one filter, increase amount if more filters will be used) and ST (1 ml for every 50 ml of HB in box). Add the filters to the box (DNA side down) removing air bubbles underneath. Incubate at 65 °C for 3 h to overnight for unused filters, 1 h or more for previously used filters.

4 When the labelled probe is ready, add it to the hybridization buffer in the corner of the box and mix well by shaking gently. Incubate for 16–48 h.

5 After hybridization, wash the membranes at 65 °C in washing solutions as follows:
- Warm the first wash solution (2×) to 65 °C. Pour into the larger washing box. Add 1 l of wash solution per box, and add no more than 15 filters per box. Remove the filters from the hybridization boxes one at a time, allowing to drip briefly. Separate with mesh screens if desired. Incubate at 65 °C shaking for 20 min.
- Pour off the first wash solution and replace with the prewarmed (65 °C) second wash solution (1×); incubate with shaking for 20 min.
- Repeat with the third (0.5×) wash solution, and incubate with shaking for 15 min. Monitor radioactivity on the blots at the end of the second wash with a Geiger counter before proceeding to the higher stringency wash.

6 Remove the membranes from the wash solution, blot slightly dry with Whatman paper. Wrap the membranes in cling film or insert into plastic sheet covers and expose to X-ray film in X-ray cassette with intensifying screens at −80 °C for 1-7 days depending on the efficiency of labelling, age and quality of filter, etc.

7 Whenever reprobing of the blots is necessary, the old signals can be removed by:
- either washing the membrane in 0.4 M NaOH for 30 min at 50 °C, followed by two 250 ml washes at room temperature with 0.1 × SSC, 0.1% (w/v) SDS, 0.2 M Tris-HCl pH 7.0;
- or incubating the membrane at ≥95 °C in 250 ml of 0.1% (w/v) SDS solution.

The efficiency of probe removal should be checked by autoradiography or with a Geiger counter prior to pre-hybridization and subsequent reprobing.

[a] HB can be made up in advance and stored at room temperature in convenient aliquots. Dissolve sodium phosphate dibasic heptahydrate in water first, warm to dissolve if necessary. Also dissolve the BSA separately in water. After mixing ingredients, adjust the buffer to pH 7.2 with 1 M NaH_2PO_4/1 M Na_2HPO_4, usually it requires 4 ml/l

5 Random amplified polymorphic DNA (RAPD) analysis

RAPD markers are generated by using single, short, random oligomers (9–12 nucleotides) as primers for the PCR amplification of targeted genomic DNA, under low stringency (see *Protocol 8*) (6). The resulting PCR product shows many bands, some of which will be polymorphic and can be scored as molecular markers in segregating populations. Amplification only occurs if priming sites homologous to the primer motif exist and are separated by 500–5000 bp in the 5′ and 3' strand. Polymorphisms between two DNA samples will occur whenever differences exist at the priming sites (mutations) leading to polymorphisms of presence/absence of bands, or when there are insertions/deletions in the intervening sequences, leading to polymorphisms in the length of amplified fragments. The latter case is less common and, therefore, RAPD markers are generally a dominant class of DNA markers. Given the length of the primers, their sequence characteristics and considering, for example, the size of the tomato genome (1000 Mbp), one would expect each primer, on average, to amplify between 5–15 independent loci throughout the genome.

Design of the oligonucleotide primers follows simple criteria: ten bases in length; G + C content of 50–80%; no palindromic motifs of six or more nucleotides. Many such primers are synthesized and tested on the parental lines to ascertain whether or not their PCR products are polymorphic.

RAPD markers do not specify sequence-tagged sites (STSs) and, therefore, they cannot immediately be used to screen a resource such as YAC or BAC libraries. For these purposes, cloning and sequencing of the RAPD band will be required in order to convert it into a conventional STS.

Protocol 8

Genotyping by use of RAPD markers

Equipment and reagents

- Thermocycler and compatible reaction tubes or microtitre plate
- Electrophoresis equipment
- RAPD primer at 10 μM (e.g. Operon)
- 10 × PCR buffer: 100 mM Tris pH 8.3, 0.5 M KCl, 15 mM $MgCl_2$, 0.1% (w/v) gelatin
- dNTP mix: 10 mM each dATP, dCTP, dGTP, dTTP (e.g. Boehringer Mannheim, PE Biosystems)
- Taq DNA polymerase, 5U/μl (e.g. Boehringer Mannheim, PE Biosystems)
- Light mineral oil (Sigma; if the thermal cycler in not equipped with a heated lid)
- Stop buffer (see *Protocol 3*)
- Genomic DNAs at 25 ng/μl
- Sterile double distilled water
- Ethidium bromide stock solution (10 mg/ml)
- High resolution agarose (e.g. Nusieve 3:1, FMC Bioproducts)[a]
- NEB buffer, 10 × stock: (see *Protocol 3*)
- Tris-acetate electrophoresis (TAE) buffer, 50 × stock: 2 M Tris-acetate, 0.05 M EDTA. To prepare 1 litre of 50 × stock solution, mix 242 g Tris-base, 57.1 ml glacial acetic acid, add 100 ml 0.5 M EDTA pH8.0, and bring to 1 l with water.

Method

1 Prepare a master mix containing all PCR components excluding the DNA template. Sufficient mix should be prepared for each genomic DNA to be analysed, plus one

Protocol 8 continued

negative control and 1–2 reactions for error. Final volume is adjustable, but a sample reaction could be:

- dH_20 (enough to make the final volume 25 μl);
- 2.5 μl 10 × PCR buffer;
- 1 μl dNTP mix;
- 1.25 μl primer;
- 0.07 μl $MgCl_2$;
- 0.32 μl Taq polymerase.

2 Aliquot 24 μl of the mixture into each PCR reaction tube (or microtitre plate well) and add 1 μl of the appropriate genomic DNA or, for the negative control, 1μl of water.

3 Vortex, and spin or tap down to the bottom of the tube.

4 Overlay each reaction with one or two drops mineral oil (if necessary).

5 Perform 45 thermal cycles as follows:
- 94 °C × 1 min;
- 35 °C × 1 min;
- 72 °C × 2 min;
- 72 °C × 7 min.

6 Add 7 μl of blue stop buffer to each reaction.

7 Analyse products by electrophoresis (typically in a 2% Agarose gel) in 1 × NEB or TAE buffer, following the recommendations of the supplier of the gel tank and of the agarose to achieve the highest resolution. See also *Protocol 3* (steps 3–13) for running and staining electrophoresis gels.

[a] We generally obtain good quality results also using regular agarose (e.g. Boehringer Mannheim, Agarose LE).

6 Principles of amplified fragment length polymorphism (AFLP) analysis

The AFLP™ approach is based on the selective PCR amplification of a subset of genomic restriction fragments (8, 9); however, unlike RFLPs, the AFLP technique results in the presence or absence of restriction fragments, rather than in length differences. Fingerprints are produced without prior sequence knowledge using a limited set of generic primers. AFLP is powerful in detecting polymorphisms originating from mutations, which abolish or create a restriction site, and inversions, insertions, or deletions between two restriction sites. AFLPs may be used for typing, identification of molecular markers, and mapping of genetic loci. AFLP fingerprints allow one to survey numerous independent loci per run and are thus especially well suited for identification of markers linked to specific phenotypes, provided one has isogenic lines available (see Section 2.5). DNA from individual bands can be removed from the gel, after linkage or polymorphism is confirmed, and cloned for use as an RFLP probe. See Chapter 9 for other comments on AFLP.

The AFLP™ technology is under patent; patent applications are owned by KeyGene N.V (http://www.keygene.com) and fingerprinting kits for research are marketed (under license) from Life Technologies (http://www.lifetech.com) and Perkin-Elmer (http://www.perkinelmer.com). The method involves several steps. Double stranded genomic DNA is cut to completion with restriction endonucleases. A single enzyme can be used, but the best results are achieved using the combination of two different enzymes, a rare-cutter and a frequent-cutter. Next, specific double stranded DNA adapters (25–30 bp) are ligated to the ends of the restriction fragments to act as PCR template DNA. A subset of adapted-restriction fragments is then amplified by PCR, using primers homologous to the adapter sequences and the adjacent restriction site, but containing additional selective nucleotides at the 3′ end. The amplified DNA fragments are resolved on 5% or 6% denaturing polyacrylamide (sequencing) gels. Finally, the PCR products are visualized by direct staining or by autoradiography.

The procedure is complex and it must be calibrated for any specific system. Many parameters can be manipulated that will affect the AFLP fingerprint: number and type of enzymes used, type, number, and combinations of primers and PCR amplification conditions.

The enzyme combinations include *Eco* RI, *Hin* dIII, *Pst* I, *Bgl* II, *Xba* I, and *Sse* 8387I (eight base cutter) as rare-cutter enzymes, together with either *Mse* I or *Taq* I, as frequent-cutter enzymes. However, the most commonly used enzyme combination is *Eco* RI (having a 6 bp recognition site) and *Mse* I (having a 4 bp recognition site). Careful primer design is crucial for successful PCR amplification (see Chapter 9). AFLP primers consist of three parts: the 5′ part corresponding to the adapter, the restriction site sequence and the 3′ selective nucleotides. The number of selective nucleotides, as well as their C and G composition, are the two major factors that determine the number of restriction fragments amplified in a single AFLP reaction and, therefore, which determine the selectivity of the reaction. The selective nucleotide extensions can vary in length from 1 to 3 bp, but are of a defined length for a given primer. In general, the more Cs and Gs used as selective nucleotides in the amplification primers, the fewer the number of fragments that will be amplified. Depending on the complexity of the genome analysed and on the sequence context of the selective nucleotides, the value can range, approximately, from 10 to 100 amplified fragments (*Figure 3*).

Commercially available kits use the AFLP procedure reported by Vos *et al.* (9) and which is based on a two-step amplification strategy including a pre-amplification with both *Eco* RI and *Mse* I primers having 1-bp, 3′-extensions, followed by AFLP-PCR with primers having 3-bp, 3′-extensions. The two-step amplification strategy has the advantages of reducing background 'smears' in the fingerprint patterns and of providing virtually unlimited amount of template DNA for AFLP

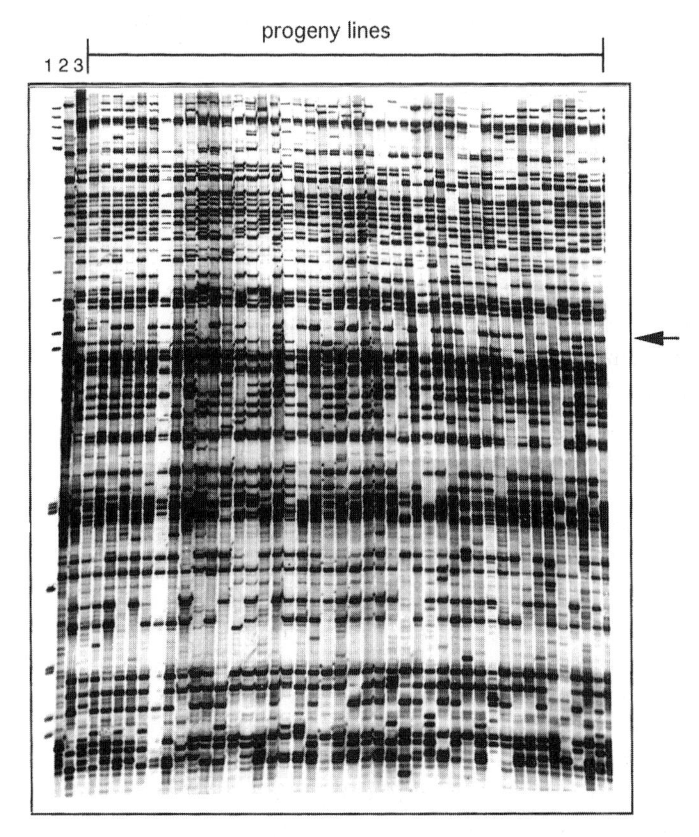

Figure 3. AFLP banding pattern of 43 progeny lines obtained by selfing a tetraploid somatic hybrid between a diploid accession of *Solanum commersonii* (CMM1T) and a *S. tuberosum* haploid (SVPII). The AFLP profiles were generated using the primer combination E + ACT/M + CAG of the GIBCO BRL AFLP Analysis System I. The arrow on the right of the picture indicates a *S. tuberosum*-specific band segregating in a 3:1 ratio. 1 = DNA size standard (Marker V of Boehringer Mannheim); 2 = *S. commersonii* parent; 3 = *S. tuberosum* parent (A. Barone *et al.* unpublished results).

reactions. However, it should be noted that it is possible to change the conditions for both steps in order to fit specific needs.

The AFLP technology predominantly produces mono-allelic markers: usually only one of the two alleles at a heterozygous locus is detected. However, it is possible to quantitatively analyse AFLP as co-dominant markers by means of an AFLP Image Analysis software developed at KeyGene (26).

7 PFGE in physical mapping of plant genomes

Identifying and mapping larger fragments of DNA such as those necessary for physical mapping requires techniques that will allow the separation and visualization of very large fragments. The development of PFGE allows the separation of DNA molecules up to approximately 10 Mb. This increase in resolution has been achieved by periodically changing the orientation of the electric field during electrophoresis by means of many different systems (10–12). Being able to resolve very large fragments of DNA also necessitates being able to digest DNA into suitably long lengths. This can be accomplished with restriction enzymes with large and therefore rare recognition sites. Since the size of the DNA fragments required for these techniques is much larger than that needed for RFLPs, for example, a specific DNA extraction technique is also required.

For small genomes, such as those of many micro-organisms and some lower eukaryotes, restriction fragment patterns can be visualized directly by gel staining. With larger genomes, the complex mixture of fragments generated after digestion and separation on pulsed field gels, needs to be blotted onto membranes and used for hybridizations similarly to blots obtained with standard gels. To establish a long-range physical map the blots are sequentially hybridized to tightly linked markers. The order of markers and of fragments is determined analysing both partial and complete digests with one or more restriction enzymes. Those markers which hybridize to the same restriction fragment in at least two enzyme digests are considered to be physically linked to each other. Double digests can then be used to determine more precisely the distance separating the markers. For a more comprehensive overview the reader is referred to recent reviews (60, 72) and to the original literature.

7.1 Preparation of large fragment DNA from plant nuclei

During standard DNA extraction techniques, the DNA is subjected to mechanical breakage and to the activity of contaminant nucleases that can digest the DNA down to fragments of less than 500 kb in length. Therefore, the preparation of high molecular weight (HMW) DNA using a method that protects the DNA from this shearing is a crucial first step for procedures requiring larger fragment sizes. In yeast and mammalian systems, this can be done by directly embedding whole cells in agarose plugs or microbeads (10, 73). Subsequent cell lysis, DNA purification, and manipulation are performed *in situ*. However, plant cells have cell walls that make it more difficult to extract high quality DNA. Therefore, in order to extract HMW DNA from plant tissue, the cell walls need to be removed before the cells are embedded in agarose. Conventional methods used for preparation of HMW DNA from plants involve the isolation of protoplasts using cell wall hydrolases (e.g. cellulase and pectinase) and subsequent embedding of the protoplasts in agarose (74–76). However, protoplast isolation is a tedious and time-consuming procedure, particularly costly on a large scale. Other disadvantages of the procedure are the frequent species specificity and the significant amount of chloroplast and mitochondrial DNA contained in HMW DNA prepared from plant protoplasts (77). For these reasons, more recently alternative techniques have been developed which allow the isolation of HMW (>2.5 Mb) DNA from plant nuclei, avoiding the protoplast isolation step (78, 79). In *Protocol 9* we describe the procedure developed by Liu and Whitter (78). To our knowledge the method is known to work in a fairly wide range of plants including *Arabidopsis*, rice, tomato, and petunia.

Protocol 9

Preparation of megabase plant DNA from plant nuclei in agarose plugs and microbeads

Equipment and reagents

- Beakers and flasks
- 50 ml plastic tubes and 1.5 ml microfugetubes
- 250, 90, 50, and 20 μm nylon mesh
- Perspex mould with slots of approximately 100 μl volume (available from BioRad and Pharmacia Biotech)
- Bench-top centrifuge
- Rocking platform
- Nuclei isolation buffer (NIB): 10 mM Tris–HCl pH 9.5, 10 mM EDTA, 100 mM KCl, 0.5 M sucrose, 4 mM spermidine, 1.0 mM spermine, 0.1% (v/v) mercaptoethanol
- Cell lysis buffer (LB): 1% (w/v) sarkosyl, 0.25 M EDTA pH 8.0, 0.2 mg/ml proteinase K
- Tris-EDTA: 10 mM Tris–HCl pH 9.5, 10 mM EDTA
- 1.2% (w/v) or 1.4 % (w/v) low-melting (LMT) agarose (SeaPlaque GTG, FMC) in Tris-EDTA
- Triton X-100
- Mineral oil

Method

1 Grind about 25 g of young plant tissue (leaves or whole plants) into powder in liquid nitrogen with a mortar and pestle, and transfer the powder to a beaker.

2 After the liquid nitrogen has evaporated completely, add 200 ml of ice-cold NIB to the powder and mix.

3 Filter the homogenate on ice sequentially through one layer each of 250, 90, 50 and 20 μm nylon mesh into 50 ml tubes.

4 Add to the filtrate one-twentieth volume of NIB supplemented with 10% Triton X-100 (this step is optional, but promotes chloroplast lysis).

5 Pellet the nuclei by centrifuging at 4 °C at 2000 g for 10 min and resuspend with NIB to a final volume of 1–1.2 ml.

6 Prewarm the nuclei at 42–45 °C for 1–2 min in a water bath before embedding them in agarose:

   (a) *Embedding the nuclei in agarose plugs*: mix gently the preheated nuclei with an equal volume of 1.4% LMT agarose in Tris-EDTA prewarmed at 45 °C. Immediately dispense the agarose-nuclei suspension mixture into slots made through a Perspex mould covered on one side with tape. Allow the samples to solidify at 4 °C or on ice for 10 min. Gently remove the solidified plugs from the mould using a sterile inoculation loop.

Protocol 9 continued

(b) *Embedding the nuclei in agarose microbeads*: mix gently the preheated nuclei in a 30 ml flask with an equal volume of 1.2% LMT agarose in Tris-EDTA (45 °C). Based on the agarose-nuclei combined volume, add 1.5 volumes of pre-warmed (45 °C) mineral oil to the flask and swirl vigorously for 10–15 sec. Immediately add another 1.5 volumes of ice cold mineral oil to the emulsion and quickly swirl the flask in an ice water bath for 2 min. Centrifuge briefly to collect the microbeads.

7 Incubate, with gentle shaking, either the agarose plugs or the agarose microbeads in LB at 50 °C for 24 h with one change of LB.

About 200–300 μg DNA should be obtained from 20–25 g of plant tissue. If the plugs or the beads are not used immediately, they can be washed once in 0.5 M EDTA, pH 9.0–9.3 for 1 h at 50 °C, once in 0.05 M EDTA, pH 8.0 for 1 h on ice, and then stored in 0.05 M EDTA, pH 8.0 at 4 °C (79).

It is advisable to clean new agarose preparations from impurities and to test them for nuclease by fractionation on pulsed field gels. Good quality HMW DNA samples should remain trapped in the wells on ethidium bromide-stained gels.

7.2 Restriction digestion of agarose-embedded DNA

The choice of rare-cutting restriction enzymes depends on the organism. While for bacteria, yeast and mammals it is possible to make reliable predictions on the basis of several criteria such as GC content, degree of methylation, and codon usage, the task becomes more difficult for invertebrates and particularly for plants for which the usefulness of restriction enzymes often needs to be verified empirically. In general, endonucleases potentially useful for PFGE in plants possess one or both of the following characteristics:

(1) Recognition sequences >6 bp which consist almost exclusively of G or C (plant genomes are on average only 40% GC). Examples of such enzymes are *Not* I and *Sfi* I.

(2) Recognition sites having one or more of the CpG and CpXpG sequence pairs. Such sites are often under-represented in the genomes of higher plants and frequently methylated at the cytosine residue, which prevents most restriction enzymes from cutting such sequences.

Examples of this type of enzymes are *Bss* HII, *Mlu* I, *Nru* I, *Sac* II, and *Sma* I. However, some plant genomes are very poorly methylated and have relatively high GC content (43–44%), which makes it very difficult to find appropriate rare-cutting enzymes.

Digestion of agarose-embedded DNA with restriction enzymes is outlined in *Protocol 10*. Digestions of agarose plugs are essentially performed following the manufacturer's recommendations, but generally using a 2–3-fold enzyme excess.

Protocol 10

Restriction enzyme digestion of DNA in agarose

An agarose plug of 100 μl volume prepared according to Protocol 9 should contain approximately 10 μg of DNA. Use between one-third and one-half of a plug for each enzyme digest to be loaded per lane on a gel.

Equipment and reagents

- Sterile scalpel
- Incubator
- 1.5 plastic microfuge tubes
- TE_{10}: 10 mM Tris–HCl pH 8.0, 10 mM EDTA
- PMSF (phenylmethylsulfonylfluoride)[a]
- Spermidine
- 10 × Restriction enzyme buffer (provided by manufacturer)
- Restriction enzyme
- Sterile distilled water

Method

1. Cut the agarose plugs in thirds or halves with a sterile scalpel.
2. Prior to digestion wash the agarose plugs or microbeads at 50 °C for 1 h in several volumes of TE_{10} containing 1 mM PMSF, then in several changes of TE_{10} buffer.
3. Equilibrate the DNA samples (40 μl, 3–5 μg DNA) for 30 min in 400 μl of the appropriate digestion buffer supplemented with 8 mM spermidine.
4. Perform restriction-enzyme digestions in 120 μl volumes with 50 units of restriction enzyme at the recommended temperature for 4–5 h (for microbeads) or overnight (for plugs).[b]
5. After digestion, chill the tubes on ice for 10 min and load the plugs directly on the gel.

[a] Prepare fresh each time by dissolving 40 mg/ml PMSF in ethanol at room temperature. (*Caution*: PMSF is highly toxic and should be handled with gloves)

[b] The enzyme can be added in two equal portions, at the beginning and half way through the digestion time.

The use of partial digests in the construction of physical maps can help to confirm linkage between DNA markers and to increase the size of the restriction map obtained around any one marker. The optimal conditions of partial digestion should be tested for each preparation of DNA. Partial digestion of DNA embedded in agarose can be produced following standard procedures of limiting enzyme concentration, or reaction time, or magnesium concentration (80). In some cases, the simultaneous addition of methylase and restriction enzyme to compete for the same recognition site has also been considered (81).

DNA embedded in agarose can be digested with multiple enzymes sequentially. If the recommended restriction buffers of the enzymes chosen are incompatible, set up digests for the restriction enzyme with the lowest salt requirement first. At the end of the first enzyme digest, remove the reaction buffer and wash the plugs twice in 1 ml of the restriction buffer appropriate for the next enzyme to be used. Finally, add the restriction buffer and the enzyme for the next digestion.

7.3 PFGE analysis

7.3.1 PFGE techniques and running conditions

Conventional agarose electrophoresis cannot resolve large molecules of DNA (>50–100 kb). This limitation in resolution has been overcome with the introduction of PFGE (10–12) systems that permit the separation of very large (up to 5–10 Mb) molecules of DNA. During PFGE, DNA molecules are subject to alternating electric fields (obtuse or octagonal), which allow DNA molecules to undergo a continuous reorientation; small DNA molecules will re-orientate faster than larger ones (60). The size range of DNA fragments that can be separated on PF gels is a function of several parameters including electric field strength, pulse time (e.g. the time between alternations of the field), pulse angle, electrophoresis run time, agarose and buffer concentration, temperature, and electrophoresis buffer; the first two being the principal parameters (60). PFGE is very sensitive to differences in the concentration of the DNA loaded resulting in slower migration for higher concentrations. It is therefore important to avoid overloading to get meaningful data from lane-to-lane comparisons or when fragment sizes are estimated.

Several modifications of the original PFGE apparatus are now available, which differ for the electrode geometry used, field homogeneity and the method of re-orientation of the electric field. The most widely used systems are transverse alternating field electrophoresis (TAFE; 82), contour-clamped homogeneous electric field electrophoresis (CHEF; 12) and the programmable autonomously-controlled electrophoresis (PACE; 83). These three systems are preferred for physical genome analysis as they ensure reasonably straight bands and lanes as well as high resolution and sharpness of bands in the entire size range from 1 to 10 000 kb. We have used the CHEF system from BioRad (known as the CHEF-DRIII). In this system, a hexagonal array of electrodes surrounds the gel and generates the electric field at angles of 60° or 120° depending on the polarity, resulting in a uniformly clamped field at all points of the gel. These features allow the resolution of molecules over 10 Mb.

Generally, the use of a single set of electrophoretic parameters does not allow analysis of the full size range of fragments produced by complete and partial digests of genomic DNA. Therefore, when starting a mapping project it is advisable to establish standard running conditions for two to four overlapping size ranges (e.g. <50 kb; 30–700 kb; 200–2000 kb; and 1000–7000 kb; 72). Information on running parameters are normally supplied by the manufacturer of the PFGE apparatus.

7.3.2 Size markers

Appropriate DNA size standards should be selected for the range of fragment sizes being resolved. Those most commonly used include concatemers of bacteriophage λ (50–600 kb) or the chromosomes of *Saccharomyces pombe* (200–1500 kb), *Kluyveromyces lactis* (900–3000 kb), and *Schizosaccharomyces pombe* (3000–6000 kb). These markers can be easily prepared or obtained directly from several commercial suppliers (e.g. BRL, BioRad, New England Biolabs; 60).

7.3.3 Gel preparation

The preparation of gels for PFGE is outlined in *Protocol 11*.

Protocol 11

Casting and loading the gel

Equipment and reagents

- 0.5 × TBE buffer: 45 mM Tris pH 8.3, 45 mM boric acid, 1 mM EDTA
- 1% (w/v) agarose (SeaKem GTG, FMC) in 0.5 × TBE buffer
- Gel casting stand and comb
- Plastic inoculation loop and sterile scalpel blade
- PFGE apparatus

Method

1. Prepare 1% (w/v) agarose in 0.5 × TBE. To separate DNA molecules greater than 2 Mb, use between 0.5% and 0.8% (w/v) agarose.
2. Assemble the gel casting stand and the comb. Ensure that it is resting on a level surface.
3. Pour agarose and leave to solidify.
4. Thoroughly remove the supernatant of the plugs containing digested DNA.
5. Transfer the plugs to the gel slots prefilled with 0.5 × TBE buffer, using a plastic inoculation loop and sterile scalpel blade.
6. Similarly, transfer the plugs containing marker DNAs to the gel slots.
7. Fix the plugs in their position by sealing the top of the gel with 0.5% low melting agarose in 0.5 × TBE.
8. Transfer the gel to the PFGE apparatus electrophoresis chamber and add running buffer (0.5 × TBE) to cover the gel. Start electrophoresis using appropriate parameters.

7.3.4 Southern hybridization of PFGE gels

After electrophoresis PF gels are blotted and hybridized according to the standard procedures described in *Protocols* 4 and 7. However, in order to fragment HMW DNA the gel is treated with short length (254 nm) UV light for 2 min on a transilluminator, and then depurinated in 0.25 M HCl for 25 min. Also in this case it is preferable to use a reusable nylon membrane (e.g. Hybond+, Amersham) so that the blot can be hybridized sequentially with multiple probes to generate a restriction map. The probes are radioactively labelled according to the random priming procedure (*Protocol 5*; 68).

7.3.5 Construction of physical maps from hybridization results

The construction of long-range restriction maps is achieved by sequentially hybridizing the same PFGE blot with neighbouring markers to establish their physical linkage. As an initial step, it is advisable to prepare PFGE blots with single restriction enzyme digests separated in different size ranges. The autoradiographs obtained from different probe hybridizations are aligned to compare the pattern of bands. The co-hybridization of two markers to a 'common' band is suggestive of physical linkage. The smallest restriction fragment to which the two DNA probes co-hybridize gives a maximum value for the physical distance between the two markers.

However, in order to exclude random co-migration of restriction fragments it is important to independently corroborate the physical linkage by verifying that the two markers also hybridize to a common band in restriction digests generated by other enzymes (particularly those with complementary properties) or in a ladder of partially digested fragments. Once physical linkage between two markers has been established, more refined restriction maps can be obtained by analysing double digests with different rare-cutting restriction enzymes.

Acknowledgements

We would like to thank Dr Steven D. Tanksley and his group for providing several of the protocols described in this chapter. Our thanks also go to Dr D. Bernacchi for helpful ideas and initial write-ups of some of the procedures, and to Dr A. Barone for allowing us to include unpublished results. This work was supported in part by the International Exchange Program of the University of Naples Federico II, Italy.

References

1. Morgan, T. H. (1911). *Science,* **34**, 384.
2. Sturtevant, A. H. (1913). *J. Exp. Zool.*, **14**, 43.
3. Botstein, D., White, R. L., Skolnick, M., and Davis, R. W. (1980). *Am. J. Hum. Genet.*, **32**, 314.
4. O'Brien, S. J. (ed.) (1993). *Genetic maps.* Cold Spring Harbour Laboratory Press, NY.
5. Saiki, R. K., Gelfland, D. H., Stoffel, S., Scharf, S. J., Higuchi, R., Horn, G. T., *et al.* (1988). *Science*, **239**, 487.
6. Williams, J. G.K., Kubelik, A. R., Livak, K. J., Rafalski, J. A., and Tingey, S. V. (1990). *Nucl. Acids Res.*, **18**, 6531.
7. Welsh, J. and McClelland, M. (1990). *Nucl. Acids Res.*, **18**, 7213.
8. Zabeau, M. and Vos, P. (1993) *Eur. Pat. Appl.*, EPO 534858.
9. Vos, P., Hogers, R., Bleeker, M., Reijans, M., Van de Lee, T., Hornes, M., *et al.* (1995). *Nucl. Acids Res.*, **23**, 4407.
10. Schwartz, D. C. and Cantor, C. R. (1984). *Cell*, **37**, 67.

11. Carle, G. F., Frank, M., and Olson, M. V. (1986). *Science*, **232**, 65.
12. Chu, G., Vollrath, D., and Davies, R. W. (1986). *Science*, **234**, 1582.
13. Burke, D. T., Carle, G. F., and Olson, M. V. (1987). *Science*, **236**, 806.
14. Schizuya, H., Birren, B., Kim, U. J., Mancino, V., Slepak, T., Tachiiri, Y., *et al.* (1992). *Proc. Natl Acad. Sci. USA*, **89**, 8794.
15. Zhang, H-B. and Wing, R. A. (1997). *Plant Mol. Biol.*, **35**, 115.
16. Mozo, T., Dewar, K., Dunn, P., Ecker, J. R., Fischer, S., Kloska, S., *et al.* (1999). *Nature Genet.*, **22**, 271.
17. Nusbaum, C., Slonim, D. K., Harris, K. L., Birren, B. W., Steen, R. G., Stein, L. D. *et al.* (1999). *Nature Genet.*, **22**, 388.
18. Allard, R. W. (1956). *Hilgardia*, **24**, 235.
19. Haldane, J. B. S. (1919). *J. Genet.*, **8**, 299.
20. Kosambi, D. D. (1944). *Annl. Eugen. Lond.*, **12**, 172.
21. Lander, E. S., Green, P., Abrahamson, J., Barlow, A., Daley, M. J., Lincoln, S. E., *et al.* (1987). *Genomics*, **1**, 174.
22. Stam, P. and van Ooijen, J. W. (1996). *Join Map(TM) version 2.0: software for the calculation of genetic linkage maps.* CPRO-DLO, Wageningen.
23. Mohan, M., Nair, S., Bhagwat, A., Krishna, T. G., Yano, M., Bhatia, C. R., *et al.* (1997). *Mol. Breeding*, **3**, 87.
24. Kurata, N., Moore, G., Nagamura, Y., Foote, T., Yano, M., Minobe, Y., *et al.* (1994). *Bio/Technology*, **12**, 276.
25. Freyre, R., Skroch, P. W., Geffroy, V., Adam-Blondon, A. F., Shirmohamadali, A., Johnson, W. C., *et al.* (1998). *Theor. Appl. Genet.* **97**, 847.
26. Haanstra, J. P.W., Wye, C., Verbakel, H., Meijer-Dekens, F., van den Berg, P., Odinot, P., *et al.* (1999). *Theor. Appl. Genet.*, **99**, 254.
27. Nilsson, N. O., Hallden, C., Hansen, M., Hjerdin, A., and Sall, T. (1997). *Genome* **40**, 644.
28. Laucou, V., Haurogne, K., Ellis, N., and Rameau, C. (1998). *Theor. Appl. Genet.*, **97**, 905.
29. Tautz, D. (1989). *Nucl. Acids Res.*, **17**, 6463.
30. Dietrich, W. F., Miller, J., Steen, R., Merchant, M. A., Damron-Boles, D., Husain, Z., *et al.* (1996) *Nature*, **380**, 149.
31. Gyapay, G., Morisette, J., Vignal, A., Dib, C., Fizames, C., Millasseau, P., *et al.* (1994). *Nature Genet.*, **7**, 246.
32. Röder, M., Plaschke, J., Konig, S. U., Borner, A., Sorrells, M. E., Tanksley, S. D., *et al.* (1995). *Mol. Gen. Genet.*, **246**, 327.
33. Wu, K. S. and Tanksley, S. D. (1993). *Mol. Gen. Genet.*, **241**, 225.
34. Moore, S. S., Sargeant, L. L., King, T. J., Mattick, J. S., Georges, M., and Hetzel, D. J. S. (1991). *Genomics*, **10**, 654.
35. Qi, X., Stam, P., and Lindhout, P. (1998). *Theor. Appl. Genet.*, **96**, 376.
36. Vuylsteke, M., Mank, R., Antonise, R., Bastiaans, E., Senior, M. L., Stuber, C. W., *et al.* (1999). *Theor. Appl. Genet.*, **99**, 921.
37. Cho, R. J., Mindrinos, M., Richards, D. R., Sapolsky, R. J., Anderson, M., Drenkard, E., *et al.* (1999). *Nature Genet*, **23**, 203.
38. Helentjaris, T. (1987). *Trends Genet.*, **3**, 217.
39. Miller, J. C. and Tanksley, S. D. (1990). *Theor. Appl. Genet.*, **80**, 437.
40. Burr, B., Burr, F. A., Thompson, K. H., Albertson, M. C., and Stuber, C. W. (1988). *Genetics*, **118**, 519.
41. Lu, C. F., Shen, L. S., Tan, Z. B., Xu, Y. B., He, P., Chen, Y., *et al.* (1997). *Theor. Appl. Genet.*, **94**, 145.
42. Reiter, R. S., Williams, J. G.K., Feldmann, K. A., Rafalski, J. A., Tingey, S. V., and Scolnik, P. A. (1992). *Proc. Natl Acad. Sci.* USA, **89**, 1477.
43. Tanksley, S. D. (1993). Mapping polygenes. *Ann. Rev. Genet.*, **27**, 205.

44. Michelmore, R. W., Paran, I., and Kesseli, R. V. (1991). *Proc. Natl Acad. Sci. USA*, **88**, 9828.
45. Giovannoni, J., Wing, R., Ganal, M. W., and Tanksley, S. D. (1992). *Nucl. Acids Res.*, **19**, 6553.
46. Tanksley, S. D., Ganal, M. W., and Martin, G. B. (1995). *TIG*, **11**, 63.
47. Poulsen, D. M. E., Henry, R. J., Johnston, R. P., Irwin, J. A. G., and Rees, R. G. (1995). *Theor. Appl. Genet.*, **91**, 270.
48. Chunwongse, J., Doganlar, S., Crossman, C., Jiang, J., and Tanksley, S. D. (1997). *Theor. Appl. Genet.*, **95**, 220.
49. Koorneef, M., Alonso-Blanco, C. and Stam, P. (1998). In Arabidopsis protocols (ed. J. M. Martínez-Zapater, and J. Salinas), p.105. Humana Press, Totowa, New Jersey.
50. Fabri, C. O. and Schäffner, A. R. (1994). *Plant J.*, **5**, 149.
51. Schäffner, A. R. (1998). In: *Arabidopsis protocols* (ed. J. M. Martínez-Zapater and J. Salinas), p. 183. Humana Press, Totowa.
52. Konieczny, A. and Ausubel, F. M. (1993). *Plant J.*, **4**, 403.
53. Bell, C. J. and Ecker, J. R. (1994). *Genomics*, **19**, 137.
54. Lister, C. and Dean, C. (1993). *Plant J.*, **4**, 745.
55. Tanksley, S. D., Ganal, M. W., Prince, J. P., deVicente, M. C., Bonierbale, M. W., Broun, P., *et al.* (1992). *Genetics*, **132**, 1141.
56. Pillen, K., Pineda, O., Lewis, C. B., and Tanksley, S. D. (1996). In: *Genome mapping in plants* (ed. A. H. Paterson). Landes Co, Austin.
57. Eshed, Y. and Zamir, D. (1995). *Genetics*, **141**, 1147.
58. Liu, Y-S. and Zamir, D. (1999). *TGC Report*, **49**, 26.
59. Ganal, M. W., Young, N. D., and Tanksley, S. D. (1989). *Mol. Gen. Genet.*, **215**, 395.
60. Monaco, A. P. (ed.) (1995). *Pulsed field gel electrophoresis*. IRL Press, Oxford.
61. Murray M. G. and Thompson, W. F. (1980). *Nucl. Acids Res.*, **8**, 4321.
62. Bernatzky, R. and Tanksley, S. D. (1986). *Plant. Mol. Biol. Rep.*, **4**, 37.
63. Fulton, T. M., Chunwongse, J., and Tanksley, S. D. (1995). *Plant Mol. Biol. Rep.*, **13**, 207.
64. Shure, M., Wessler, S., and Federoff, N. (1983). *Cell*, **35**, 225.
65. Tai, T. H. and Tanksley, S. D. (1990). *Plant Mol. Biol. Rep.*, **8**, 297.
66. Hall, L. (1995). In: *Gene probes 2* (ed. B. D. Hames and S. J. Higgings), p. 119. IRL Press, Oxford.
67. Hames, B. D. and Higgins, S. J. (ed.) (1995). *Gene Probes 1*. IRL Press, Oxford.
68. Feinberg, A. P. and Vogelstein, B. (1984). *Anal. Biochem.*, **137**, 266.
69. Rigby, P. W. I., Dieckmann, M., Rhodes, C., and Berg, P. (1977). *J. Mol. Biol.*, **113**, 237.
70. Anderson, M. L.M. and Young, B. D. (1985). In: *Nucleic acid hybridization* (ed. B. D. Hames, and S. J. Higgins), p. 73. IRL Press, Oxford.
71. Church, G. and Gilbert, W. (1984). *Proc. Natl Acad. Sci.*, **81**,1991.
72. Bautsch, W., Römling, U., Schmidt, K. D., Samad, A., Schwartz, D. C., and Tümmler, B. (1997). In: *Genome mapping* (ed. P. H. Dear), p. 281. IRL Press, Oxford.
73. Overhauser, J. and Radic, M. Z. (1987). *Focus*, **9**, 8.
74. Ganal, M. W. and Tanksley, S. D. (1989). *Plant Mol. Biol. Rep.*, **7**, 17.
75. Cheung, W. Y. and Gale, M. D. (1990). *Plant Mol. Biol.*, **14**, 881.
76. Wing, R. A., Zhang, H-B., and Tanksley, S. D. (1994). *Mol. Gen. Genet.*, **242**, 681.
77. Woo, S-S., Jiang, J. M., Gill, B. S., Paterson, A. H., and Wing, R. A. (1994). *Nucl. Acid. Res.* **22**, 4922.
78. Liu, Y. G. and Whittier, R. F. (1994). *Nucl. Acids Res.*, **22**, 2168.
79. Zhang, H. B., Zhao, X., Ding, X., Paterson, A. H., and Wing, R. A. (1995). *Plant J.*, **7**, 175.
80. Hoheisel, J.D ., Nizetic, D., and Lehrach, H. (1989). *Nucl. Acids Res.*, **17**, 9571.
81. Hanish, J. and Mc Clelland, M. (1989). *Anal. Biochem.*, **179**, 357.
82. Gardiner, K., Laas, W., and Patterson, D. (1986). *Som. Cell. Mol. Genet.*, **12**, 185.
83. Clark, S. M., Lai, E., Birren, B. W., and Hood, L. (1988). *Science*, **241**, 1203.

Chapter 7
Arabidopsis YAC, BAC and cosmid libraries

Paul Muskett, Ian Bancroft* and Caroline Dean
Departments of Molecular Genetics, and *Brassica and Oilseeds,
John Innes Centre, Norwich, UK

1 Introduction

Yeast artificial chromosome (YAC), bacterial artificial chromosome (BAC), and cosmid libraries are central to gene cloning, the generation of high resolution physical maps (see Chapter 6) and large-scale genome sequencing experiments. This has been clearly demonstrated in *Arabidopsis thaliana*, for which detailed YAC, BAC, and cosmid-based physical maps have been generated (1–8), some of which now cover the entire genome (9, 10). In this chapter, we provide protocols of how to construct and screen these libraries, and discuss the uses of these different clone types.

2 YAC libraries

YACs are the vector system currently capable of maintaining the largest inserts, and clones that harbour 1 Mb or more have been constructed (11, 12). The ability to clone large DNA fragments has made YAC vectors a very attractive cloning system for genome analysis, and has contributed greatly to the use of YAC vectors in physical mapping and positional cloning experiments in many organisms. Despite these advantages it is important to be aware of the limitations of the YAC cloning system. YAC libraries often contain high proportions of chimaeric clones and YACs do not maintain all inserts stably (repeated sequences especially can be deleted out; 13). Therefore, care must be taken to confirm that YACs faithfully represent the genomic region. Chimaeric clones can make chromosome walking experiments using YAC end-probes very difficult, and all walking steps must be confirmed using overlapping clones. YAC DNA is also difficult to purify in large amounts and to isolate free of contamination by endogenous yeast DNA. Therefore, YACs are rarely used for sequencing and if gel-purified YACs are used in hybridization experiments it is necessary to carry out control experiments using yeast DNA.

The four Arabidopsis YAC libraries containing genomic DNA from the Columbia ecotype, yUP (14), EW (15), EG (16), and CIC (17), have formed the basis for systematic genome-scale mapping in Arabidopsis. The CIC library has been the most useful for mapping, due to its large insert size (average 420 kb) and low level of chimaeric clones (less than 10%). All four libraries and colony filter sets are available from the Arabidopsis Biological Resource Centre (ABRC; http://aims.cps.msu.edu/aims/). The current YAC-based physical maps of the five Arabidopsis chromosomes (3–7, 18–20) are available through the internet (see Table 1).

In addition to physical mapping, YACs have been used extensively in map-based cloning experiments. The locus is mapped relative to YAC end fragments (21) until it has been localized to a single YAC clone. DNA from the YAC clone can be subcloned as smaller fragments into cosmid vectors suitable for *Agrobacterium tumefaciens*-mediated transformation (22; see Chapters 3) and used for complementation tests. Alternatively, the YAC can be used as a probe to screen a cosmid vector library containing plant genomic DNA from another Arabidopsis ecotype (see Section 4). The hybridizing cosmid vectors are then used for transformation experiments to look for complementation of the mutant phenotype (23). More recently, it has been reported that YACs retro-fitted with plant selectable markers can be used for the biolistic transformation of plants (24) and these may become useful directly for complementation analysis.

Arabidopsis YAC clones can now also be used to rapidly map new sequences onto the genome by hybridization using YAC library filters. For example, expressed sequence tag (EST) sequences (M. Stammers, Z. Lenehan, K. Love and C. Dean, unpublished data: http://genome-www.stanford.edu/Arabidopsis/

Table 1 Internet sites for YAC-based Arabidopsis physical maps

URL (http://)	Information available
cbil.humgen.upenn.edu/~atgc/physical-mapping/ch1-graphics-all-legend.html	YAC map for chromosome 1
//weeds.mgh.harvard.edu/goodman/c2. html	YAC map for chromosome 2
genome-www.stanford.edu/Arabidopsis/Chr3-INRA/	YAC map for chromosome 3
genome-www.stanford.edu/Arabidopsis/JIC-contigs/Chr4_YACcontigs.html	YAC map for chromosome 4
genome-www.stanford.edu/Arabidopsis/JIC-contigs/Chr5_YACcontigs.html	YAC map for chromosome 5
genome-www3.stanford.edu/cgi-bin/AtDB/Pchrom	Arabidopsis Physical Map Overview, various types of physical maps displayed together and linked by common markers.
genome-www3.stanford.edu/Arabidopsis/chromosomes/	Arabidopsis Genomic View, single entry point for all AtDB's maps.
genome-www.stanford.edu/Arabidopsis/	*Arabidopsis thaliana* Database (AtDB), provides access to all aspects of Arabidopsis information, including the above sites.

EST2CIC.html; 25) and T-DNA flanking sequences (P. Muskett and C. Dean, unpublished data: http://www.jic.bbsrc.ac.uk/staff/caroline-dean/dslaunch.htm) have been quickly and accurately mapped using YAC CIC library filters. As with physical mapping, the CIC library is most useful due to its low content of chimaeric clones. Even so, hybridization data should only be trusted when two or more overlapping YAC clones are identified.

Below we describe techniques for efficient screening of YAC libraries by using colony hybridization experiments and the preparation of intact yeast chromosomes. For methods describing the preparation of YAC colony filters and the subsequent analysis of YAC clones, such as preparation of yeast genomic DNA and preparation of YAC end fragments see Schmidt and Dean (26).

2.1 Screening of YAC libraries by colony hybridization

Hybridization-based screening of YAC libraries is carried out using membranes carrying yeast colonies (27). To help scoring of positively-hybridizing YAC colonies, hybridization and washing conditions should be established so that a low level of non-specific background hybridization remains on all colonies (*Figure 1a*). Alternatively, ^{35}S-labelled YAC vector can be included in the hybridization buffer, which will hybridize to all YAC clones. Colonies hybridizing to a ^{32}P-labelled specific DNA fragment can be clearly scored against the uniform background of the ^{35}S-labelled colonies. *Protocol 1* describes the method we use to screen YAC library filters.

Protocol 1

Hybridization analysis of YAC library filters

Equipment and reagents

- 0.5 M phosphate buffer pH 7.2: 0.5 M Na_2HPO_4, add 4 ml 85% (v/v) H_3PO_4 per l
- Hybridization solution: 0.25 M NaCl, 0.25 M phosphate buffer, 10% (w/v) PEG 6000, 7% (w/v) SDS, 1 mM EDTA
- Hybridization boxes (8 × 12 cm; e.g. Bunzl 6064)
- 20 × SSC: 3 M NaCl, 0.3 M tri-sodium citrate
- Solution A: 3 × SSC, 0.1% (w/v) SDS
- Solution B: 0.1 × SSC, 0.1% (w/v) SDS

Method

1 Label 50 ng of gel purified probe (PCR product, cDNA clone/EST, genomic DNA clone or YAC DNA) using a random primers labelling kit (see Chapter 6). We usually use α-[^{32}P]dCTP, but other labels could be used. There is no requirement to remove unincorporated [^{32}P]-labelled nucleotides prior to hybridization, in fact these help to provide background, which aids scoring of the filters.

Protocol 1 continued

2 Prehybridize membranes at 62 °C for at least 1 h with 10 ml of hybridization solution in a small hybridization box.

3 Remove filter from the hybridization solution and add boiled denatured probe. Mix well and place filter back into the hybridization solution.

4 Hybridize overnight at 62 °C with gentle shaking.

5 Transfer the filter into a large box and rinse briefly with cold solution A.

6 Add 250 ml of solution A and wash for 20–30 min at 62 °C with gentle shaking.

7 Remove solution A and replace with 250 ml of solution B, incubate for 20–30 min at 62 °C with gentle shaking.

8 Finally, rinse the filter in 3 × SSC, blot dry to remove excess liquid, seal in a plastic bag, and expose to film. Adjust the length of exposure depending on the strength of the signal. An ideal autoradiograph should show a just visible background of most of the colonies on the filter, with the positive signals clearly stronger (*Figure 1a*).

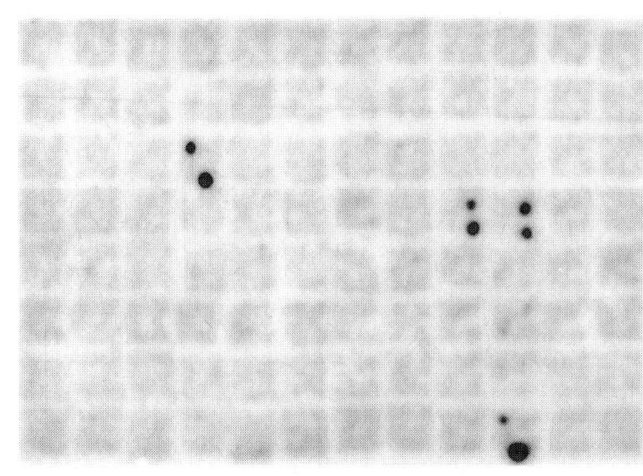

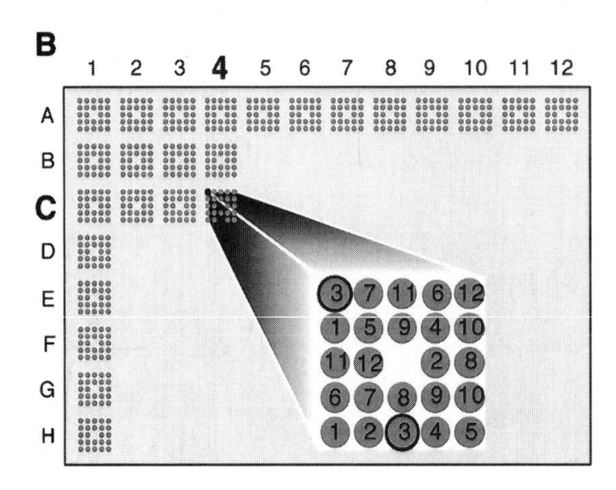

Figure 1 Autoradiograph of a CIC YAC library filter probed with a T-DNA flanking fragment (A), and a diagram showing the layout of the filter (B). The CIC YAC library has been arrayed in duplicate by using 24 offsets of a 96-prong replicator. By arraying the library in duplicate, identification of positively hybridizing clones is easily achieved as each pair of duplicated clones has its own spatial arrangement in the offset square of 24 clones (see B). The identity of the duplicate positive determines which of the twelve 96-well microtitre plates holding the library contains the clone, and the plate row and column coordinate is determined by the coordinate of the offset square containing the duplicate positive on the filter. For example, (B) shows the scoring of one of the duplicate positives from the autoradiograph (A). The duplicate hybridizing positives represent clone 3 in the offset square of 24 clones, and the square is in row C and column 4 of the filter. The co-ordinate of this hybridizing positive would therefore be 3C4. The other hybridizing positives on the autoradiograph (A) are 4D9, 4D10, and 3H10, and together with 3C4 represent overlapping YACs on chromosome 3.

An example of a CIC YAC library filter hybridization is shown in *Figure 1*, together with details to help the scoring of positive hybridizing clones. Once positive hybridizing clone co-ordinates have been identified, the next step is to determine the position of these clones on the physical map. There are several Internet sites showing YAC-based maps of the five Arabidopsis chromosomes and sites such as 'Arabidopsis Physical Map Overview' have a search facility (co-ordinates should be prefixed by the name of the library, e.g. CIC4C3). These sites are listed in Table 1.

2.2 Preparation of intact yeast chromosomes

For many of the uses of individual YAC clones, such as for hybridization probes or for subcloning, the ability to prepare high molecular weight DNA from YAC-containing yeast strains is required. The method we routinely use, together with a method for the gel-purification of YAC clones to be used as hybridization probes, is given in *Protocol 2*.

Protocol 2

Preparation and gel purification of YAC chromosomal DNA[a]

Equipment and reagents

- YEPD medium: 1% (w/v) yeast extract, 2% (w/v) peptone, 2% (w/v) glucose; autoclave
- 1 M sorbitol
- InCert agarose (FMC)
- 100 μl block formers (Pharmacia)
- SCEM: 1 M sorbitol, 100 mM sodium citrate pH 5.8, 10 mM EDTA, 30 mM 2-mercaptoethanol
- 100 mg/ml novozyme 234 in SCE: 1 M sorbitol, 100 mM sodium citrate pH 5.8, 10 mM EDTA; filter sterilized
- ESP(1.0): 100 mM EDTA pH 8.0, 1% (w/v) sarkosyl, 1 mg/ml proteinase K
- ES: 100 mM EDTA pH 8.0, 1% (w/v) sarkosyl
- ESP(0.5): 100 mM EDTA pH 8.0, 1% (w/v) sarkosyl, 0.5 mg/ml proteinase K
- 1 mM phenlymethylsulphonyl fluoride (PMSF) in 1 mM EDTA pH 8.0, freshly made up using a stock solution of 100 mM PMSF in isopropanol
- TE: 10 mM Tris-HCl, 1 mM EDTA, pH 8.0
- 2 × CA buffer: 40 mM Tris pH 8.0, 14 mM $MgCl_2$, 200 mM KCl, 400 μg/ml BSA
- β-agarase I (New England Biolabs)
- 10 mg/ml yeast RNA
- SeaKem GTG agarose (FMC)
- Phenol:chloroform:iso-amyl alcohol (24:25:1)

Methods

A. Preparation of intact yeast chromosomes

1 Transfer a single yeast colony into a plastic centrifuge tube containing 50 ml YEPD medium and grow to stationary phase (2 days) in an orbital shaker at 30 °C.

Protocol 2 continued

2 Pellet cells by centrifugation at approximately 2600 g for 10 min at room temperature.

3 Resuspend cells in 10 ml 1 M sorbitol and transfer to a 15-ml screw-capped centrifuge tube.

4 Centrifuge at approximately 2600 g for 10 min at room temperature.

5 Resuspend cells in 1 M sorbitol to a final volume of 0.6 ml.

6 Make up 1% (w/v) InCert agarose in 1 M sorbitol, cool to 50 °C, and add 0.6 ml to the resuspended cells.

7 Mix thoroughly and quickly dispense into 100 μl block formers. Set on ice for 10 min.

8 Press blocks out into 13 ml SCEM in a 15 ml screw-capped centrifuge tube and add 120 μl 100 mg/ml novozyme 234 solution. Screw on cap and place on cell mixer at room temperature for 4 h.

9 Transfer blocks to a 15 ml screw-capped centrifuge tube containing 7.5 ml ESP(1.0), fill tube with ES, and incubate at 50 °C overnight.

10 Drain blocks and add fresh ESP (0.5) and incubate at 50 °C for a further 24 h.

11 Transfer blocks to a 15 ml screw-capped centrifuge tube and wash twice for 1 h with 1 mM PMSF in 1 mM EDTA pH 8.0, then six times for 20 min each with 1 mM EDTA pH 8.0, and finally for 20 min with TE. Store the blocks (containing intact YAC chromosomes) in TE at 4 °C.

B. Gel purification of YAC clones to be used as hybridization probes

1 Prepare a 1% (w/v) SeaKem GTG agarose gel using 0.5 × TBE buffer. Cast gel without wells and reserve some molten agarose.

2 Cut a trough near one end of the gel and insert the blocks containing intact YAC chromosomes. Insert the blocks side-by-side and place in contact with the 'front' face of the trough. Place lambda ladder markers in the trough alongside the plant DNA blocks. Fill in the remainder of the trough with the reserved molten agarose to seal in the blocks, and let the agarose set.

3 Resolve intact yeast chromosomes by Pulsed Field Gel Electrophoresis (PFGE; see Chapter 6) using pulse parameters to achieve maximum resolution of the YAC, visualize under long-wave UV, and excise the minimum sized gel slice containing the YAC.

4 Wash gel slice in at least 10 volumes of water in an appropriate sized tube on a cell mixer.

5 Transfer gel slice to a 1.5 ml microfuge tube and place tube in a 65 °C water bath for 10 min.

Protocol 2 continued

6 Add molten gel slice to an approximately equal volume of 2 × CA buffer (pre-warmed to 65 °C) and mix.

7 Add 10 U *Taq*I and incubate at 65 °C for 30 min.

8 Add 0.05 volumes of EDTA, transfer tube to a 37 °C water bath, and equilibrate for 10 min.

9 Add 1 U β-agarase I per 150 μl original 1% gel slice and incubate at 37 °C overnight.

10 Centrifuge tube briefly and heat to 65 °C for 20 min.

11 Equilibrate tube at 37 °C, add 0.5 U β-agarase I per 150 μl original 1% gel slice and incubate at 37 °C for 4 h.

12 Heat to 65 °C and transfer to a new microcentrifuge tube containing 0.1 vols of 3 M sodium acetate, 2 μl yeast RNA carrier, and an equal volume of phenol. Invert to mix and centrifuge at 12 000 g for 10 min at room temperature.

13 Transfer aqueous phase to a new tube and extract with an equal volume of phenol:chloroform:iso-amyl alcohol. Centrifuge at 12 000 g for 2 min.

14 Transfer aqueous phase to a new microcentrifuge tube and add 2.5 vols of 100% (v/v) ethanol to precipitate DNA. Precipitate at −20 °C overnight.

[a] Methods reproduced from ref. 28, with permission.

3 BAC libraries

In recent years, two cloning systems have been developed, BACs (29) and vectors based on bacteriophage P1 (30), that combine the advantages of large inserts with ease of handling. Both use *Escherichia coli* as host, so DNA is easily purified and, because clones are maintained at a very low copy number per cell, the problem of instability is reduced. The use of the BAC system has become prevalent as clones can carry larger inserts (up to 300 kb) than P1 clones (up to 95 kb).

BAC libraries have been constructed from many plant species, including maize, tomato, rice, and Arabidopsis (http://hbz.tamu.edu). Two BAC libraries, TAMU BAC (31) and IGF BAC (32), and one P1 library, Mitsui P1 (33), constructed from genomic DNA of the Columbia ecotype have been used extensively in Arabidopsis physical mapping and genome sequencing projects. These libraries are available from ABRC (http://aims.cps.msu.edu/aims/) as whole libraries, colony filter sets, or individual clones. The average insert sizes are approximately 100 kb for the BAC libraries and approximately 80 kb for the P1 library. The two BAC libraries together have been shown to provide excellent coverage of the genome, though the redundancy of coverage varies widely (1).

Recently two complete BAC-based physical maps of the Arabidopsis genome have been reported (9, 10) which were assembled using TAMU and IFG BACs by iterative hybridization of end probes to colony filters, or by generating restriction 'fingerprints'. These physical maps are extremely valuable for studies regarding genome organization, sequencing and map-based cloning in Arabidopsis, and are available via the internet (see Table 2).

3.1 Construction of BAC libraries

In many circumstances BAC libraries of non-standard Arabidopsis ecotypes or other plant species would be useful. For example, in the analysis of plant/pathogen interactions the cloning of complex disease resistance loci from the different ecotypes may require large insert clones.

Two vectors have been developed for large-insert library construction, pBeloBAC11 (34) and pECBAC1 (35), and one for large-insert library construction and plant transformation, BIBAC2 (36). These vectors are present at 1–2 copies per cell resulting in only a low production of beta-galactosidase and, therefore, only pale blue colonies and the yield of insert DNA is limited. Also, there are limited restriction sites available for cloning in these vectors, BIBAC2 only has a single cloning site. Recently, it has been reported that the cosmid vector pCLD04541 (designed as a binary vector for *A. tumefaciens*-mediated plant transformation; see Section 4) has been used to clone large DNA fragments (40–310 kb), and is well suited for large-insert library construction (37). pCLD04541 is present at a higher copy number than BAC vectors, four or five copies per cell, and the clones are highly stable. This vector also has a range of cloning sites available (*Figure 3*). pCLD04541 has been used for the construction of large-insert DNA libraries in a range of plants, including sorghum (37), Arabidopsis (K. Meksem *et al.*, unpublished results; Y-L. Chang *et al.*, unpublished results), soybean (K. Meksem *et al.*, in preparation), *Triticum tauschii* (O. Moullet *et al.*, in preparation) and maize (http://hbz.tamu.edu; V41 = pCLD04541).

Table 2 Internet sites for BAC-based Arabidopsis physical mapping tools

URL (http://)	Information available
genome.wustl.edu/gsc/arab/arabsearch.html	*Arabidopsis thaliana* BAC fingerprint database (TAMU clones = M; IGF clones = I)
www.mpimp-golm.mpg.de/101/mpi_mp_map/bac.html	BAC-based physical map of the *Arabidopsis thaliana* genome essentially represented by overlapping IGF-BAC clones.
www.tigr.org/tdb/at/atgenome/bac_end_search/bac_end_search.html	BAC end-sequence database
www.kasuza.or.jp/arabi/	Integrated physical maps of chromosomes 3 and 5
synteny.nott.ac.uk/agr/agr.html	Arabidopsis Genomic Resource (AGR), integrated genetic, physical and sequence maps of chromosomes 4 and 5

Protocol 3 describes a generalized method for the preparation of a BAC library.

Protocol 3

Construction of a BAC library[a]

Equipment and reagents

- Nucleobond AX 10 000 kit (CLONTECH)
- 1 × CA buffer: 20 mM Tris pH 8.0, 7 mM $MgCl_2$, 100 mM KCl, 200 μg/ml BSA
- Calf intestinal alkaline phosphatase (CIAP; New England Biolabs)
- Phenol:chloroform:iso-amyl alcohol (24:25:1)
- ESP(0.5): 100 mM EDTA pH 8.0, 1% (w/v) sarkosyl, 0.5 mg/ml proteinase K
- 1 mM phenlymethylsulphonyl fluoride (PMSF) in 1 mM EDTA pH 8.0, freshly made up using a stock solution of 100 mM PMSF in isopropanol
- 2 × CA buffer: 40 mM Tris pH 8.0, 14 mM $MgCl_2$, 200 mM KCl, 400 μg/ml BSA
- SeaKem GTG agarose (FMC)
- GeneCapsules (Geno Technology Inc.)
- 0.025 μm VS filters (Millipore)
- T4 DNA ligase, 2000 U/μl (New England Biolabs)
- Electromax DH10B cells (Gibco BRL)
- Cell-porator with voltage booster (Gibco BRL) or equivalent device
- Isopropyl-1-thio-galactopyranoside (IPTG)
- 5-bromo-4-chloro-3-indolyl-galactopyranoside (X-Gal)
- LB: 1% (w/v) tryptone, 0.5% (w/v) yeast extract, 1% (w/v) NaCl, pH 7.2
- Freezing broth: 1% tryptone, 0.5% yeast extract, 36 mM K_2HPO_4, 1.5 mM tri-sodium citrate, 0.75 mM $MgSO_4$, 7 mM $(NH_4)_2SO_4$, 13 mM KH_2PO_4, 0.48 M glycerol, pH 7.4, autoclaved

Methods

A. Preparation of the vector

1. Inoculate 4 l of LB medium (with appropriate selection) with the bacterial culture harbouring the BAC vector plasmid, and grow overnight at 37 °C with shaking.
2. Isolate plasmid DNA using a Nucleobond AX 10 000 kit according to the manufacturers instructions, except elute the DNA from the column with four 25 ml aliquots of buffer N5, rather than with one 100 ml aliquot, and discard the first 10 ml of buffer off the column.
3. Linearize 30 μg of plasmid DNA with 100 U of appropriate restriction endonuclease (e.g. *Hin* dIII for pBeloBAC11, *Bam* HI for BIBAC2) in a total volume of 400 μl of 1 × CA buffer at 37 °C overnight.
4. Check linearization of the plasmid by analysing a 3 μl aliquot of the digest by agarose gel electrophoresis.
5. Dephosphorylate by adding 30 U CIAP to the linearized plasmid and incubate at 37 °C for 1 h.
6. Inactivate the CIAP by incubation at 65 °C for 20 min.

Protocol 3 continued

7 Extract once with an equal volume of phenol:chloroform:iso-amyl alcohol. Add 40 μl of 3 M sodium acetate and 1 ml 100% (v/v) ethanol to the aqueous phase, and precipitate for 30 min on ice.

8 Pellet plasmid DNA by centrifugation for 10 min at 12 000 g and resuspend vector DNA in 30 μl 0.1 × TE.

B. Preparation of the insert DNA

1 Prepare plant cell nuclei and embed in InCert agarose blocks as described in the Texas A&M Training Manual (http://www.tamu. edu:8000/~creel/TOC.html; see also Chapter 6).

2 Transfer the blocks to a 15 ml screw-capped centrifuge tube containing 13 ml ESP (0.5), and incubate at 50 °C overnight.

3 Replace buffer with fresh ESP (0.5) and incubate at 50 °C for another 24 h.

4 Transfer the blocks to a new 15 ml screw-capped centrifuge tube and wash twice for 1 h each with 1 mM PMSF in TE, and then six times for 20 min with TE. Store blocks at 4 °C in TE.

C. Calibration of restriction endonuclease for partial digestion of the DNA preparation

1 Transfer 4 blocks to a 15 ml screw-capped centrifuge tube containing sterile water and place the tube on a cell mixer for 30 min.

2 Cut the blocks in half and place each into a microfuge tube.

3 To each tube add 50 μl 2 × CA buffer and incubate at 37 °C for at least 2 h.

4 Exchange the buffer with fresh 1 × CA buffer and incubate on ice for 20 min.

5 Add 5 μl of the appropriate restriction endonuclease diluted in 1 × CA buffer (e.g. *Hin* dIII for pBeloBAC11, *Sau* 3AI for BIBAC2) and incubate on ice for 30 min. Use a 2-fold dilution series to add 0, 0.125, 0.5, 1.0, 2.0, 4.0, and 8.0 units to each half block.

6 Transfer the tubes to a 37 °C water bath for exactly 1 h.

7 Transfer the tubes to ice and immediately replace the buffer with 50 μl of ice-cold 50 mM EDTA pH 9.0. Incubate on ice for 1 h.

8 Resolve the samples by PFGE (see Chapter 6), using lambda ladder markers and appropriate pulse parameters, in order to identify the partial digests that produce the maximum amount of DNA in the desired 130–200 kb size range.

D. Size fractionation and purification of the insert DNA

1 Carry out a large scale digest on 12–16 full blocks, using the procedure described above, but with one whole block per tube, double volumes, and with the optimum

Protocol 3 continued

quantity of restriction endonuclease added (remembering to take account of the doubling of solution volumes and the quantity of DNA in each tube).

2 Prepare a 1% (w/v) SeaKem GTG agarose gel using 0.5 × TBE buffer. Cast gel without wells and reserve some molten agarose.

3 Cut a trough near one end of the gel and insert the blocks of the partially digested plant DNA. Insert the blocks in two rows of 6 or 8 blocks, placed in contact with each other and the 'front' face of the trough. Place lambda ladder markers in the trough alongside the plant DNA blocks. Fill in the remainder of the trough with the reserved molten agarose to seal in the blocks, and let set.

4 Run PFGE for 18 h using pulse parameters selected so as to resolve to approximately 250 kb.

5 Cut off the gel outer edges containing the lambda markers and the edges of the plant DNA smear. Stain the gel edges with ethidium bromide and visualize on a UV transilluminator.

6 Measure the exact migration distance of DNA in the size range 130–200 kb.

7 Guided by the measurement taken, excise the slice of gel containing 130–200 kb DNA from the unstained portion of the gel.

8 Rotate the gel slice through 180° and load onto a fresh 1% (w/v) SeaKem GTG agarose gel, such that the largest DNA is now furthest along the gel.

9 Run the second pulse field gel, use exactly the same parameters as used in step 4.

10 Analyse gel as described in step 5.

11 The majority of the DNA should be concentrated into a sharp band. Measure the exact migration distance of the band.

12 Based on the migration distance, recover the DNA from the unstained portion of the gel by electroelution. We use GeneCapsules, following the manufacturers instructions.

13 Pool eluted DNA and analyse a 10 μl sample by PFGE to check integrity and estimate yield.

14 Float a 0.025 μm VS filter on 50 ml 0.1 × TE in a 50-ml screw-capped centrifuge tube. Gently pipette the eluted DNA solution onto the membrane and place at 4 °C overnight to dialyse. Store dialysed DNA at 4 °C.

E. Ligation, electroporation and arraying the library

1 Prepare ligation reactions using 200 ng of insert DNA, a 10-fold molar excess of vector DNA, 0.1 volume 10 × ligation buffer (as supplied by the manufacturer), 2000 U (1 μl) T4 DNA ligase in a final volume of 100 μl. Incubate at 4 °C overnight.

Protocol 3 continued

Multiple 100 μl ligations should be set up in parallel. As a control, set up one ligation lacking insert DNA.

2. Float dialyse the ligations as described in step D14. Store at 4 °C.
3. Electroporate 1 μl aliquots of the ligation into 20 μl Electromax DH10B cells, using an electroporator with voltage booster and the protocol supplied by the manufacturer. Electroporation parameters: Fast charge, low Ohms, 330 μF, voltage booster set to 4000 Ohms. Plate cells on appropriate selective medium (LB agar plus 0.5 mM IPTG, 40 mg/ml X-Gal, and 12.5 μg/ml chloramphenicol for pBeloBAC11; LB agar plus 0.5 mM IPTG, 40 mg/ml X-Gal, 40 μg/ml kanamycin, and 5 % sucrose for BIBAC2).
4. Check that the transformation efficiency is adequate for library construction and that the ligations including insert DNA result in at least a 50-fold increase in transformants over the control ligation without insert DNA. If the background of non-recombinants is too high (indicated by a lower ratio), a new preparation of vector must be made.
5. Plates of transformed cells can be stored for several days at 4 °C prior to picking. Constructs in pBeloBAC11 must be stored at 4 °C for several days to allow the blue colour of non-recombinant clones to fully develop.
6. Use toothpicks to pick colonies and inoculate 384-well microtitre plates containing 50 μl per well of freezing broth and appropriate selection. Grow at 37 °C for 2 days, stir and freeze on dry ice. Store at −80 °C.

[a] Methods reproduced from ref. 28, with permission.

3.2 Screening of BAC libraries using colony hybridizations

BAC libraries can be used for whole genome mapping (9, 10) or for the assembly of contigs in specific regions of the genome. For physical mapping in Arabidopsis the extensive YAC-based physical maps provide a valuable resource to help assemble BAC contigs in defined genomic regions. The YAC maps are well integrated with the genetic maps of Arabidopsis. Chosen YACs can be used to identify clones in the BAC library by preparing gel-purified YAC chromosomal DNA (*Protocol 2*) and hybridizing the isolated DNA to BAC colony filters. It is very important that the prepared yeast chromosomes are largely intact, as contamination by large quantities of degraded yeast DNA results in strong non-specific hybridization, which can mask true positive signals on autoradiographs. BAC contigs can be assembled by restriction fingerprinting of the positive hybridizing clones, followed by iterative hybridization (as detailed in Bent *et al.*; 1). Additionally, as with the YAC colony filters, the BAC filters can be used to readily map ESTs or T-DNA/transposon flanking sequences (see Chapter 4), and with higher resolution.

Hybridization-based screening is performed using membranes carrying bacterial colonies. To maximize the number of colonies on each membrane a multiprong replicator (or 'hedgehog') is used. This procedure can be carried out manually or by a robot. Economical handling of large numbers of colonies can be achieved by slightly offsetting each colony print, and up to 3 × 3 × 384 colonies (*Figure 2*) can be transferred to a single 11 × 7.5 cm membrane. To make the identification of positively hybridizing colonies even more reliable, the colonies can be duplicated. The preparation of BAC colony filters by hand is described in *Protocol 4*. The screening of BAC filters by hybridization is described in *Protocol 5*.

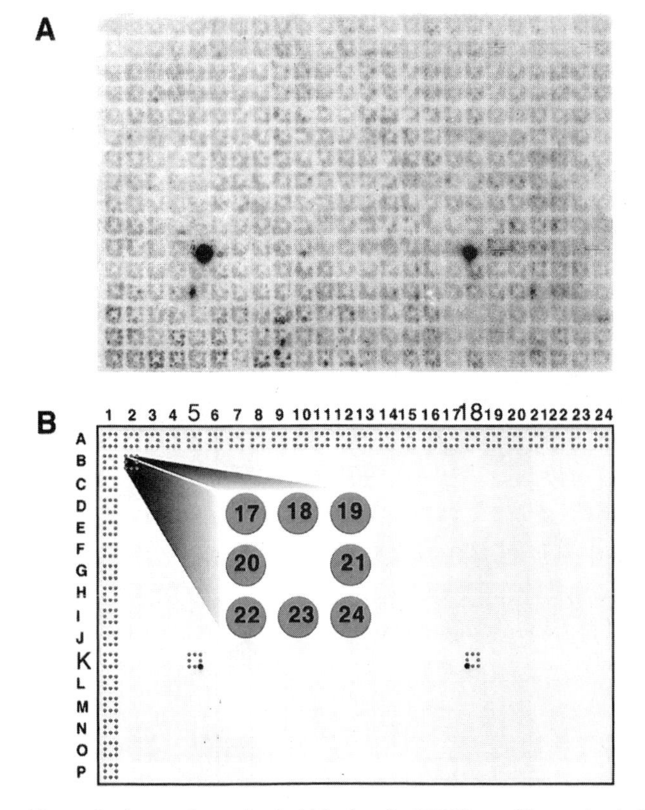

Figure 2. Autoradiograph of a high density BAC library filter probed with a T-DNA flanking fragment (A), and a diagram showing the filter layout (B). The BAC library filter has been arrayed by using eight offsets (3 × 3, centre position blank) of a 384-prong replicator (see B). This filter contains colonies of the IGF BAC library (plates 17–24), and is one of a set of four that represent the entire library. The scoring of the filter is carried out in an analogous way to scoring the YAC library filter (*Figure 1*). Unlike the YAC library filter, the BAC clones have not been arrayed in duplicate, however, the positive clones can be clearly scored against the background hybridization. The positively hybridizing clones from the autoradiograph (A) are illustrated in (B), and the coordinates of the two positive clones are F24K5 and F22K18. The prefix 'F' denotes a clone from the IGF library, for the TAMU library the co-ordinate would be prefixed by 'T'.

Protocol 4

Preparation of colony filters for hybridization analysis[a]

Equipment and reagents

- LB: 1% (w/v) tryptone, 0.5% (w/v) yeast extract, 1% (w/v) NaCl, pH7.2 (add 1.5% bacto-agar for LB agar)
- 8.6 × 12.8 cm OmniTrays (Nalge Nunc International)
- Hybond N^+ membrane (Amersham)
- Whatman 3MM paper
- Denaturation solution: 0.5 M NaOH, 1.5 M NaCl
- Neutralization buffer: 1 M Tris pH 7.5, 1.5 M NaCl
- Scrape solution: 5 × SSC, 0.5% (w/v) SDS, 1 mM EDTA pH 8.0

Method

1. Pour LB agar plates containing appropriate selection using 8.6 × 12.8 cm OmniTrays, and allow to dry.
2. Cut 11.7 × 7.6 cm rectangles of Hybond N^+ membrane. Label using a pencil or fine, permanent marker pen (in position A1) and place on the surface of an LB agar plate.
3. Using a 384-prong replicator, spot the colonies of one 384-well microtitre plate onto each filter (up to nine plates may be spotted if an offsetting strategy is used). Prepare several replicates for simultaneous hybridization analysis.
4. Invert the plates, and incubate at 37 °C overnight.
5. Carefully lift the filters from the agar plates, place on a sheet of 3 MM paper soaked in denaturation solution, and leave for 3 min.
6. Transfer the filters to a fresh sheet of 3 MM paper soaked in denaturation solution, and leave for another 3 min.
7. Transfer filters to a sheet of 3 MM paper soaked in neutralization solution, and leave for 6 min.
8. Rinse the filters in 2 × SSC and air dry. Place filters in sleeves of 3MM paper when they begin to curl.
9. Bake at 80 °C for 2 h.
10. Soak the filters for at least 2 h at 42 °C in scrape solution.
11. Using a paper towel, scrape off the bacterial debris. Rinse filters in 2 × SSC. The wet filters can be used immediately for hybridization, or dried for use later.

[a] Method reproduced from ref. 28, with permission.

Protocol 5

Hybridization analysis of colony filters[a]

Equipment and reagents

- Random prime labelling kit (Gibco BRL)
- 20 × SSPE: 3 M NaCl, 0.2 M NaH_2PO_4, 25 mM EDTA, pH 7.4
- 50 × Denhardts: 1% (w/v) Ficoll, 1% (w/v) polyvinylpyrrolide, 1% (w/v) BSA
- Hybridization boxes (8 × 12 cm; e.g. Bunzl 6064)
- Hybridization buffer: 5 × SSPE, 5 × Denhardts, 0.5% (w/v) SDS, 5 μg/ml salmon sperm DNA
- Hybridization wash buffer: 0.1 × SSC, 1% (w/v) SDS
- Membrane neutralization buffer: 0.1 × SSC, 0.1% (w/v) SDS, 0.2 M Tris pH 7.5

Method

This protocol is designed for use with sets of 8 membranes of size 11.7 × 7.6 cm, and should be scaled up or down as appropriate.

1. Label 50 ng of gel purified probe (PCR product, cDNA clone/EST, genomic DNA clone or YAC DNA) using a random primers labelling kit (see Chapter 6). We usually use α-[^{32}P]dCTP, but other labels could be used. There is no requirement to remove unincorporated [^{32}P]-labelled nucleotides prior to hybridization; in fact, these help to provide background which aids scoring of the filters.
2. Rinse the filters in 2 × SSC and then place in a hybridization box containing 30 ml of hybridization buffer. Add filters one-by-one, making sure each is fully submerged before adding the next.
3. Prehybridize at 65 °C for at least 4 h with gentle shaking.
4. Add denatured probe to a fresh hybridization box containing 10 ml hybridization buffer, and transfer the filters one-by-one as previously.
5. Hybridize overnight at 65 °C with gentle shaking.
6. Transfer the filters, one-by-one, to a larger box containing 250 ml hybridization wash buffer. Incubate at 65 °C for 20 min with gentle shaking.
7. Discard the wash buffer and replace with fresh solution. Incubate as in step 6.
8. Repeat step 7 to give a third wash.
9. Blot dry, seal filters in a plastic bag and expose to film, with the film to the DNA side of the filters. Adjust the length of exposure depending on the strength of the signal.
10. It may be necessary to re-wash the filters to reduce excessive background. An ideal autoradiograph should show most of the colonies on each filter as a just visible background, with the positive signals clearly stronger (*Figure 2a*).
11. Filters can be stripped by incubation in 0.4 M NaOH twice for 20 min each at 45 °C with gentle shaking, followed by a 15-min wash in membrane neutralization buffer.

[a] Method reproduced from ref. 28, with permission.

An autoradiograph of a high-density BAC library filter is shown in *Figure 2*. Internet sites that have a search facility for positioning clones on the Arabidopsis physical map, and other useful web sites including BAC end-sequence databases, BAC and P1 clones used in the Arabidopsis genome sequencing project, and sites integrating BAC and P1 maps with other data such as genetic maps, YAC maps and nucleotide sequences, are listed in *Table 2*.

4 Cosmid libraries

Cosmid clones are extremely useful for high resolution mapping of defined regions and have played essential roles in positional cloning experiments. Recently, the generation of transformation-competent, or 'binary', cosmids that are capable of replication in both *E. coli* and *A. tumefaciens* and contain sequences necessary for the integration of the insert into the plant genome has enabled clones to be directly used in complementation tests. One binary cosmid vector, pCLD04541 (*Figure 3*; http://www. jic.bbsrc.ac.uk/staff/caroline-dean/04541.htm), has been used widely for gene identification in Arabidopsis (22, 23) and tomato (38), and is available from ABRC.

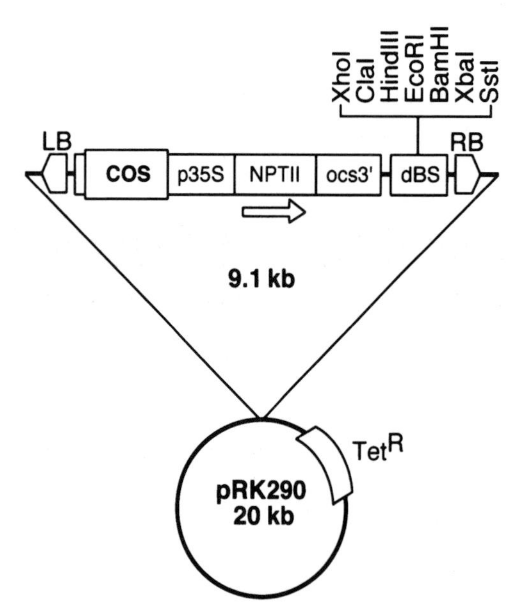

Figure 3. Binary cosmid vector pCLD04541 consists of a 9.1 kb insert into the broad host range cosmid pRK290 (40). The insert carries the *cos* site (for lambda packaging), a 35S-NPTII gene (conferring kanamycin resistance in plants), and a dark Bluescript (dBS) polylinker (41) between the T-DNA borders. The dBS polylinker provides a number of unique restriction enzyme sites for cloning and gives a 'bluer' colour to *E. coli* colonies on X-Gal and IPTG, which is an advantage in a low copy binary vector. Only unique restriction enzyme sites are shown. pRK290 carries a bacterial tetracycline resistance gene (selection is on 10 μg/ml in *E. coli*, but 1 μg/ml in *Agrobacterium*). For more information about pCLD04541 visit 'http://www.jic.bbsrc.ac.uk/staff/caroline-dean/04541.htm'. pCLD04541 is available from ABRC (http://aims.cps.msu.edu/aims/).

Cosmid libraries made using pCLD04541 carry inserts of 15–25 kb (22, 23), and each clone therefore contains on average three or four genes. Screening of cosmid libraries is carried out in the same way as BAC libraries and contig assembly approaches are also similar. Contigs are constructed by digesting clones with a combination of restriction endonucleases and then matching restriction fragment patterns. The small size of the inserts and the construction of restriction maps leads to very high resolution physical maps.

4.1 Construction of a cosmid library using pCLD04541

Cosmid libraries can be easily constructed from a range of source DNA, such as genomic DNA, YAC inserts or BAC inserts. We have routinely used the binary cosmid vector pCLD04541 to construct cosmid libraries and a reliable method for construction involves the partial digestion of source DNA with *Sau* 3A, size selection to 15–25 kb, phosphatase treatment of the insert and ligation, at high concentration, into *Bam* HI-cut vector. The ligated DNA is then packaged into lambda phage using commercial packaging extracts, and these are used to infect *E. coli*, in which the DNA is circularized and propagated as a plasmid. *Protocol 6* describes a detailed method for the construction of a library using Arabidopsis genomic DNA. This method can also be used for subcloning BAC DNA, and can be adapted for subcloning a YAC, as the source DNA in this case is limiting. For methods describing partial digestion and size fractionation of a gel-purified YAC see Bancroft *et al.* (22). After the library has been constructed and arrayed, colony filters for hybridization screening can be prepared as described in *Protocol 4*. Hybridization of colony filters is carried out as described in *Protocol 5*.

Protocol 6

Construction of a binary cosmid library from Arabidopsis genomic DNA

Equipment and reagents

- Cosmid vector pCLD04541 (available from ABRC)
- 100–150 μg CsCl-pure Arabidopsis genomic DNA
- 10 μg CsCl-pure vector pCLD04541
- LB: 1% (w/v) tryptone, 0.5% (w/v) yeast extract, 1% (w/v) NaCl, pH7.2
- Phenol:chloroform:iso-amyl alcohol (24:25:1)
- CIAP (New England Biolabs)
- T4 DNA ligase and ligation buffer (New England Biolabs)
- Gigapack II XL or Gold lambda packaging extracts (Stratagene)
- *E. coli* Sure Tet^S strain (Stratagene)
- IPTG
- X-Gal
- Freezing broth: 1% tryptone, 0.5% yeast extract, 36 mM K_2HPO_4, 1.5 mM tri-sodium citrate, 0.75 mM $MgSO_4$, 7 mM $(NH_4)_2SO_4$, 13 mM KH_2PO_4, 0.48 M glycerol, pH 7.4, autoclaved.

Protocol 6 continued

Methods

A. Calibration of restriction endonuclease for partial digestion of plant genomic DNA

1. To 10 μg genomic DNA add restriction buffer (1×), spermidine (4 mM), BSA (1×) and water to a final volume of 150 μl, mix well.
2. Label 9 microfuge tubes 1–9, add 30 μl of the above mix to tube 1, 15 μl to tubes 2–8, and the remainder of the mix to tube 9.
3. Add 1 U *Sau* 3A to tube 1 and mix well.
4. Transfer 15 μl from tube 1 to tube 2 and mix well. Continue the two-fold dilution until tube 8. Do not add enzyme to tube 9.
5. Incubate at 37 °C for 1 h.
6. Stop the reaction on ice and add 1 μl of 0.5 M EDTA (pH 8.0).
7. Resolve the digests by 0.4% (w/v) agarose gel electrophoresis. NB. Make a 1–2% (w/v) agarose base and then put 0.4% (w/v) agarose on top, the 1–2% gel supports the low percentage gel during transfer to the UV transilluminator. Use DNA size markers that have fragments between 9 and 30 kb.
8. Stain the agarose gel with ethidium bromide solution and determine which concentration of enzyme appears to best produce DNA of the required size, 15–25 kb. Half this amount of enzyme will give the maximum amount of molecules of the required size (39).

B. Large scale partial digestion of genomic DNA

1. Using the same concentration of DNA (1 μg/15 μl) and components as in the calibration experiment, conduct a series of 10 μg partial digests on the remaining genomic DNA using half the concentration of enzyme that appeared to produce most amount of DNA in the desired 15–25 kb range.
2. Incubate at 37 °C for 1 h.
3. Remove 1 μg aliquots from the digestion mixes and add 1 μl 0.5 M EDTA (pH 8.0).
4. Stop the rest of the reaction by incubation at 65 °C for 20 min.
5. Add 0.1 vols of 3 M sodium acetate and an equal volume of phenol: chloroform: isoamyl alcohol and mix gently. Centrifuge at 12 000 g for 5 min.
6. Transfer the aqueous phase to a new microfuge tube and precipitate the partially digested DNA with 2 vols of 100% (v/v) ethanol.
7. Pellet the DNA in a microcentrifuge at 12 000 g for 10 min.
8. Wash the pellets with 70% (v/v) ethanol, and centrifuge briefly at 12 000 g.
9. Resuspend the pellets in water to produce a pool of 270 μl.
10. Resolve 1 μg aliquots by agarose gel electrophoresis (0.4% gel on a 1–2% base, as in step A7) to confirm that the digested fragments are of the correct size range.

Protocol 6 continued

C. Treatment of partially digested genomic DNA with CIAP

1. To the 270 μl of partially digested DNA add 30 μl 10 × CIAP buffer and 2.5 U CIAP.
2. Incubate at 37 °C for 20 min and then at 68 °C for 15 min.
3. Add 0.1 vols of 3 M sodium acetate and an equal volume of phenol: chloroform: iso-amyl alcohol and mix gently. Centrifuge at 12 000 *g* for 5 min.
4. Transfer the aqueous phase to a new microfuge tube and add 2 vols of 100% (v/v) ethanol to precipitate the partially digested DNA.
5. Pellet the DNA in a microcentrifuge at 12 000 g for 10 min.
6. Wash the pellet with 70% (v/v) ethanol and centrifuge briefly at 12 000 g. Air dry.
7. Resuspend the DNA in 500 μl TE.

D. Size fractionation and purification of genomic DNA

1. Prepare a sucrose gradient with 10–40% (w/v) sucrose solutions (diluted in: 1 M NaCl, 20 mM Tris pH 8.0, 5 mM EDTA).
2. Load the 500 μl of DNA obtained from step C on to the top of the gradient, and centrifuge at approximately 50 000 g (e.g. 20 000 rpm in an SW41 rotor) for 18 h.
3. Following centrifugation, collect 20 fractions of 300–500 μl from the bottom of the tube.
4. Analyse 10 μl of each fraction by agarose gel electrophoresis (0.4% gel on a 1–2% base, as in step A7) and identify fractions containing genomic DNA fragments between 15 and 20 kb (NB: it is necessary to add NaCl to the size markers so that the NaCl concentration is the same as in the fractions).
5. Dilute the fractions that contain DNA fragments 15–25 kb 1 in 4 with water.
6. Precipitate the DNA by adding 1/30 volume of 3 M sodium acetate (pH 5.5), 1 μl tRNA, and 0.7 volumes of isopropanol, incubate on ice for 60 min.
7. Pellet the DNA in a microcentrifuge for 20 min at 12 000 g.
8. Wash and dry the pellet as described in step C6.
9. Resuspend each fraction in 5 μl of water.
10. Check the size of the purified fractions by resolving 0.5 μl from each on a 0.4% (w/v) agarose gel (0.4% gel on a 1–2% base, as in step A7).
11. Pool fractions of the required size into a single tube.

E. Preparation of vector

1. Digest 10 μg of pCLD04541 vector to completion with *Bam* HI. Use at least 5 U *Bam* HI/μg vector DNA and incubate at 37 °C for 2–3 h.
2. Remove a 100 ng aliquot, and add an equal volume of phenol: chloroform:iso-amyl alcohol. Centrifuge at 12 000 g for 5 min.
3. Ethanol precipitate the vector DNA and wash with 70% (v/v) ethanol as in steps C4, 5, and 6.

Protocol 6 continued

4 Resolve the 100 ng aliquot by agarose gel electrophoresis to check that the vector is completely linearized.

F. Ligation of vector and insert DNA

1 For the ligation combine 1 μg dephosphorylated, purified genomic DNA fragments, and 0.25 μg vector DNA in a 7.5 μl volume.
2 Remove a 0.5 μl aliquot and store at −20 °C (unligated control).
3 Incubate the mix at 65 °C for 5 min, at 42 °C for 20 min, and finally at room temperature for 2 h.
4 To the tube add 2 μl of ligation buffer and 1 μl (2000 units) of T4 DNA ligase.
5 Incubate at 12 °C for 18 h.
6 To check the ligation analyse 0.5 μl of the ligation mix against the unligated control, from step 2, by agarose gel electrophoresis.

G. Packaging and plating the library

1 Perform a test packaging with 2 μl of ligation using Gigapack II XL or GOLD lambda packaging extracts. Perform according to the manufacturers protocol.
2 Combine 10 μl of phage stock, 100 μl phage buffer, and 200 μl plating cells (*E. coli* Sure Tet^S strain).
3 Incubate at 37 °C for 20 min.
4 Add 1 ml LB medium and incubate at 37 °C for 45 min.
5 Spread 0.1 and 0.5 ml aliquots on selective plates (LB agar + 10 μg/ml tetracycline) including 0.5 mM IPTG and 40 mg/ml X-Gal (NB: spin down cells and resuspend in fresh LB medium before plating to avoid adding $MgCl_2$ to tetracycline plates, as this inhibits tetracycline selection).
6 Incubate at 37 °C overnight.
7 Count the colonies and calculate the titre of the library.
8 Plate out the library, as described above, at a density such that single colonies can easily be picked.
9. Pick colonies into 384-well microtitre plates containing 50 μl freezing broth plus 10 μg/ml tetracycline per well. Grow at 37 °C for 2 days, and freeze at −80 °C.

References

1. Bent, E., Johnson, S., and Bancroft, I. (1998) *Plant J.*, **13**, 849.
2. Wang, M. L., Huang, L., Bongard-Pierce, D. K., Belmonte, S., Zachgo, E. A., Morris, J. W., Dolan, M., and Goodman, H. M. (1997) *Plant J.*, **12**, 711.
3. Zachgo, E. A., Wang, M. L., Dewdney, J., Bouchez, D., Camilleri, C., Belmonte, S., Huang, L., Dolan, M., and Goodman, H. M. (1996) *Genome Res.*, **6**, 19.
4. Schmidt, R., West, J., Love, K., Lenehan, Z., Lister, C., Thompson, H., Bouchez, D., and Dean, C. (1995) *Science*, **270**, 480.

5. Schmidt, R., Love, K., West, K., Lenehan, Z., and Dean, C. (1997) *Plant J.*, **11**, 563.
6. Kotani, H., Nakamura, Y., Sato, S., Asamizu, E., Kaneko, T., Miyajima, N., and Tabata, S. (1998) *DNA Res.*, **5**, 203.
7. Camilleri, C., Lafleuriel, J., Macadre, C., Varoquaux, F., Parmentier, Y., Picard, G., Caboche, M., and Bouchez, D. (1998) *Plant J.*, **14**, 633.
8. Sato, S., Kotani, H., Hayashi, R., Liu, Y. G., Shibata, D., and Tabata, S. (1998) *DNA Res.*, **5**, 163.
9. Mozo, T., Dewar, K., Dunn, P., Ecker, J. R., Fischer, S., Kloska, S., Lehrach, H., Marra, M., Martienssen, R., Meier-Ewert, S., and Altmann, T. (1999) *Nature Genet.*, **22**, 271.
10. Marra, M., Kucaba, T., Sekhon, M., Hillier, L., Martienssen, R., Chinwalla, A., Crockett, J., Fedele, J., Grover, H., Gund, C., McCombie, W. R., McDonald, K., McPherson, J., Mudd, N., Parnell, L., Schein, J., Seim, R., Shelby, P., Waterston, R., and Wilson, R. (1999) *Nature Genet.*, **22**, 265.
11. Burke, D. T., Carle, G. F., and Olson, M. V. (1987) *Science*, **236**, 806.
12. McCormick, M. K., Shero, J. H., Cheung, M. C., Kan, Y. W., Hieter, P. A., and Antonarakis, S. E. (1989) *Proc. Natl Acad. Sci. U S A*, **86**, 9991.
13. Schmidt, R., Putterill, J., West, J., Cnops, G., Robson, F., Coupland, G., and Dean, C. (1994) *Plant J.*, **5**, 735.
14. Ecker, J. R. (1990) *Methods*, **1**, 186.
15. Ward, E. R. and Jen, G. C. (1990) *Plant Mol. Biol.*, **14**, 561.
16. Grill, E. and Somerville, C. (1991) *Mol. Gen. Genet.*, **226**, 484.
17. Creusot, F., Fouilloux, E., Dron, M., Lafleuriel, J., Picard, G., Billault, A., Le Paslier, D., Cohen, D., Chaboute, M. E., Durr, A., Fleck, J., Gigot, C., Camilleri, C., Bellini, C., Caboche, M., and Bouchez, D. (1995) *Plant J.*, **8**, 763.
18. Schmidt, R., West, J., Cnops, G., Love, K., Balestrazzi, A., and Dean, C. (1996) *Plant J.*, **9**, 755.
19. Tutois, S., Cloix, C., Cuvillier, C., Espagnol, M. C., Lafleuriel, J., Picard, G., and Tourmente, S. (1999) *Chromosome Res.*, **7**, 143.
20. Dewar, K., Dunn, P. J., Li, Y.-P., Kim, C. J., Bouchez, D., and Ecker, J. R. (1996) *7th International Conference on Arabidopsis Research, Norwich, UK.*, Abstract S45.
21. Schmidt, R. and Dean, C. (1992) In: *Genome analysis volume 4: strategies for physical mapping* (ed. K. E. Davies and S. M. Tilghman) p. 71–9. Cold Spring Harbor Laboratory Press, NY.
22. Bancroft, I., Love, K., Bent, E., Sherson, S., Lister, C., Cobbett, C., Goodman, H. M., and Dean, C. (1997) *Weeds World*, **4ii**.
23. Macknight, R., Bancroft, I., Page, T., Lister, C., Schmidt, R., Love, K., Westphal, L., Murphy, G., Sherson, S., Cobbett, C., and Dean, C. (1997) *Cell*, **89**, 737.
24. Adam, G., Mullen, J. A., and Kindle, K. L. (1997) *Plant J.*, **11**, 1349.
25. Taji, T., Seki, M., Yamaguchi-Shinozaki, K., Kamada, H., Giraudat, J., and Shinozaki, K. (1999) *Plant Cell Physiol.*, **40**, 119.
26. Schmidt, R. and Dean, C. (1995) *Methods Mol. Cell. Biol.*, **5**, 309.
27. Coulson, A., Sulston, R., Brenner, J., and Karn, J. (1986) *Proc. Natl Acad. Sci. USA*, **83**, 7821.
28. Bancroft, I. (2000) In: *Arabidopsis: a practical approach* (ed. Z. Wilson). Oxford University Press, Oxford.
29. Shizuya, H., Birren, B., Kim, U.-J., Mancino, V., Slepak, T., Tachiiri, Y., and Simon, M. (1992) *Proc. Natl Acad. Sci. USA*, **89**, 8794.
30. Ioannou, P. A., Amemiya, C. T., Garnes, J., Kroisel, P. M., Shizuya, H., Chen, C., Batzer, M. A., and de Jong, P. J. (1994) *Nature Genet.*, **6**, 84.
31. Choi, S. D., Creelman, R., Mullet, J., and Wing, R. A. (1995) *Weeds World*, **2**, 17.
32. Mozo, T., Fischer, S., Meier-Ewert, S., Lehrach, H., and Altmann, T. (1998) *Plant J.*, **16**, 377.

33. Lui, Y-G., Mitsukawa, N., Vasquez-Tello, A., and Whittier, R. F. (1995) *Plant J.*, **7**, 351.
34. Kim, U. J., Birren, B. W., Slepak, T., Mancino, V., Boysen, C., Kang, H. L., Simon, M. I., and Shizuya, H. (1996) *Genomics*, **34**, 213.
35. Frijters, A. C. J., Zhang, Z., van Damme, M., Wang, G.-L., Ronald, P. C., and Michelnore, R. W. (1997) *Theor. Appl. Genet.*, **94**, 390.
36. Hamilton, C. M., Frary, A., Lewis, C., and Tanksley, S. D. (1996) *Proc. Natl Acad. Sci. USA*, **93**, 9975.
37. Tao, Q. and Zhang, H. B. (1998) *Nucl. Acids Res.*, **26**, 4901.
38. Brommonschenkel, S. H. and Tanksley, S. D. (1997) *Mol. Gen. Genet.*, **256**, 121.
39. Sambrook, J., Fritsch, E. F., and Maniatis, T. (ed.) (1989) *Molecular cloning: a laboratory manual* (2nd edn). Cold Spring Harbor Laboratory Press, NY.
40. Ditta, G., Stanfield, S., Corbin, D., and Helinski, D. R. (1980) *Proc. Natl Acad. Sci. USA*, **77**, 7347.
41. Jones, J. D. G., Shlumukov, L., Carland, F., English, J., Scofield, S. R., Bishop, G. J., and Harrison, K. (1992) *Transgenic Res.*, **1**, 285.

Chapter 8

In situ hybridization to plant chromosomes and DNA fibres

J. S. Heslop-Harrison and Trude Schwarzacher
Department of Biology, University of Leicester, Leicester LE1 7RH, UK

1 Introduction

In situ hybridization of labelled probes to microscope preparations of chromosomal DNA is a powerful method to investigate the physical organization of DNA sequences along the chromosome (*Plates 1–4*) and in the three dimensions of the interphase nucleus. In plant molecular biology, the methods have proved particularly valuable to localize and characterize the organization of repetitive sequences along chromosomes. Major uses of *in situ* hybridization to chromosomal targets include the physical localization of genes to chromosome arms or along chromosomes, the determination of order of genes with respect to centromere/telomere, and the determination of locations of repetitive sequences. Many important questions about the genomic distribution of repetitive sequences in clusters or as dispersed sites along chromosomes cannot be answered in other ways. In mammalian, and particularly human, research, many papers describing gene mapping complement genetic data with *in situ* hybridization results to show the physical location of a gene. In plants, single copy *in situ* hybridization remains a specialized technique—few laboratories have substantial numbers of publications using the technique, despite the many questions that the method could answer. In plants, well characterized single copy sequences as short as 800 bp have been localized, but 2.5 kb or longer sequences are easier to work with, and more than 20 kb of homologous sequence is required for reliable hybridization to both chromatids of both chromosomes in a diploid metaphase.

As with many molecular biology techniques, the techniques for *in situ* hybridization are demanding and require some specialized equipment (particularly a well set-up fluorescence microscope). Setting up a new laboratory requires extensive resources, so it is important to consider the option of collaboration with a laboratory carrying out experiments routinely, although a working knowledge of the steps involved is valuable when involved in collaboration. Knowledge of the methods involved and their limitations also helps interpretation of published data.

For research applications, detection of hybridization sites by fluorescence methods is strongly recommended and is described here. Although there is no real evidence that fluorescence detection is more sensitive than enzymatic or radioactive methods, problems of background signal, safety, and low contrast are less, while there is the critical ability to localize two probes simultaneously using different labels and fluorescent colours for detection.

The resolution of *in situ* hybridization to chromosomes depends on the condensation of the chromosomes analysed. In general, the spindle microtubule inhibitor colchicine (widely used for chromosome counting) gives heavily condensed chromosomes, where the distribution of hybridization sites along chromosome arms cannot be measured accurately. Other chromosome condensation agents such as iced-water or 8-hydroxyquinoline give more extended chromosomes, where the location of signals can be estimated down to a few Mb or μm along the chromosome. Since the early 1990s, the application of *in situ* hybridization to extended chromatin fibres (*Plate* 5) has enabled pairs of probes to be located with an accuracy of a few hundred base pairs.

The basic procedures for *in situ* hybridization are analogous to those used for Southern hybridization. The target is a chromosome preparation or DNA spread onto a microscope slide, while the probe DNA is the sequence of interest labelled with a radio nucleotide, biotin, fluorescent molecule, or a hapten. Probe and target are denatured and allowed to form hybrid molecules. The stringency of hybridization—the similarity between probe and target DNA sequences that is required before they can remain stably hybridized in a double-stranded DNA helix—is controlled by the temperature, formamide (a helix de-stabilizing agent), and sodium ion concentration in the hybridization and washing solutions. In publications, it is essential to present the composition and temperature of the most stringent solutions because this is required to interpret the results. The protocols here are based around those presented by Schwarzacher and Heslop-Harrison (5). More extensive details of the methods and variations for chromosome preparation, the composition of the hybridization solution, hybridization, detection, imaging and troubleshooting are given there, while the authors' website gives additional material and frequently asked questions (http://www.heslop-harrison.com under methods).

2 Preparation of chromosome targets

2.1 Sources of material

For most questions about the location of DNA sequences along chromosomes, metaphase chromosome and interphase nucleus preparations are made on glass microscope slides. The quality of the fixation and chromosome preparation is critical to the success of the *in situ* hybridization experiment. Low numbers of metaphases often indicate unhealthy plants or seedlings, while overlapping chromosomes will be hard to analyse, and surface films of cytoplasm will prevent penetration of the probe and detection reagents. Use of dirty or contaminated reagents will give rise to precipitates onto slides and must be avoided. Interphase cytogenetics can be attempted when metaphases cannot be obtained, but results are usually poor and difficult to interpret.

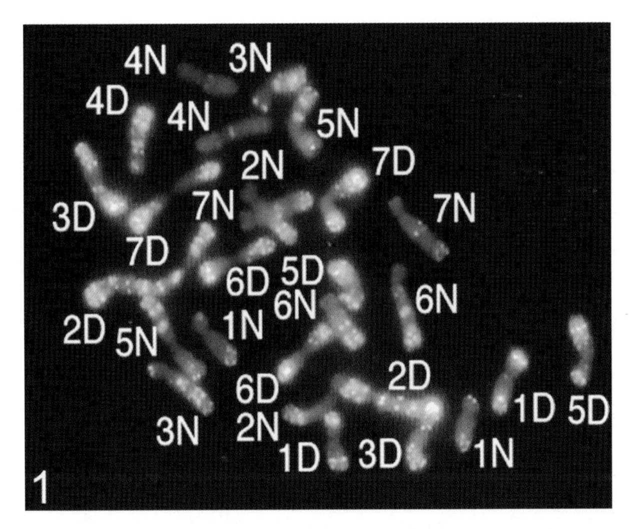

Plate 1 *In situ* hybridization of 45S rDNA (red) and a repetitive DNA sequence dpTa1 (green) to DAPI counterstained metaphase chromosomes of the wild tetraploid ($2n = 4x = 28$) wheat *Aegilops ventricosa*, which as D and N genomes. Using the probes, all individual chromosomes, although morphologically very similar, can be identified. Furthermore, notable features of genome evolution are apparent: the major rDNA loci are present on the N genome chromosomes, while the repetitive dpTa1 sequence is much more abundant on the D genome chromosomes, although both genomes presumably originated from a common ancestor (see ref. 1 for further details).

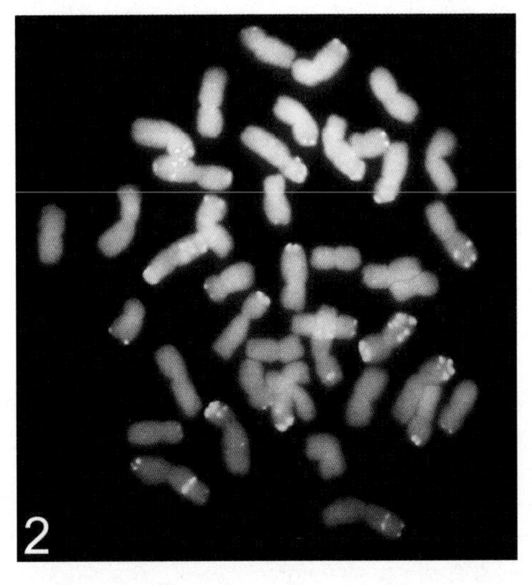

Plate 2 Genomic DNA used as a probe to identify the alien chromosomes in a metaphase of bread wheat (*Triticum aestivum*) incorporating a pair of chromosomes originating from *Aegilops umbellulata*. The *Aegilops* genomic DNA labels the alien chromosome pair in red, while a repetitive DNA sequence originally isolated from rye, pSc119.2, labels sites on many chromosomes allowing their identification (see ref. 2).

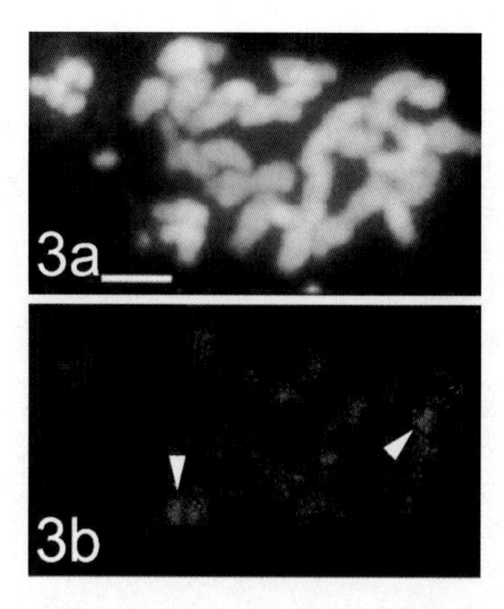

Plate 3 *In situ* hybridization of banana streak virus (a pararetrovirus) DNA to chromosomes of a banana cultivar ($2n = 3x = 33$). DAPI stained chromosomes are seen in (a), while *in situ* hybridization shows the major sites of integration of virus-homologous DNA into the chromosomes (red, b; see ref. 3).

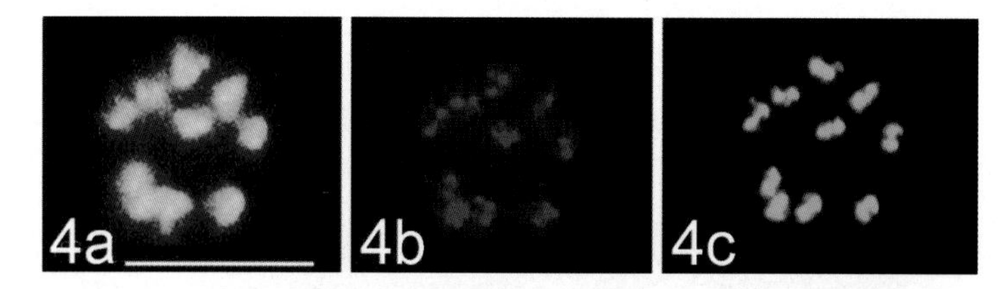

Plate 4 Chromosomes of *Arabidopsis thaliana* ($2n = 10$) (a) counterstained with DAPI; (b) showing sites of *copia*-like retro-elements with concentration around the centromere and dispersed distribution along chromosome arms (red), while (c) the 180bp repeat around the centromeres is only present in this chromosomal region (see ref. 4).

Plate 5 DNA fibre *in situ* hybridization to stretched nuclei of *Arabidopsis thaliana* labelled with the probes as *Figure 4*. The interspersion, at the molecular level, of the red (*copia*) and green (180 bp repeat) sequences is clear (see ref. 4).

Protocol 1 describes the pretreatments required for obtaining plant material containing suitable numbers of slightly condensed chromosomes for analysis by *in situ* hybridization. More detailed methods for fixation and chromosome preparations from *Arabidopsis thaliana* are given elsewhere in the Practical Approach series (6). Because of the very small chromosomes, minor modifications are required to the more universal methods presented here.

Protocol 1

Selection, pre-treatment, and fixation of plant material for chromosome preparations

Equipment and reagents

- Clean glassware and plasticware[a]
- Drinking-quality water[b]
- Filter paper
- Metaphase arresting agent:
  Ice-water (typically for seedlings of temperate plants): put 1–2 ml water in a small tube, shake vigorously to aerate, add ice, and store in an ice bucket
  8-hydroxyquinoline (for many species): make a 2 mM solution (pale yellow, taking 12 h to dissolve; store for up to 6 months at 4 °C)
  Colchicine (for most plants, but may give over-condensed chromosomes with poor morphology): use an 0.05% solution (store in dark at 4 °C)
- Fresh (less than 30 min old) fixative consisting of 1 part glacial acetic acid to 3 parts 96% ethanol

Method

1. Choose suitable plant material from which to obtain metaphases. Seedling root tips are often most convenient, but apical meristems, roots coming from the sides of root-balls in pots, roots growing into compost placed around trees or large plants, floral bud parts, or tissue culture material can be used successfully.
2. For seedling germination, place seeds on moist filter paper and allow to germinate. Cycles of 3 days at 4 °C and 4 days at 20–25 °C may encourage recalcitrant seeds to germinate. Roots about 10–15 mm are suitable for pretreatment from cereals and many other species, but they may show a rhythm in division, so time of fixation may matter. For reproducibility, it is best if seeds are germinated under standard conditions, typically 25 °C.
3. Transfer seedlings, roots from pots, apical meristems, or buds to an ample volume of metaphase arresting solution:
   - ice water: leave material in ice water for 24 h;
   - 8-hydroxyquionoline: leave material in solution at temperature of plant growth for 30–45 min, then place tubes in melting ice for 30–45 min;
   - colchicine: leave material in solution for 2–4 h.

Protocol 1 continued

4 Fix material by plunging into fresh fixative. Do not cross-contaminate other plant material with the fixative or fixative vapours.

[a] Traces of fixatives on glassware, tools used to handle plant material, or in the atmosphere of, for example, refrigerators will reduce greatly the metaphase index.

[b] The water used for seed germination can make a large difference to the chromosome morphology and metaphase index. Water with chlorine or amines (from sterilization) or heavy metals (e.g. from pipes) will reduce the quality for fixations, but tap water may be better than purified water.

2.2 Chromosome spread preparation

The chromosome preparation aims to give flat preparations of chromosomes and interphase nuclei, separated from each other and free from cytoplasm (*Protocol 2*). Unlike cytological preparations for chromosome counting, acid hydrolysis is not usually suitable, so the plant material is digested using enzymes that allow the material to be spread onto the slide.

Protocol 2

Preparation of metaphase spreads on glass slides

Equipment and reagents

- Fixed plant material (from *Protocol 1*)
- 45% or 60% acetic acid in water
- Enzyme buffer: 0.1 M citric acid, 0.1 M tri-sodium citrate; make a pH 4.6 stock solution by mixing 4 parts citric acid with 6 parts tri-sodium citrate. Can be stored at 4 °C unless it is contaminated with fungi
- Enzyme solution: 2% (w/v) cellulase from *Aspergillus niger* (Calbiochem, 21947, 4000 units/g), or a mixture of 1.8% Calbiochem and 0.2% 'Onozuka' RS cellulase (5000 units/g), and 3% (v/v) pectinase from *Aspergillus niger* (solution in 40% glycerol, Sigma P4716, 450 units/ml). Make up in 1× enzyme buffer. Store in aliquots at −20 °C. Modifications such as addition of pectolyase (1% solution) or alteration of ratios of different enzymes may be necessary to obtain optimised preparations from different species
- Microscope slides soaked in chromium trioxide solution in 80% sulphuric acid for 3 h or more
- Glass microscope coverslips, 18 × 18-mm size[a]
- Dry ice or liquid nitrogen
- 96% ethanol
- Dissecting microscope
- Dissecting instruments
- Diamond pencil

Method

1 Make enzyme buffer by diluting 1 part stock solution with 9 parts water.

2 Wash fixative from plant material twice in enzyme buffer for 10 min.

Protocol 2 continued

3 Transfer plant material to 1–2 ml enzyme solution and incubate at 37 °C for 30 min to 2 h, depending on the material.

4 Remove microscope slides from chromium trioxide solution and rinse in running tap water for 5 min, rinse in distilled water, air dry, and immediately transfer dry slides to 96% ethanol.

5 Transfer plant material from enzyme solution to enzyme buffer.

6 Transfer material for two or three preparations to 0.5 ml 45% or 60% acetic acid and leave for a few min.

7 Dry a microscope slide with a lint-free tissue, place a drop of 45% or 60% acetic acid on the slide, and put a small piece of root tip, anther, or other material into the drop.

8 Under a dissecting microscope, tease apart the tissue or tap with a glass rod, and remove any particles. Metaphases tend to float free.

9 Wipe a cover slip with a tissue and place onto the preparation. Examine with a phase-contrast microscope. Squashing with light to very heavy hand pressure might be required or the cover slip can be tapped to give the optimum preparations.

10 Put slide on a metal plate on a dry ice block for 5–10 min (not more), then flick off coverslip with a razor blade edge. Allow preparation to air dry.

11 Screen slides, without mounting, using a phase contrast microscope. Look for well spread nuclei with no clumps, metaphases, lack of surface film. Take notes about each slide to refine the preparation method after hybridization.

12 Mark the position of the preparation by scratching lines underneath the glass slide. The position of the preparation will not be visible during later stages on the wet slides.

13 Suitable slides can be stored desiccated at 4 °C for a few days or for months at −20 °C.

[a] The optimum preparations are often near the periphery of the cover slip. Therefore, use a small (e.g. 18 × 18 mm) coverslip for making preparations, and larger coverslips for hybridization and mounting (e.g. 40 × 24 mm).

2.3 Extended DNA fibres

Protocol 3 is suitable for plant DNA spreading, but alternatives using nuclei embedded in agarose (as for pulse-field electrophoresis), different lysis, and spreading methods or 'combing' as droplets dry out are also possible. The extraction of nuclei follows Liu and Whittier (7), and the spreading method is modified from Fransz *et al.* (8, 10).

Protocol 3

Preparation of extended DNA fibres

Equipment and reagents

- Mortar and pestle
- Liquid nitrogen
- Ice-cold nuclear isolation buffer (NIB): 10 mM Tris–HCl pH 9.5, 10 mM EDTA, 100 mM KCl, 500 mM sucrose, 4 mM spermidine, 1 mM spermine, 0.1% v/v mercaptoethanol
- Nylon cloth with 50 μm pore size
- Cooled centrifuge
- DAPI in McIlvaine's buffer (see *Protocol 9*)
- Phosphate buffered saline (PBS): 130 mM NaCl, 7 mM Na_2HPO_4, 3 mM NaH_2PO_4
- STE lysis buffer: 0.5% sodium dodecyl sulphate, 100 mM Tris–HCl pH7, 50 mM EDTA
- Acid-washed microscope slides (see *Protocol 2*)
- 3:1 fixative (ethanol: acetic acid, made freshly)
- Microscope cover slips

Material

- Young leaves for nuclear extraction

Method

1. Grind 1–2 g of young leaves to a fine powder in liquid nitrogen with a mortar and pestle.
2. Transfer leaf powder to a 15 ml centrifuge tube and add 10–15 ml ice-cold NIB.
3. Mix thoroughly by inversion.
4. Filter through nylon cloth with 50 μm pore size.
5. Centrifuge filtrate at 4 °C, 2000 *g* for 10 min[a]
6. Resuspend pellet including nuclei in 0.5 ml NIB.
7. Check nuclei by microscopy; stain by mixing filtrate with 10% v/v of the DAPI in McIlvaine's buffer solution. A typical concentration of 10^4–10^5 nuclei/ml is suitable.
8. Centrifuge 25–100 μl of nuclear suspension in a microcentrifuge tube at 1100 *g* for 5 min at 4 °C.
9. Discard supernatant and resuspend pellet in 25–50 μl PBS.
10. Place 1–5 μl of nuclear suspension towards the end of a microscope slide and allow to air-dry.
11. Add 10 μl of STE lysis buffer to dried down nuclei and leave for 5 min at room temperature.
12. Tilt slide slowly so buffer flows down; the flow must be slow to extend DNA fibres.
13. Allow to air-dry, place in 3:1 fixative for 2 min, and then air-dry.
14. To examine the slides, place a drop of DAPI in McIlvaine's buffer onto the slide and put a cover slip on top. Examine with a UV fluorescent microscope. Cell wall debris

Protocol 3 continued

will be seen, along with bright blue nuclei. From many of these nuclei, blue DNA fibres should be seen running in the direction of stretch. Nuclei should be well separated; if not, grind leaf tissue more and adjust volumes of solutions and suspensions.

15 The slides are then used for *in situ* hybridization with the same probe mixtures as are used for chromosomal preparations (*Protocol* 5). Denaturation, hybridization, washing, and visualization follow subsequent protocols (*Protocols* 6–8), without further pretreatment. Because there is no structure to the chromatin, denaturation temperatures up to 90 °C can be used.

[a] Much cell debris will remain in the pellet; however, if nuclei are not free from cells, then Triton X-100 may be added to the filtrate to a final concentration of 0.5% with a 10% stock solution in NIB.

3 Probes and Labels

3.1 The nature of probes

Probes for *in situ* hybridization can be any sequence homologous to a DNA sequence target on the chromosomes or DNA fibres, and are chosen to answer particular questions about genome organization or evolution. Any DNA sequence can be labelled with a modified nucleotide and used as a probe for *in situ* hybridization, as is the case with Southern hybridization. Many applications use cloned sequences, but PCR products, synthetic oligonucleotides, total genomic DNA, and restriction or centrifugation satellite DNA fragments are all widely used as probes. Two, or increasingly three, different types of sequences are often used simultaneously for *in situ* hybridization: for example, a cloned repeat might be used to allow identification of chromosome arms, while a sequence of interest is localized simultaneously.

3.2 Cloned DNA sequences

3.2.1 rDNA

Perhaps the most widely used probes for *in situ* hybridization are the ribosomal RNA gene sequences and the intergenic spacers, the 45S (*Plate 1*) and 5S rDNA. These sequences are present as long, tandemly repeated arrays, at one or more chromosomal sites in each haploid chromosome set of all plants. Because of the high conservation of the sequences between species, probes will work from any species, and the wheat sequences are particularly widely used. They provide a reliable and universal probe, and the results are often interesting for examination of chromosome evolution and for chromosome identification. For example, in hexaploid bread wheat, the major loci are on two chromosomes of one of the constituent genomes, chromosomes 1B and 6B, while loci on chromosomes of the A and D genome have much reduced loci compared to the ancestral diploid species.

3.2.2 Repetitive DNA sequences

Repetitive sequences represent 50% or more of most plant genomes, and their organization and evolution has been widely examined by *in situ* hybridization (e.g. *Plates 1, 2, and 4*). Many sequences are tandemly organized in long arrays and have a characteristic length of about 180 bp, related to the packaging around nucleosomes. Such sequences are often seen as ladders of fragments in ethidium bromide-stained size separations of restriction digests (frequently with the enzyme *Hae* III). Other repetitive sequences are dispersed with more-or-less uniform distribution over the chromosomes of a genome. Like total genomic DNA (see below), such sequences may be useful to investigate genome constitutions in hybrids or the introgression of chromosome segments.

3.2.3 Single copy and low copy sequences

The concept of using single-copy, short cloned DNA fragments, including those mapped by RFLP or other methods, as probes for *in situ* hybridization is very attractive. However, the technique required is difficult: short probes may not find the single complementary target sequences on the chromosomes, or may incorporate so few label molecules that detection is extremely difficult. Frequently, clones may include gene domains common to several sites or other types of repetitive DNA, which will be found to hybridize to multiple sites on chromosomes. Nevertheless, in plants, there are many reports of successful hybridization of sequences of a few hundred to few thousand base pairs, including to short gene clusters.

3.3 PCR products

PCR products from genomic DNA can be used as probes for *in situ* hybridization. With some repetitive sequences, this method can give a much better representation of a family of sequences than the use of individual clones. This is particularly important for retro-elements, where conserved domains flank much more degenerate regions of the elements, but researchers are often interested in the distribution of all sequences in a family, rather than single, cloned, family members.

PCR labelling methods are also frequently used with cloned DNA sequences to amplify specific inserts and incorporate label molecules.

3.4 Total genomic DNA for chromosome and genome painting

Because the repetitive DNA sequences in plant genomes evolve relatively rapidly and, hence, have many sequences that differ, labelled total genomic DNA can be a very valuable probe for distinguishing different ancestral genomes in polyploid species, parental genomes in interspecific hybrids, or alien chromosomes (*Plate 2*) or chromosome segments in lines of plants derived from hybrids.

3.5 Synthetic oligonucleotide sequences

Synthetic DNA provides a well-defined probe for *in situ* hybridization, and are increasingly used for hybridization. The telomeric sequence, TTTAGGG, is similar in most plant species so far examined and enables the labelling of the termini of their chromosomes. Synthetic simple sequence repeats, of DNA motifs 2–4 bp long, have proved to give valuable information about chromosome structure, genome evolution and are useful for chromosome identification by *in situ* hybridization.

3.6 Probe labelling and purification

Probes for *in situ* hybridization should be labelled with either fluorescent-conjugated nucleotides (direct labels) or with nucleotides conjugated to other molecules, such as biotin or digoxigenin that are detected by secondary, fluorescent antibodies, or other molecules. Some experiments use radioactive labels, where hybridization sites are detected by coating the slide with photographic emulsion. Other use colorimetric detection systems: sites of probe hybridization are detected with an enzymatic conjugate attached to the label, where the enzyme causes an insoluble, coloured precipitate. Colorimetric systems are widely used for *in situ* hybridization to RNA targets (see Volume 2, Chapter 2 or ref. 5).

The probe must incorporate a suitable proportion of label molecules to allow detection, must be divided into suitable fragments (typically 200–600 bp) after labelling, and be purified from unincorporated nucleotides. In a typical labelling reaction, the labelled nucleotide (most frequently dUTP) is mixed in a 1:2 ratio with unlabelled TTP (or suitable nucleotide). Manufacturers of the labelled nucleotides usually make recommendations of the concentrations and dilutions that give optimal incorporation. These are a good starting point, but the absolute amounts of very expensive labelled nucleotides can often be reduced by 50–70%, particularly with fresh reagents that have not been through multiple freeze-thaw cycles.

Using fluorescent *in situ* hybridization methods, it is almost always worth using two different probes simultaneously on each slide. This provides a control that the experiment has worked, and always gives additional information about the material, although there are possible increase in background hybridization.

In practice, for probes for *in situ* hybridization, most DNA labelling protocols can be used. In low-volume applications, it is most convenient to use random primer labelling (oligolabelling) or nick translation kits from molecular biology supply companies. With suitable probes, PCR labelling can also be used and oligonucleotides can be labelled by end tailing. Suitable protocols are available from any of the manufactures of labels (e.g. Molecular Probes, Roche, Amersham), with labelling enzymes, from other chapters in this volume (e.g. Chapters 6 and 7), or in Schwarzacher and Heslop-Harrison (5).

After labelling, it is important to purify probes to remove proteins (enzymes) and unincorporated nucleotides, which will otherwise contribute to background hybridization. Ethanol precipitation is widely used for this, although column purification, using commercial columns available for cleaning PCR products, for example, may be preferable because of increased recovery of probe and speed. Probes should be handled in dim light because many, particularly when labelled with fluorochromes, are light sensitive. Typical levels of bright artificial lighting in laboratories are satisfactory, but bright daylight where electric lighting is not needed is probably too bright.

3.7 Testing label incorporation and quality control of probes

Because of the great effort involved in making chromosome preparations and in the hybridization and detection protocols, and the lack of reliability of most enzymatic labelling reactions, it is worth testing the incorporation of label into the probe by making and detecting the probe on a test blot using protocols for non-radioactive Southern hybridization which detect the same label molecules. If the probe is labelled directly with a fluorochrome, it can be checked by placing a small drop on a microscope slide and examining with a fluorescence microscope and the appropriate filter set. The most frequent reasons for failure of labelling are impurity of the source DNA; it is worthwhile using a commercial DNA purification column (e.g. Promega or Qiagen) and trying an alternative labelling protocol (e.g. random primer labelling rather than nick translation).

The length of probe after labelling is also an important parameter. If probes are too long, they will give little hybridization signal, either because of poor penetration or limited diffusion to homologous target sites. PCR products can be sonicated or sheared to break them to the ideal length. Probe length can be measured by electrophoresis, and the labelled probe should ideally give a smear between 200 and 600 bp in length. If it is shorter, enzyme and primer concentrations can be adjusted in the labelling reactions.

3.8 Fluorochromes

In the most frequent fluorescent *in situ* hybridization experiments, three different fluorochromes are used simultaneously: DAPI to stain the chromosomes, giving blue fluorescence after excitation with UV light, and two fluorochromes to detect probe hybridization sites, one excited by blue light giving green fluorescence (e.g. fluorescein, Alexa 488, or Cy2) and one excited by green light giving red fluorescence (e.g. Texas Red, Cy3, or Alexa 594). A fourth fluorochrome can be added with fluorescence in the near-infra-red (e.g. Cy5 or Bodipy630/650), although this cannot be visualized by eye or most films. The fluorochromes are selected so they can be excited by the major emission lines of the mercury vapour lamps used in most fluorescence microscopes and emit light that does not overlap with the major mercury lines nor other fluorochromes used in the experiment.

If the labels are not themselves the fluorochrome conjugated to a nucleotide such as dUTP, the hybridization sites are detected by an antibody (or other molecule that binds to the label) conjugated to a fluorochrome. The major labels used are the vitamin biotin, which is detected with avidin, streptavidin, or with an antibody and digoxigenin, detected with an antibody. Fluorescein can also be detected with an antibody-fluorochrome conjugate, enabling amplification of the hybridization signal.

4 The hybridization mixture, stringency and denaturation

After making chromosome preparations, they may be stored for a few days at room temperature, or much longer, preferably dessicated at −20 °C. The preparations are then pretreated to improve probe hybridization and retain preparations on the slides (*Protocol* 4). The labelled probe is prepared in the hybridization mixture (*Protocol* 5) and applied to the slides. Both are then denatured and allowed to hybridize together (*Protocol* 6), usually overnight, before washing (*Protocol* 7) and detection (*Protocol* 8).

The stringency of hybridization and washes—how similar a probe must be to the target to remain stably hybridized as a double helix—must be carefully considered in any *in situ* hybridization experiment, and the information to enable its calculation should always be given in publications. Tables and a detailed consideration of the factors affecting stringency, including temperature, and ion and formamide concentrations, is given by Schwarzacher and Heslop-Harrison (5). Typical stringencies allowing probes with 80–90% similarity to the target are given in the examples below. At higher stringency, only probe sequences with high homology to the target (typically 90% or more) will form stable hybrids, while at low stringencies (70–80%) 'families' of sequences related to the probe will be detected.

It is often convenient to use eight slides in an experiment; with more than this, slides can be left behind in solutions, the preparation is often scratched with forceps or the edges of other slides, and it is easy to confuse the preparation side of slides or the probe mixture to be applied.

4.1 Pretreatments of chromosome preparations

Before the *in situ* hybridization procedure, the slide preparations are pretreated (*Protocol* 4) to enhance probe access to the target sites (e.g. by removing surface proteins with pepsin or protease treatments), and to reduce non-specific probe and detection reagent binding (e.g. RNase treatment and pepsin/protease treatments). The preparation is then stabilized by refixation in paraformaldehyde and alcohol, which also helps retain the preparation during the many washes.

Protocol 4

Pre-treatment of slide preparations

Equipment and reagents

- 20× SSC: 3 M NaCl, 0.3 M sodium citrate, adjusted to pH 7
- 2× SSC, diluted from 20x stock solution
- RNase solution; 10 mg/ml in 10 mM Tris-HCl, pH8, diluted to 100 μg/ml before use in 2× SSC.
- 10 mM HCl
- Pepsin solution. Make up stock of 500 μg/ml in 10 mM HCl (assuming activity is *c.* 4000 U/mg)
- Ethanol series: 96, 90, and 70% in water
- Paraformaldehyde (electron microscope grade)
- 10 M NaOH and 1 N H_2SO_4
- Coverslips (20 × 25mm) cut from pieces of autoclavable plastic bags of the type used for contaminated waste.
- Coplin jars holding eight slides and 50–100 ml solution

Method

1. Add 200 μl RNase solution to the marked area on each slide. Cover with a plastic coverslip and incubate for 1 h in a humid chamber.
2. Prepare paraformaldehyde fixative. In the fume hood, add 4 g paraformaldehyde to 80 ml water and heat to 60 °C for about 10 min, clear the solution with a few drops of concentrated (10 M) NaOH, let cool down to below 30 °C before use, and adjust pH to 8 with 1 N H_2SO_4.
3. Dip slides into coplin jar of 2× SSC, allowing cover slips to float away. Wash for 5 min each in two further jars of 2× SSC.
4. Dilute pepsin to 5 μg/μl in 10 mM HCl.
5. Place slides in 10 mM HCl for 1 min.
6. Add 200 μl pepsin solution to each slide, cover with a plastic coverslip, and incubate for 10 min 37 °C.
7. Wash off coverslips in distilled water and then wash twice for 5 min in 2× SSC.
8. Place slides in coplin jar with paraformaldehyde fixative for 10 min in a fume hood.
9. Wash twice for 5 min in 2× SSC.
10. Dehydrate through an ethanol series (70, 90, and 96%, 2 min each) and allow to air dry.

4.2 The hybridization mixture

The labelled probe, which may be stored for many months at −20 °C, is prepared in a hybridization mixture for application to the preparations (*Protocol* 5). The composition of the mixture includes sodium ions and formamide; together with the temperature, these control the stringency of hybridization. A typical experiment with eight slides will use three to six different probe mixtures, with different probes and combinations of label.

Many grades of formamide, a chemical that destabilizes the DNA double-helix, are available for molecular biology. For this application, a good grade, but not necessarily the highest is required. Purchase a medium-priced grade from a supplier with high turnover and immediately separate to aliquots of 40 ml before freezing at −20 °C. If the aliquots do not freeze completely, then there are impurities and another grade or fresher batch should be used.

Protocol 5

Preparation of hybridization mixture

Equipment and reagents

- Formamide
- Dextran sulphate (50% solution in water—takes several days to dissolve, then sterilize by forcing through a 0.22 μm filter)
- SDS solution (10% in water)
- 20× SSC: 3 M NaCl, 0.3 M sodium citrate, adjusted to pH 7
- Labelled probe DNA (50–100 ng/slide)
- Plastic coverslips cut from autoclavable plastic bags
- Salmon sperm DNA (1 μg/μl)

Method

1. In a microcentrifuge tube prepare the hybridization solution by adding, in order, the following volumes per slide[a] for a low-stringency mixture (50% formamide, 2× SSC)[b]:

formamide	20 μl	10% SDS	0.5 μl
water	0.5–4.5 μl	salmon sperm DNA	2 μl
20× SSC	4 μl	probe DNA[c]	1–5 μl
50% dextran sulphate	8 μl	Total volume	40 μl

2. Vortex the probe mixture.
3. Denature probe in a water bath at 70 °C for 10 min[d].
4. Place in ice until ready to put onto slide preparation (Protocol 6).

[a] Adjust volume to be suitable for the area on the slide. 20–40 μl can be used depending on the area of the coverslip.

[b] For a high-stringency hybridization solution, the SSC concentration could be reduced to 0.4× (0.8 μl 20× SSC per 40 μl hybridization solution); alternatively formamide concentration could be increased.

[c] Add the probe DNA last to the mixture; in particular do not add to either concentrated solutions or pure water.

[d] Although it is denatured again with the preparation, the probe can be denatured under stronger conditions than the chromosome preparation, since there is no need for preservation of morphology.

4.3 Setting up the hybridization

Following denaturation of the probe mixture, the simplest procedure involves application to the slide preparation, covering with a plastic coverslip, and denaturation of probe and target together on a heated plate, before transfer to a heated oven for overnight hybridization (*Protocol 6*). The temperature and time for target denaturation must be determined carefully, to ensure relatively complete DNA denaturation, while maintaining the morphological integrity and attachment to the slide of the chromosome preparation. Temperature changes of 1–2 °C may give differences in morphology and hybridization results. Many makes of PCR machine have modifications for holding slides (see ref. 9), and these are particularly good at enabling temperature and time to be controlled accurately. Alternatives include thermostatic heated plates or placing slides in a tray heated in a water bath. In both cases the slide temperature must be accurately monitored by placing a thermometer next to the preparations. Some protocols denature the slide preparations separately from the probe by dipping into a 70% formamide, 2× SSC mixture at 60–80 °C for 6–10 min, before applying the hybridization mixture and coverslip.

Protocol 6

Denaturation and hybridization

Equipment and Materials

- Denatured probe mixture (*Protocol 5*)
- Re-fixed preparation on microscope slide (*Protocol 4*)
- Plastic cover slips cut from autoclavable plastic bags (or commercially available)
- Heated block or water bath for denaturation
- Heated block or oven at 37 °C for hybridization

Method

1. Place 30–50 μl of probe mixture over area of preparation on the slide.
2. Cover preparations with a plastic cover slip, being careful to remove any bubbles by lifting and replacing coverslip.
3. Place slides onto heated block and raise temperature to that required for denaturation. A typical time and temperature is 80 °C for 8 min. Temperatures between 70 and 95 °C and times from 6 to 12 min may be used depending on the species, exact method of preparation (depends on person as well), and storage time of material before and after making the preparation.
4. Lower temperature to 37 °C over 10–20 min, and maintain temperature at 37 °C for the hybridization.

Protocol 6 continued

5 Leave slides on the heated block if they can be maintained in a moist atmosphere, or transfer to a humid box in an oven for hybridization. Most experiments use overnight (16 h) hybridization, but times from 2 to 96 h are used in some applications. Be careful that preparations do not dry-out or accumulate drops of condensation on them.

4.4 Stringent washes

Following hybridization, preparations must be washed to remove the hybridization solution, and unhybridized or weakly hybridized probe (i.e. probe with low homology to the target sequences). The most stringent is usually carried out at the same or slightly higher stringency than the hybridization mixture.

Protocol 7

Stringent washes after hybridization

Equipment and reagents

- 20× SSC: 3M NaCl, 0.3 M Na citrate: dilute to concentrations required—for a typical experiment you will need 500 ml of 0.1× SSC and 500 ml of 2× SSC pre-heated to 42 °C
- Formamide (typically 40 ml)
- Stringent wash solutions: 20% formamide in 2× SSC for low stringency or 0.1× SSC for high stringency
- Water bath at 45 °C
- Staining jars for washing slides, containing about 100 ml of solution

Method

1 Remove slides from hybridization chamber and check that they have neither dried out nor become wet from condensation.

2 Float off coverslips in 2× SSC at 42 °C; allow coverslips to fall away[a].

3 Wash in 2× SSC at 42 °C for 2 min.

4 Incubate slides twice for 5 min in the stringent wash solution at 42 °C. Measure the temperature of the wash solution with slides accurately and record[b].

5 Wash slides in 0.1× SSC (high stringency) or 2× SSC (low stringency) at 42 °C twice for 5 min.

6 Wash slides three times in 2× SSC at 42 °C.

Protocol 7 continued

7 Take slides in last wash out of water bath and allow to cool for 10–15 min before detecting hybridization sites with indirect labels (*Protocol 8*), or counterstaining and mounting if fluorescent labels were used (*Protocol 9*).

[a] Be careful not to scratch slides against each other; avoid leaving one slide behind when transferring between solutions.

[b] Temperatures must be accurately maintained; too high temperatures may remove all hybridized probe, destroy chromosome morphology, or remove preparations from the slide.

5 Fluorescent detection and visualization of hybridization sites

Where directly labelled probes have been used (e.g. fluorescein-11-dUTP), the preparation can be counterstained and mounted immediately (*Protocol 9*). With probes labelled with biotin or digoxigenin, hybridization sites must be detected with fluorochrome-conjugates of antibodies or avidin (*Protocol 8*); fluorescein antibodies are also available and can be used to amplify fluorescein signals.

5.1 Detection of hybridization sites of indirect labels

If biotin or digoxigenin have been used to label probes, then these must be detected at the hybridization sites using a molecule that binds to them. It is widely considered that such indirect labels give greater sensitivity, since multiple fluorescent molecules may be attached to one of the binding molecules. However, any non-specific binding will increase background noise. Secondary antibodies and extra amplification steps were extensively used in the 1990s, but improvements in microscopy and fluorochromes have made these methods unnecessary in most applications.

Note that procedures involving fluorochromes should be carried out in subdued light.

Protocol 8

Indirect detection of hybridization sites with fluorochromes

Equipment and reagents

- Detection buffer: 4× SSC containing 0.2% (v/v) Tween 20
- BSA block: 5% bovine serum albumin in detection buffer
- Detection reagent for digoxigenin labels, anti-digoxigenin (FAB-fragment) conjugated to a suitable fluorochrome (Roche): typically 1:250 dilution in detection buffer[a]
- Detection reagent for biotin labels, avidin, streptavidin, extra-avidin, or anti-biotin conjugated to suitable fluorochrome (e.g. Vector, Sigma, Molecular Probes, Roche); typically 1:250 dilution in detection buffer[a]
- Where two labels are being detected, the two reagents are mixed in the detection buffer

Protocol 8 continued

Method

1. Place slides in coplin jars in detection buffer for 5 min.
2. Apply 200 μl BSA block to the marked area of each slide, cover preparation with a plastic cover slip, and leave for 5–30 min.
3. Shake off BSA block and place 30–50 μl of detection solution onto the slides, cover with a plastic cover slip and incubate at 37 °C for 60 min.
4. Remove coverslips by floating off in detection buffer and wash in detection buffer for 3 × 6 min at 40 °C.

[a] Dilutions of antibody vary widely between batches even from one manufacturer and must be optimized in any application. Typical dilutions vary from 1:50 to 1:600.

5.2 Counterstaining and mounting

Most preparations require a DNA counterstain to visualize the chromosomes (*Protocol 9*). However, with some probes, there may be enough cross-hybridization that counterstaining is unnecessary. It is important that counterstaining is not too bright or hybridization sites may be obscured; hence, we recommend staining and then mounting. Some protocols mix the counterstain with the mountant, which saves one stage of preparation, but can lead to over staining and high backgrounds.

Most fluorochromes fade very rapidly in light (the energy destroys the molecule). Thus, slides must be kept away from bright lights (in particular, not left on sunny benches), and must be mounted with an anti-fading solution. Glycerol is quite effective, but other compounds are available commercially. Fluorochromes stabilities may differ in different anti-fade reagents, so if fading is a problem it is helpful to check with the manufacturer.

Protocol 9

Counterstaining and mounting of slide preparations

Equipment and reagents

- McIlvaine's buffer pH7: 18 ml of 100 mM citric acid and 82 ml 200 mM Na_2HPO_4
- 4× SSC-Tween: 0.2% Tween 20 in 4× SSC (20×: 3M NaCl, 0.3 M Na citrate)
- Anti-fade reagent[a]
- 24 × 40 or suitable size of thin (No. 0) coverslip
- DAPI (2′,6-diamidino-2-phenylindole)[b] staining solution: stock 0.1 mg/ml in water, stored at −20 °C, diluted to 2 μg/ml in McIlvaine's buffer for use

Protocol 9 continued

Method

1. Place a drop (50–100 μl) of staining solution onto the slide and cover with a plastic coverslip.
2. Wash slide in 4× SSC-Tween until coverslip falls off.
3. Shake and blot edges of slide with tissue, leaving surface moist.
4. Place a drop of anti-fade mountant onto the slide surface and apply a coverslip. It is easy to incorporate bubbles at this stage and these may be hard to remove.
5. Squash coverslip with high hand pressure between sheets of filter paper to squeeze out excess mountant. It is often worth doing this each time a slide is put onto the microscope.
6. Observe slide. The anti-fade may take a few hours to penetrate preparations, so wait overnight before detailed observation and photography.

[a] Fluorochromes will be destroyed rapidly (a few seconds) under the microscope unless they are stabilized with an anti-fade reagent.

[b] DAPI is used most often, since fluorochromes with UV emission are not as easy to use to detect hybridization sites (although they are available). The emission of DAPI shows minimal interference with other fluorochromes. Propidium iodide may be used as a fluorescent counterstain with fluorescein or other yellow-green fluorochromes, although frequently it obscures minor hybridization sites and a red fluorochrome cannot be used to detect hybridization sites.

5.3 Visualization and imaging

To visualize sites of probe hybridization detected with fluorochromes, an epi-fluorescence microscope is used. In this chapter, we do not aim to give instructions for the use of the microscope, but below outline problems we have encountered frequently in laboratories setting up fluorescent *in situ* hybridization experiments.

If DAPI staining is used, the optics and immersion oil used for visualization must be specified for UV, and the oil must be specified as being for fluorescent applications. Oil must be reasonably fresh and stored at room temperature (heating in the sun in a salesman's car will make some oils autofluorescent or UV opaque). Most makes of lenses or at least the glues used in their manufacture, and filter coatings deteriorate or become autofluorescent after 1000–2000 h of use with high-power UV bulbs, so these must be replaced regularly. Fluorescence filter technology improved markedly (with better transmission and sharper wavelength cut-offs) in the last decade of the 20th century and it is worth replacing filters older than a few years. The most useful filter sets in most applications are those for individual fluorochromes; multi-bandpass sets are available, but often the brightnesses of the different fluorochromes are not matched enough to make them useful, and cross-excitation of the different fluorochromes will often make analysis of results difficult.

Even with strong fluorochromes and good anti-fade mountants, *in situ* hybridization signals may be very weak and fade rapidly. Thus, the location of the microscope in a completely dark room—no bright computer screens, and nobody opening doors to light or turning on lights—is critical to obtaining good results. The microscope must also be on a stable surface, although it is hard to predict which rooms in a building will be most suitable. Remember that the microscope operator will be sitting with the microscope for many hours so a fully adjustable chair, good ventilation, and low-wattage, dimmable lamps controlled next to the microscope are vital. If the light goes on in a room, the operator must wait at least 10 min, and perhaps half an hour, to regain dark adaptation. We emphasize these points because we have seen many microscopes, costing several times a researcher's salary, located in positions where they will never be able to give state-of-the-art results.

The microscope must be equipped with a suitable film or digital camera system. Now, both give resolutions suitable for publication. However, it must be noted that a digital system, no matter how expensive nor how persuasive the salesman, cannot make up for deficiencies in the set-up of the microscope. Because of the fading problems noted above, it is essential that images can be acquired with minimal time (typically less than 5 s) spent focusing and adjusting the camera, although exposure times of 30 s or more may be used with both digital and film cameras. When taking sequential photographs at different wavelengths, the longest (reddest) wavelength should be photographed first, with the shortest, highest energy (usually UV with DAPI) being photographed last since this fades the preparations fastest. Where film is used, colour print film is usually preferable to slide (reversal) film since it has higher exposure latitude and is cheaper to buy and print than slide films.

In most cases, the final image must be presented in a form for publication. Whether digital cameras or film is used, it is now most convenient to use a microcomputer-based image-processing program (e.g. Adobe Photoshop, Corel Draw, or one of the equivalent packages; there is a huge investment of time in learning to use these programs so it is usually worth purchasing a fully-featured version). However, care must be taken not to over process images and supervisors must be well aware that they are responsible for the supervision of image processing by their laboratory members!

6 Conclusions

In situ hybridization provides a powerful method to determine and study the organization of DNA sequences along metaphase chromosomes and within the interphase nucleus. Future developments with increased sensitivity, allowing location of shorter probes, and better spatial resolution of signal to allow both discrimination of location of adjacent probes and their localization in the nucleus or chromosomes are likely to increase the power of the methods. *In vivo* probing methods, using such methods as triple helixes or nucleotide substitutions are conceivable, while increased numbers of fluorescent colours will also be used more extensively. Because individual cells are observed and, hence, no averaging is used, the methods along with other tools of immunocytochemistry will answer key questions in cell and molecular biology about genome organization, behaviour, segregation, recombination, and evolution.

References

1. Bardsley, D., Cuadrado, A., Jack, P., Harrison, G., Castilho, A., and Heslop-Harrison, J. S. (1999). *Theor. Appl. Genet.*, **99,** 300–304.
2. Castilho, A., Miller, T. E., and Heslop-Harrison, J. S. (1996). *Theor. Appl. Genet.*, **93,** 816–825.
3. Harper, G., Osuji, J. O., Heslop-Harrison, J. S., Hull, R. (1999). *Virology*, **255,** 207–213.
4. Brandes, A., Thompson, H., Dean, C., and Heslop-Harrison, J. S. (1997). *Chromosome Res.*, **5,** 238–246.
5. Schwarzacher, T. and Heslop-Harrison, J. S. (2000). *Practical in situ hybridization*. Oxford: Bios.
6. Jones, G. H. and Heslop-Harrison, J. S. (2000). In: B. Mulligan, ed., *Arabidopsis: a practical approach*. Oxford: IRL Press.
7. Liu, Y.G. and Whittier, R. F. (1994). *Nucl. Acids Res.*, **22,** 2168–2169.
8. Fransz, P. F., de Jong, J. H., and Zabel, P. (1998). *Plant Mol. Biol. Manual*, **G5,** 1–18.
9. Heslop-Harrison, J. S., Schwarzacher, T., Anamthawat-Jónsson, K., Leitch, A. R., Shi, M., and Leitch, I. J. (1991). *Technique*, **3,** 109–115.
10. Fransz, P. F., Alonso-Blanco, C., Liharska, T. B., Peeters, A. J. M., Zabel, P., and de Jong, H. (1996). *Plant J.*, **9,** 421–430.

Chapter 9
PCR cloning approaches

Michiel Vandenbussche
Department of Genetics, University of Gent, K Lledeeganckstraat 35, BE9000, Gent, Belgium

Tom Gerats
Department of Experimental Plant Sciences, University of Nijmegen, Toernooiveld 1, B525ED Nijmegen, The Netherlands

1 Introduction

The polymerase chain reaction has become one of the most widely used techniques in molecular biology due to its remarkable speed, specificity, sensitivity, and flexibility. The latter is reflected by the fact that ever since its implementation, new PCR applications keep emerging. Nowadays, the cloning of a gene or any other defined sequence does not represent a major time-consuming task. Thus, more effort can be directed towards the analysis of its biological function. Moreover, thanks to massive sequencing efforts of increasingly more plant genomes and the development of a range of new techniques, we enter an exciting era in which genome-wide or system-wide approaches are becoming an integral part of plant research. We thus witness a shift from single gene approaches (that might appear to have provided anecdotal viewpoints) towards more holistic approaches that aim at identifying 'all' genes. In this chapter, we will focus on PCR-based techniques, which facilitate genome-wide or system-wide approaches to the elucidation of biological questions.

In Section 2 of this chapter, we will discuss differential display (1) and cDNA-AFLP (2) methods, two RNA fingerprinting techniques, which allow the simultaneous visualization of the expression patterns of a large number of genes. These approaches make it possible to clone and identify 'all' genes involved in a particular process. Alternative procedures are based on micro-array and DNA-chip technology, and are discussed elsewhere (see Volume 2, Chapter 3).

Large-scale sequencing programmes show that eukaryotic genomes are quite redundant, with many genes occurring as multi-gene families. This redundancy is considered to be one of the major reasons why often no phenotype is observed when loss-of-function mutants are studied. The functional analysis of each member of a gene family may require the construction of lines carrying mutations in a number of family members. To be able to do this, the full extent of such a gene family has to be known. In Section 3 of this chapter, we will describe in detail a highly efficient method for cloning gene families from genomic DNA.

2 RNA fingerprinting methods

2.1 Differential display

Isolating differentially expressed genes using differential cDNA library screening methods is labour-intensive and is likely to yield clones derived mainly from highly differentially expressed genes. Therefore, more sensitive methods for RNA fingerprinting have been developed, which allow the detection of DNA fragments derived from RNA using cDNA synthesis and subsequent PCR amplification. One of the most popular RNA fingerprinting methods is differential display (1). In this method, total RNA is reverse transcribed using a poly-dT primer with two selective nucleotides at the 3′-end, followed by PCR amplification using the same poly-dT primer and an arbitrary decamer primer. In an alternative protocol (3), arbitrary primers are used both for cDNA synthesis and PCR amplification. Both variants of RNA fingerprinting require a low annealing temperature (T_{ann}) during thermocycling in order to achieve visualizable products. The methodologies involved in these approaches are not discussed in this volume, but interested readers are referred to references 1 and 3 for detailed descriptions.

2.2 cDNA-AFLP

2.2.1 Introduction

More recently, a new RNA fingerprinting technique has been developed, based on AFLP (4) (amplified fragment length polymorphism; see also Chapter 6). The cDNA-AFLP method is based on the use of highly stringent PCR conditions, facilitated by adding double-stranded adaptors at the ends of cDNA restriction fragments which serve as annealing sites for amplification (*Figure 1*). Selective amplification is achieved by adding one or more bases at the 3′-end of PCR primers, which will then be extended successfully only if the complementary sequence is present in the fragment flanking the restriction site, thereby reducing the number of visualized bands. The sensitivity of detection increases by adding more selective nucleotides (2). The highly stringent PCR conditions applied in the cDNA-AFLP technique offer some important advantages over differential display. First, the quantity of individual amplification products for a given mRNA is a function only of its initial concentration, whereas in differential display, this also depends on the quality of the match between primer and a particular template (5). Secondly, the higher specificity of the amplification produces much clearer and more reproducible fingerprints due to a significant background reduction. An example of a cDNA-AFLP gel is given in *Figure 2*.

2.2.2 General methodology

In our laboratory, template preparation for cDNA-AFLP is routinely performed on poly(A)$^+$RNA derived from total RNA using biotinylated poly-d[T]$_{25}$V oligonucleotides coupled to paramagnetic beads (Dynal A.S. Oslo, Norway) following

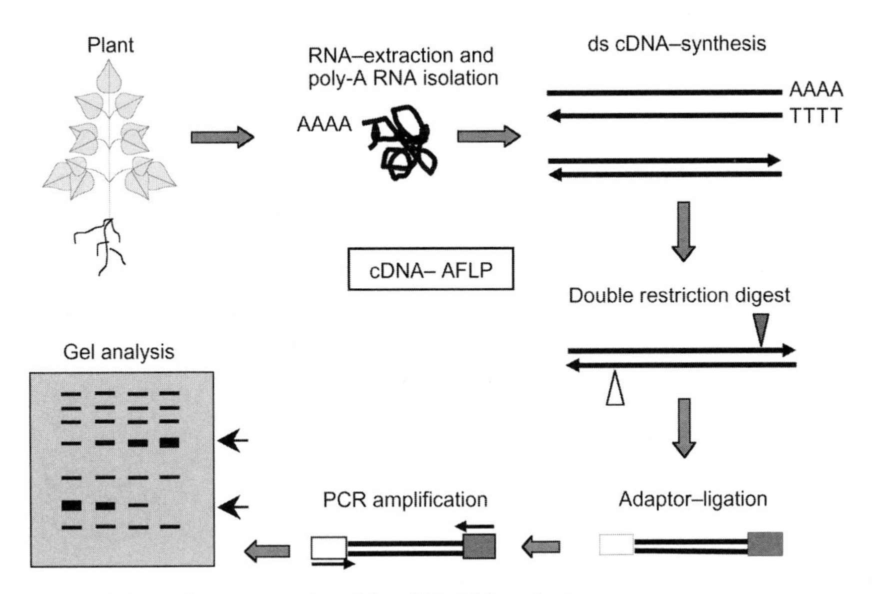

Figure 1 Schematic representation of the cDNA-AFLP method.

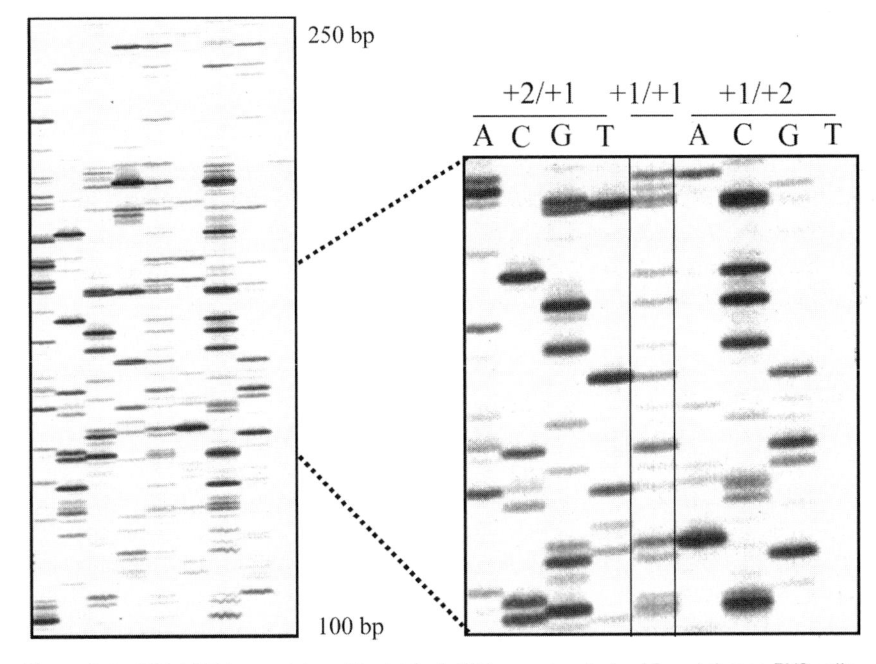

Figure 2 A cDNA-AFLP transcript profiling gel of cDNA samples derived from tobacco BY2 cells (courtesy P. Breyne). Central lane (+1/+1): amplicons obtained with two adaptor primers, A and B, each having one selective nucleotide at the 3′ end. (+2/+1): amplicons obtained with adaptor primer A with two selective nucleotides (last nucleotide A, C, G, or T) in combination with adaptor primer B with one selective nucleotide. (+1/+2): amplicons obtained with adaptor primer A with one selective nucleotide in combination with adaptor primer B with two selective nucleotides (last nucleotide A, C, G, or T).

the supplier's recommendations. Total RNA is extracted according to the LiCl method as described by Goormachtig *et al.* (6; see also other protocols for RNA extraction in Volume 2, Chapter 2). Double-stranded cDNA synthesis is done as described in *Protocol 1*. The choice of the enzymes used for creating the restriction fragments is of critical importance in the cDNA-AFLP method and is determined by the frequency of cleavage within cDNA sequences. To generate transcript-derived fragments (TDFs) with different adapters on both sides (as in standard AFLP), a rare cutter and a frequent cutter are used. Ideally, the restriction enzymes should cut every cDNA species once (rare cutter) and cut the cDNA on one or both adjacent sides (frequent cutter) to yield TDFs of between 100 and 1000 bp in size. For some plant species, the enzyme combination that most closely meets these criteria can be determined by an *in silico* restriction analysis of a large number of mRNA sequences present in sequence databases. Different enzyme combinations can be used to improve coverage of the transcripts. We are presently refining our cDNA-AFLP protocols such as to maximize the number of transcripts captured in the analysis and to limit the number of amplified fragments to one per transcript (M. Zabeau, personal communication; contact Dr Peter Breyne (email: pebre@gengenp.rug.ac.be) for more information and a detailed protocol). The original cDNA-AFLP methodology, as described by Bachem *et al.* (2), can be found at http://www.spg.wau.nl/pv/staff/aflp.htm#1. More efficient protocols for the isolation and sequencing of the bands of interest are described in Section 3.6.4 of this chapter.

Protocol 1

First and second strand cDNA synthesis

Reagents

- 5× first strand buffer: 250 mM Tris.HCl pH 8.3, 15 mM $MgCl_2$, 375 mM KCl
- 5× second strand buffer: 94 mM Tris.HCl pH 7.0, 23 mM $MgCl_2$, 453 mM KCl, 750 μM NAD^+, 50 mM $(NH_4)_2SO_4$
- Biotinylated oligo-dT_{25}
- DTT (dithiothreitol, 0.1 M)
- dNTPs (10 mM each)
- Superscript II (200 U/μl; Gibco-BRL)
- *Escherichia coli* DNA ligase (Gibco-BRL)
- *Escherichia coli* DNA polymerase I (Pharmacia)
- RNase-H (Pharmacia)
- S-300 spin columns equilibrated in 10 mM Tris.HCl pH 8.0, 0.1 mM EDTA (Pharmacia)

Methods

A. First strand cDNA synthesis

1 Mix together 20 μl poly$(A)^+$RNA (± 1.0 μg), 1μl biotinylated oligo-dT_{25} (700 ng/μl) and 4 μl H_2O.

Protocol 1 continued

2 Then add 8 μl 5× First Strand buffer, 4 μl 0.1 M DTT, 2 μl 10 mM dNTPs, and 1 μl Superscript II (200 U/μl).

3 Incubate the mixture for 2 h at 42 °C.

B. Second strand synthesis

1 Set up a reaction containing 40 μl first strand reaction mixture, 32 μl 5× second strand buffer, 3 μl 10 mM dNTPs, 6 μl 0.1 M DTT, 15 units *E. coli* DNA ligase, 50 units *E. coli* DNA polymerase I, 1.6 units RNase-H, and add H_2O to a final volume of 160 μl.

2 Incubate the reaction for 1 h at 12 °C and 1 h at 22 °C.

3 Extract the reaction mixture once with phenol/chloroform and purify the cDNA using a Pharmacia S-300 column according to the manufacturers instructions.

4 Check the quality and yield of cDNA by agarose gel electrophoresis. Usually, 30 μl of the double stranded cDNA preparation is used for AFLP (± 400 ng when starting with 1 μg mRNA).

2.3 Obtaining full length cDNA clones: RACE-PCR

Incomplete cDNAs are frequently obtained from RNA fingerprinting techniques, such as differential display and cDNA-AFLP, and from the screening of libraries by PCR using degenerate primers. Rapid amplification of cDNA ends (RACE) allows the addition of either the 5′ or 3′ end of a specific cDNA starting from an mRNA population (5′-RACE and 3′-RACE, respectively; 7, 8). A schematic representation of the RACE strategy is outlined in *Figure 3*. Several modifications and variants of RACE have been published; for an overview, we refer to Bertioli (9). One of the critical steps in this procedure is obviously the choice of tissue type for RNA isolation and subsequent cDNA synthesis. Any available information about expression pattern (tissue specificity and timing) is useful for selecting material enriched in the mRNA of interest. When using RNA fingerprinting methods, such as differential display and cDNA-AFLP, it is advisable to isolate an excess of starting material (RNA) in the beginning. Once cDNA fragments are isolated via these methods, the same RNA source can be used as starting material for RACE. This guarantees that the mRNA of interest will, indeed, be present in the RNA pool. Nowadays, the majority of researchers use commercially available kits to perform RACE experiments. Frequently used is the Marathon™ cDNA Amplification Kit (Clontech).

3 Multigene family cloning by the FamilyWalker method

3.1 Introduction

One of the basic goals in molecular genetics is to define the biological function of an isolated gene. A direct way to obtain functional information is to create loss-

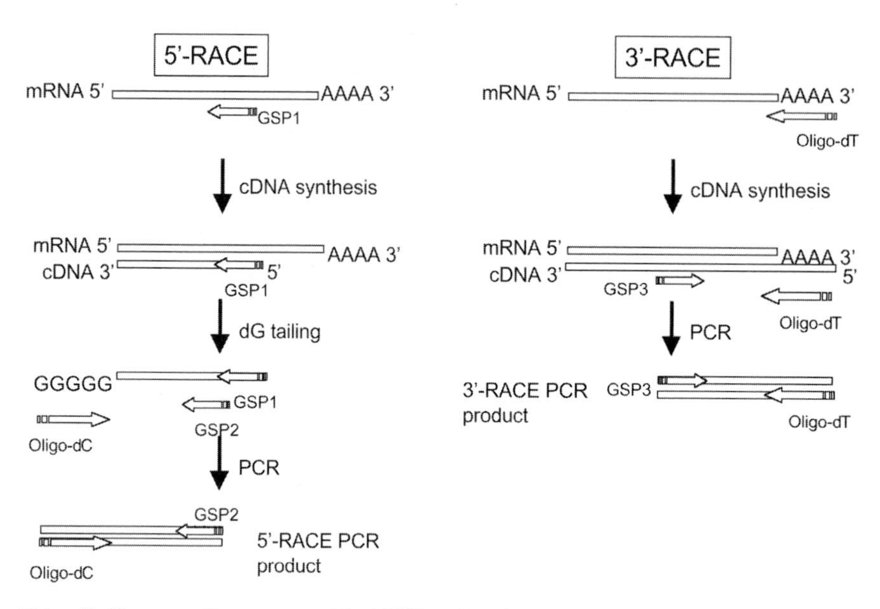

Figure 3 Diagrammatic summary of the RACE method. The key feature of the method is the use in PCR of a gene specific primer (GSP1–3) designed using the known sequence within the mRNA together with a 'general' primer complementary either to the mRNA poly-A tail (for 3′-RACE) or to a homopolymer tail that has been added to the 3′- end of the cDNA (for 5′-RACE).

of-function mutations and study the resulting phenotypes. Since homologous recombination is not yet a feasible approach in plants other than *Physcomitrella patens* (see Volume 2, Chapter 14), to target mutations into specific genes, large collections of plants mutagenized with an insertion element (T-DNA or transposons, see Chapters 3 and 4, respectively) are screened for insertions in genes of interest. Although this is a straightforward approach, a high percentage of the isolated knock-out lines do not show any visible phenotype when grown under standard conditions (10). The redundancy of eukaryotic genomes is probably one of the major reasons why so many of the mutant lines in which a single gene is disrupted do not show a visible phenotype (other reasons may be that the gene has a function under specific environmental conditions or that the mutation induces only subtle effects which are easily overlooked). The functional analysis of individual members of such gene families may require the construction of lines carrying mutations in different family members. To do this, the full extent of the family has to be identified and characterized.

3.2 The FamilyWalker strategy

Gene families characteristically exhibit at least one conserved amino acid domain. PCR provides a powerful tool for cloning new members on the basis of such a signature. Because the genetic code is degenerate, PCR has to be performed with degenerate primers (11). Conventionally, amplification is performed using a forward and a reverse degenerate primer, chosen in two conserved regions. This kind of approach is already adequately documented elsewhere (12–14). In this chapter, we will present a method called 'family walking', which requires the design of only one degenerate primer. The FamilyWalker method is based on the Universal GenomeWalker™ Kit (Clontech), originally designed for finding unknown genomic DNA sequences adjacent to a known sequence such as a cDNA (15). We adapted the original protocol so as to allow amplification with degenerate primers (*Figures 4 and 5*). The FamilyWalker method offers several advantages that make it highly suitable for the rapid characterization of large gene families:

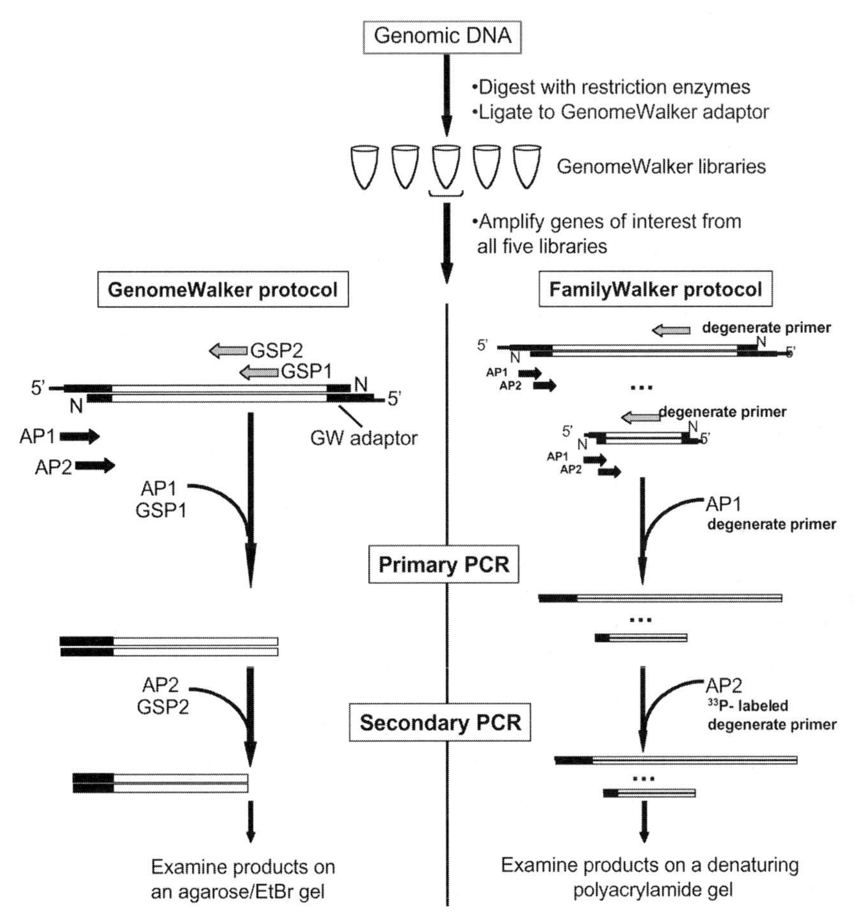

Figure 4 A comparison of the GenomeWalker™ (GW) and the FamilyWalker approaches. In the GW method, genomic DNA is digested with five different hexacutters and ligated to adaptors. Primary amplification products are obtained with an outer, gene specific primer (GSP1) and an outer adaptor primer (AP1), followed by a nested PCR with a nested GSP2 and nested AP2. In the FamilyWalker approach, primary PCR is performed using a degenerate primer (which can potentially anneal to a range of sequences) and the AP1 primer. Secondary amplification products are obtained with the same degenerate primer (^{33}P-labelled) and the AP2 primer. Products are analysed on a 3% denaturing polyacrylamide gel.

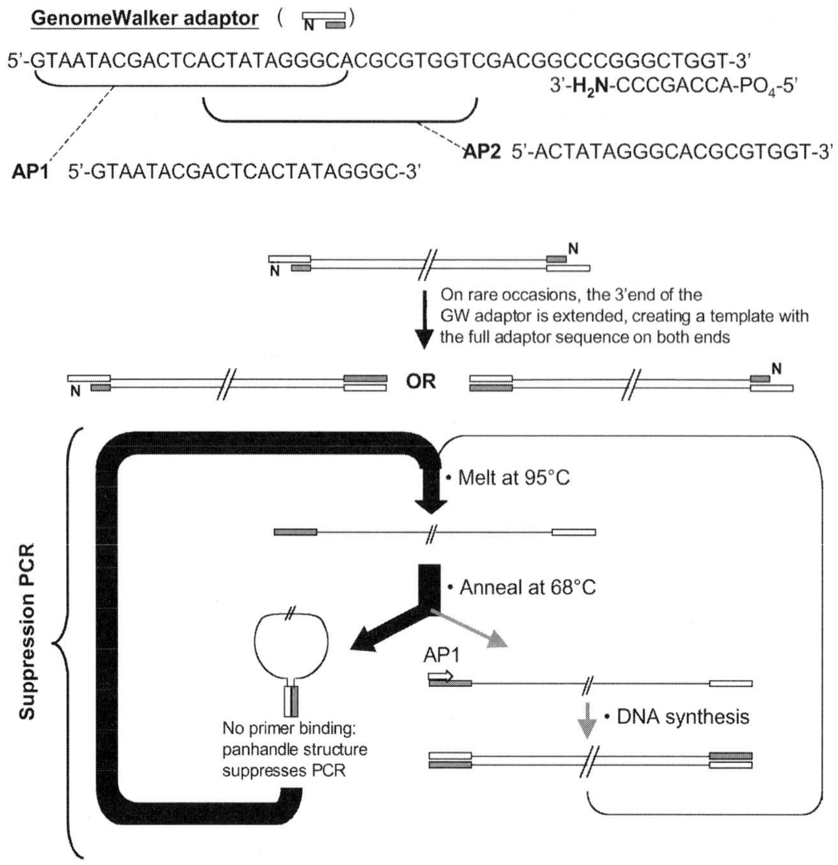

GenomeWalker adaptor

5'-GTAATACGACTCACTATAGGGCACGCGTGGTCGACGGCCCGGGCTGGT-3'

3'-H_2N-CCCGACCA-PO_4-5'

AP2 5'-ACTATAGGGCACGCGTGGT-3'

AP1 5'-GTAATACGACTCACTATAGGGC-3'

On rare occasions, the 3'end of the GW adaptor is extended, creating a template with the full adaptor sequence on both ends

OR

Suppression PCR

• Melt at 95°C

• Anneal at 68°C

AP1

No primer binding: panhandle structure suppresses PCR

• DNA synthesis

Figure 5 Structure of the GW adaptor and the suppression PCR effect. The amine group on the lower strand of the adaptor blocks extension of the 3′-end of the adaptor-ligated genomic fragments, and thus prevents formation of an AP1 binding site on the general population of fragments. In rare cases, the 3′-end gets extended, thereby creating a molecule that has the full-length adaptor sequence on both ends and can serve as a template for end-to-end amplification. However, in suppression PCR, the adaptor primer is much shorter than the adaptor itself. Thus, during subsequent thermal cycling, nearly all the DNA strands will form the 'panhandle' structure shown, which cannot be extended. At the appropriate T_{ann}, this intra-molecular annealing event is strongly favoured over the inter-molecular annealing of the much shorter adaptor primer to the adaptor (figure redrawn after figure 5 in Genome Walker™ user manual, Clontech).

1 Only one degenerate primer needs to be designed. Therefore, gene families, which are characterized by a limited conserved region, can also be amplified by this method. Additionally, more distantly related sequences can potentially be amplified if only one degenerate primer is used (since only one conserved region has to be recognized by the primers in order to allow amplification).

2 Different family members can be amplified simultaneously.

3 Genomic DNA is used as a template. When the aim is to amplify as many new members of a multigene family as possible, the use of genomic DNA as a template offers an important advantage over cDNA: a cDNA population is a representation of an mRNA population at a given developmental stage in specific cells or tissues and under specific environmental conditions. This implies that the number of mRNA copies of different members of the same gene family may be highly different and even that some mRNAs can be completely absent. Since PCR is a competitive process, it is likely that rare transcripts will be very hard to visualize when present together with more abundant transcripts that are co-amplified with the same degenerate primer. We therefore prefer to use genomic DNA as a substrate since every gene is present in equimolar amounts and the experimenter does not need to choose the tissue from which the RNAs will be extracted.

In this section, we will start with an extensive discussion about the design of degenerate primers and how to optimize PCR conditions, as these are the two most critical steps in the FamilyWalker method. Subsequently, the complete FamilyWalker protocol is described in detail and illustrated with the amplification of MADS-box genes (MBGs) in *Petunia hybrida*.

3.3 Design of degenerate primers

3.3.1 Identification of the conserved domains

Most often, only a few sequences of a specific gene family are available to start with. Querying sequence databases with these sequences can identify other members. The search should be repeated with every newly discovered member until no more sequences are found. BLAST (16, 17) is a good choice for sequence-to-sequence searches, but a keyword search of the database can also be useful when the name of the protein is known. Scientific literature can also be a source for identifying related sequences, especially for very divergent members that might be missed in a single sequence search. Deciding if a sequence should be included in a family relies on the significance of its similarity to other members and, more importantly, on the researcher's knowledge of the functions of the protein(s) being studied.

Once all related sequences have been collected, the conserved domains can be identified by aligning the sequences. Several multiple alignment tools are available. Examples are the pile up function in the GCG software package [Wisconsin Package Version 10.0, Genetics Computer Group (GCG), Madison, Wisc.], ClustalW (18), and BlockMaker (19). Both ClustalW and BlockMaker can be found at http://dot.imgen.bcm.tmc.edu:9331/multi-align/multi-align.html. Again, it must be stressed that experimental data can be used in evaluating and even in constructing multiple alignments. This can help to enhance the biological significance of the alignment. For example, if catalytic sites within the proteins are known, these sites will be aligned together and may 'push' the alignment in that direction. Such manual alignments can be good start points to align more sequences. Finally, an excellent database containing already made multiple alignments of a number of gene families can be found at http://www.sanger.ac.uk/Software/Pfam/.

3.4 Primer design

Design of degenerate primers involves finding the right balance between conflicting parameters. Moreover, a successful PCR amplification with degenerate primers can be achieved in several ways and each case will be different. It is therefore useful to discuss the most critical parameters and how they can be manipulated. This should allow the experimenter to create a mental picture of the process, which can be helpful when decisions have to be taken that may seem rather intuitive.

3.4.1 Degeneracy and complexity

We define degeneracy of a primer as the theoretical number of potential target sequences in the template to which the primer pool can anneal. The complexity of a primer is the number of different primer sequences present in the primer pool. The values for degeneracy and complexity will be different if inosines are included in the primer (see *Figure* 6). In practical terms, the relationship between degeneracy and complexity, on the one hand, and specificity and yield of the PCR amplification, on the other hand, can be explained as follows:

1 Specificity—when a primer is highly degenerate, more background amplification can occur. Therefore, primers with the smallest possible degeneracy should be chosen to obtain the highest specificity. This can be done as follows:

(a) Back-translate the conserved domains and choose the area of minimum ambiguity. Degeneracy can be further reduced by using codon usage data tables (20) (website: http://www.kazusa.or.jp/codon/) and inter-codon dinucleotide frequency tables (21)

(b) When working with gene families for which a large number of members in a range of species has been identified (e.g. the MBG family with >150 known genes), degeneracy can be further reduced by analysing DNA sequences

		Degeneracy	Complexity	T_m range
Primer 1:	5' TGY GTN GGN TTY GGN TAY AC 3'	512	512	54–66°C
Primer 2:	5' TGY GT I GG I TTY GG I TAY AC 3'	512	8	48–54°C

Y = C or T;
N = A, C, T or G;
I = Inosine

Figure 6 Comparison of two nearly identical primers. Primer 2 differs from primer 1 in the N's, which are substituted by inosines in primer 2. The two primers have the same degeneracy but different complexity values. The difference in complexity is reflected by a different T_m range. The T_m range is calculated with the formula $T_m = 2 (A + T) + 4 (G + C)$, and of the primer sequences with lowest and highest GC content present in the degenerate primer pool. Inosines are not taken into account for calculating the T_m, since their contribution to overall primer-template stability can be neglected.

(c) Gene families can often be divided in subfamilies that have by definition a higher degree of sequence similarity than the family as a whole.

2 Yield—the complexity of a degenerate primer directly affects the T_m range of the primer pool (*Figure 6*). When applying stringent conditions (e. g. a high T_{ann}) during PCR amplification with very complex primers, only a small part of the primer pool (the primers that fit exactly) will contribute to the amplification. As a result, yield will be so poor that no products will be visible. On the other hand, when applying less stringent conditions, background amplification can get too high. This problem can be overcome by using a touchdown PCR profile (22, 23; see Section 3.5.2) combined with the incorporation of inosines (24) in the primers (see Section 3.5.2). Using higher primer concentrations (up to 10-fold higher than in standard PCR reactions) has also proven useful to increase yield.

3.4.2 Other remarks on primer design

1 It is useful to have some conserved residues downstream of the degenerate primer in order to facilitate the identification of the selected fragments as new members of the family

2 Conventionally, degenerate primers are chosen with the 3′ end being the most conserved part of the primer because a good annealing of the 3′ end is most important for primer extension. However, recently a fully automatic method for degenerate primer design was described, called Codehop (25; website: http://www.blocks.fhcrc.org/codehop.html). Interestingly, these primers are degenerate at the 3′ end and conserved at the 5′ end. The authors claim that such primers are more effective for amplifying distantly related sequences

3 Try to design a primer that is at least 30 nucleotides long. This will allow the use of initially high annealing temperatures, which favours the suppression PCR effect of the Genomewalker™ kit (see *Figure 5*)

4 Keep in mind the general rules for designing primers such as avoiding self-hybridization and avoiding a high GC content (i.e. > 50%) of the last six nucleotides at the 3′ end.

3.4.3 Database-aided primer design

Sequence databases can be searched with degenerate primer sequences, which provides a quick and elegant method to test the specificity of a degenerate primer designed with the above guidelines. The GCG tool 'findpatterns' is well suited for this approach because the number of allowed mismatches between primer and template can be set before each search. We suggest searching nucleotide databases with your degenerate primer, and allow zero or one mismatch to start with. When analysing the results, two questions should be considered:

1 Do the majority of the identified sequences belong to the gene family or subfamily you are interested in?

If many sequences do not belong to the right gene family, try to reduce the degeneracy of the primer. If this is not possible without losing the starter sequences, then the conserved part you chose is probably not unique for that particular gene family. If possible, choose another conserved region and design a new primer

2 Do you find back the majority of previously identified members of the particular family or subfamily you want to amplify? If not, gradually increase the number of allowed mismatches after each search; analyse the positions where the mismatches are situated and change your primer based on that information.

Ideally, the primer should be able to recognize the majority of the already known family members when allowing a maximum of two or three mismatches. If you have chosen a region that is really family specific, then no other sequences will be recognized even if five or six mismatches are allowed (when using a 30-mer).

When EST (expressed sequence tag) databases and non-annotated genomic sequences are searched by this approach, new family members can be identified that would be missed by conventional homology searches. In this way, we have identified eight Arabidopsis and two rice MADS-box containing sequences which had not yet been recognized as being MBGs. Interestingly, degenerate primer sequences designed for amplifying Petunia MBGs were used for this screen.

3.5 PCR optimization

3.5.1 Hot start PCR

We highly recommend using some form of hot start PCR (26) when optimizing PCR conditions for degenerate primers. In a hot start PCR, DNA synthesis is prevented during warming up to 94 °C in the first cycle. This can be accomplished mechanically by adding the enzyme after the denaturation temperature is reached or by using special *Taq* polymerases (see *Protocol 2*) that become active only after the first heating step. This provides a tighter control over the conditions that allow primer-template annealing. In many cases, hot start also reduces background amplification significantly.

3.5.2 Touchdown PCR

In a touchdown (TD) PCR (22, 23), a high annealing temperature (T_{ann}) is used during the first cycles to enhance specificity of the amplification, followed by a gradual decrease in the T_{ann} to increase yield. Typically, a TD PCR programme is run with two cycles at each temperature with annealing temperatures declining over a 10–20 °C range at 1 °C intervals, followed by 30–40 cycles at the lowest T_{ann}. Although the T_{ann} drops to levels that normally would promote spurious amplification, the desired product, having already undergone several amplification cycles, will be in a position to out-compete most lower T_m (spurious) amplicons. Specific guidelines on how to determine a touchdown profile based on the T_m range of a degenerate primer pool will be given in the next section. However, when the simultaneous amplification of different family members with one degenerate primer is desired (which is the aim of the family walking protocol), a very large T_m range of the degenerate primer pool and a subsequent long touchdown PCR profile would strongly favour the amplification of family members with a higher GC content. In such cases, we recommend reducing the complexity of the primer by incorporating inosines on the highly degenerate positions in the primer. Although incorporating inosines lowers the average T_m of the primer pool, the difference in T_m of primer-template duplexes of sequences with different GC contents will be reduced (see *Figure 6*).

3.5.3 Test experiments

Determining an optimal T_{ann} for a degenerate primer depends on finding the right balance between specificity and yield. Unfortunately, even with the most sophisticated programs such as OLIGO Primer Analysis Software (National Biosciences, Inc., Plymouth, MN; 27, 28), this optimal T_{ann} is difficult to predict. On the other hand, an already cloned member of the gene family of interest (which will in many cases be available) used as a template in an optimization experiment can yield valuable empirical information for determining a PCR profile for the FamilyWalker method (see *Protocol 2*). The goal of such an optimization experiment is to calculate by extrapolation the optimal annealing temperatures (which is a combination of yield and specificity) of the primers with highest and lowest GC content present in the primer pool. This extrapolation will be based on the empirically defined optimal T_{ann} and the GC content of the primer that is complementary to the test template. For our purposes, PCR profiles based on these assumptions have proven to be successful for family walking. A practical example of these calculations for a degenerate MBG primer is given in *Figure 7*.

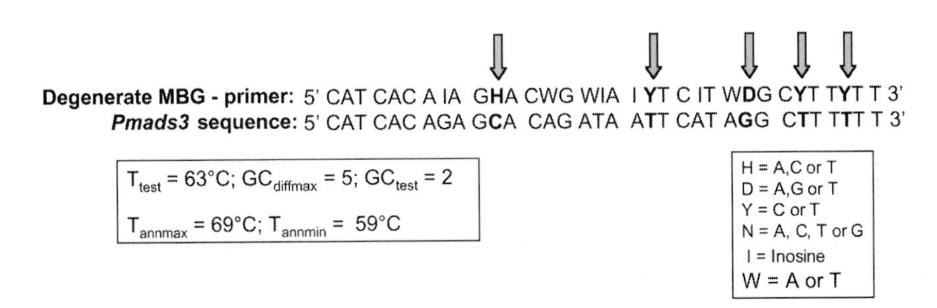

Figure 7 A practical example of a degenerate primer, as explained in Section 3.5.3. *Pmads3* is a *Petunia hybrida* MADS-box gene used as template in a FamilyWalker PCR optimization experiment for the degenerate MBG primer. All represented values are calculated according to the method described in *Protocol 2*. The positions in the degenerate primer that can contain either a purine or a pyrimidine are indicated by arrows.

Protocol 2

PCR optimization experiment for family walking

Equipment and reagents

- A long degenerate primer (30-mer or more, we typically use 33–35mers)
- An already cloned and sequenced member[a] of the gene family in question, which can be amplified with the degenerate primer that will be used for family walking
- A reverse primer[b] to be used in combination with the degenerate primer
- A hot start polymerase (in our hands, Platinum *Taq*[c] DNA polymerase (GibcoBRL) gave the best results)
- dNTPs (10 mM each)
- deionized H_20 (dH_2O)
- 10 × PCR reaction buffer containing 15 mM $MgCl_2$ (Boehringer)
- A thermal cycler
- Equipment for 1.5 % agarose gel electrophoresis

Method

1. For a single PCR reaction, mix[d]: 5 μl 10× PCR buffer, 1 μl dNTPs (10 mM), 2 μl degenerate primer (10 μM)[e], 1 μl reverse primer (10 μM), 0.25 μl Platinum *Taq* (5u/μl), 1 μl template (5–100ng/μl)[a], and 39.75 μl dH_2O[f].
2. Run the following programme: 2 min denaturation at 94 °C, 38 cycles (94 °C for 30 s, 68 °C for 30 s, 72 °C for one min/kb), 10 min at 72 °C.
3. Analyse 25 μl of the PCR products on an agarose gel. If no clear band of the expected size can be seen, repeat steps 1 and 2, and reduce the T_{ann} in step 2 by 2 °C. Repeat[g] this until the desired product is obtained[h]. Call this the optimal T_{ann} for the test template T_{test}.
4. Count the number of degenerate positions in your primer that leave the choice between a purine or a pyrimidine[i]. This number ($GC_{diffmax}$) is the maximum difference in the number of GC's between the primer with lowest GC content and highest GC content present in the degenerate primer pool.
5. Count the number of G's and C's in the template sequence on the positions identified in step 4 and call this value GC_{test}.
6. Calculate the optimal annealing temperatures for the primers with highest GC content (T_{annmax}) and lowest GC content (T_{annmin}) present in the degenerate primer pool with the following formulas[j]:
   - $T_{annmax} = T_{test} + 2(GC_{diffmax} - GC_{test})$
   - $T_{annmin} = T_{test} - 2(GC_{test})$
7. Use these values to determine the touchdown profile described in *protocol 3* of the FamilyWalker method (see Section 3.6.1).

Protocol 2 continued

[a] The gene fragment may either be cloned in a plasmid or be a crude PCR fragment. Family members can be used from any species as long as the degenerate primer can recognize the template (check the sequence!).

[b] This primer can be a gene specific primer or a primer based on plasmid sequence (in the case that a plasmid is used as a template) and should be situated not further than 2 kb from the degenerate primer. Importantly, the reverse primer should have a higher T_m than the degenerate primer in order to prevent the selection of a too low T_{ann} based on a short reverse primer.

[c] Advantage *Tth* Polymerase Mix (Clontech)—the hot start enzyme mix used in the GenomeWalker protocol—works equally well, but we prefer Platinum *Taq* for the optimization experiments.

[d] The PCR reactions can be assembled at room temperature since Platinum *Taq* only becomes activated after heating to 94 °C.

[e] The amount of degenerate primer used is twice the amount of a standard PCR reaction in order to increase yield.

[f] If a thermal cycler without heated lid is used, then 100 μl mineral oil should be added.

[g] If a thermal cycler with a gradient heating block is available, this optimization experiment can be performed in one single run.

[h] When the degenerate primer is at least a 30-mer and not too complex, this T_{ann} should be above 60 °C.

[i] For example, M(A,C), R(A,G), Y(C,T), K(G,T), V(A,C,G), H(A,C,T), D(A,G,T), B(C,G,T), N(A,C,G,T), but not W(A,T) and inosines.

[j] The formulas are deduced as follows: if $T_m = 2(A + T) + 4(G + C)$ then $T_m = 2(G + C + \text{primer length})$. The difference in T_m of 2 primers A and B with the same primer length is then $T_{mA} - T_{mB} = 2[(G + C)_A - (G + C)_B]$. If two primers of the same degenerate primer pool are compared, then only the positions that give a difference in GC content should be considered (see step 4). Optimal annealing temperatures calculated by this method are not completely correct because the T_m depends not only on the overall GC content but also on where exactly the GC's are situated in the primer.

3.6 Amplifying new *Petunia hybrida* MADS-box genes by the FamilyWalker method

MADS-box genes (MBGs) form a family of regulatory transcription factors whose name derives from *MCM1* (*Saccharomyces cerivisae*), *AGAMOUS* (*Arabidopsis thaliana*), *DEFICIENS* (*Antirrhinum majus*), and *SRF* (*Homo sapiens*; 29). These genes are involved in developmental decisions in all eukaryotes in which they have been found so far, including plants, animals, yeast and fungi. They all possess a DNA-binding domain, encoded by the 180 bp MADS-box, which is conserved across the different species. The majority of defined plant MBGs function as floral meristem or organ identity genes (30). Whether this reflects a natural phenomenon or perhaps a bias in research preferences remains to be seen.

The FamilyWalker protocol is a modified version of the protocol accompanying the Universal GenomeWalker™ Kit (Clontech). Genomic DNA preparation and construction of the GenomeWalker™ (GW) libraries is done as described in the protocol included in the kit and by using the kit components. Here, we will only describe the PCR amplification steps and subsequent analysis of the amplified fragments, which have been optimized for degenerate primer amplification.

3.6.1 Primary PCR for the FamilyWalker protocol

Use *Protocol 2* to calculate T_{annmax} and T_{annmin}, which will be needed in *Protocol 3* for determining a TD profile.

A picture of the resulting primary PCR products, using the degenerate MBG primer described in *Figure 7*, is shown in *Figure 8*.

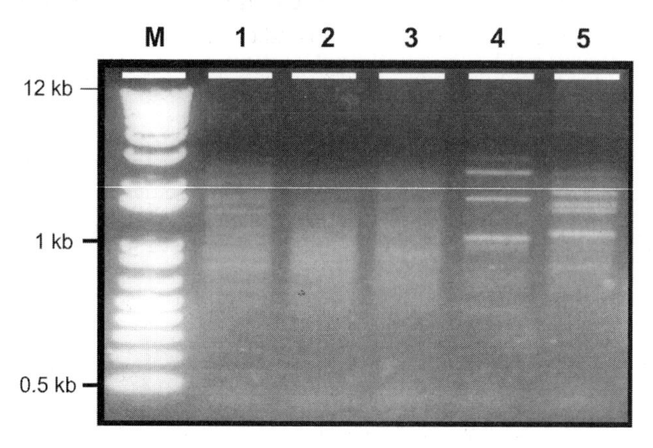

Figure 8 Primary PCR products of the FamilyWalker protocol run on a 1% agarose gel. Amplification was done as described in *protocol 3* with the degenerate MBG primer described in *Figure 7* on *Petunia hybrida* GW libraries. M = 1 kb Plus DNA ladder (GibcoBRL). Lanes 1–5: 10 μl of primary PCR products amplified from *Sca* I, *Dra* I, *Eco* RV, *Pvu* II, and *Stu* I GW libraries, respectively.

Protocol 3

Primary PCR for the FamilyWalker method

Equipment and reagents

- Universal Genome Walker™ Kit (Clontech)
- 5 GenomeWalker (GW) libraries constructed according to the suppliers' instructions
- Advantage *Tth* Polymerase Mix (Clontech)
- A thermal cycler that can perform complex programmes (e.g. Perkin Elmer GeneAmp PCR System 9600)
- Thin-walled PCR tubes
- dNTPs (10mM)
- dH_2O
- Degenerate primer

Protocol 3 continued

Method

1 Add 1 μl of each DNA library to a PCR tube.

2 Combine the following reagents in a 1.5-ml microfuge tube (enough for the 5 GW libraries): 25 μl 10× *Tth* PCR reaction buffer[a], 11 μl $Mg(OAc)_2$[a] (25 mM), 5 μl dNTPs (10 mM), 10 μl degenerate primer (10 μM), 5 μl primer AP1[a] (10 μM), 5 μl Advantage *Tth* Polymerase Mix (50×), and 179 μl dH_2O.

3 Mix well and briefly spin the tube in a microcentrifuge to collect the contents at the bottom.

4 Add 49 μl mix to the tubes containing the five different GW libraries and mix very carefully by pipetting the mixture two or three times up and down.

5 Commence cycling in a Perkin-Elmer GeneAmp PCR system 9600 or equivalent using the following parameters[b]:

94 °C for 2 s, (T_{annmax} + 2 °C) for 4 min 30 s and declining 1 °C per cycle until (T_{annmin} + 3 °C) is reached[c];

Thirty-seven cycles [94 °C for 2 s, (T_{annmin} + 2 °C) for 4 min 30 s];

67 °C for 7 min.

6 Analyse 10 μl of the PCR products on a 1.5% agarose gel (see *Figure 8*)[d].

[a] These components are provided with the Universal Genome Walker™ Kit.

[b] We optimized PCR conditions using a Perkin-Elmer GeneAmp PCR system 9600. If other systems are used, cycling parameters should be optimized again. In particular the length of the denaturation and elongation steps may need optimization. Guidelines for other thermocyclers are described in the GW protocol.

[c] We also tried a three-step cycling programme, but this resulted in higher background amplification. We strongly recommend design of the primers in such a way that (T_{annmin} + 3 °C) is higher than 60 °C. If (T_{annmax} + 2 °C) is higher than 71°C, start the TD profile at 71 °C. Importantly, the TD profile should not be longer than 10 cycles.

[d] If multigene families are amplified, a smear is expected in all five lanes, but several distinct bands should be visible in some lanes.

3.6.2 Secondary PCR for the FamilyWalker method

We have found it necessary to perform two subsequent PCRs before fragments become clearly visible and can be cloned. In the secondary PCR (see *Protocol 5*), a ^{33}P-labelled degenerate primer (see *Protocol 4*) is used together with the nested AP2 adaptor primer from the GW kit. The degenerate primer is labelled for two reasons:

1 To allow direct visualization on polyacrylamide gels, which are used for analysis because of their high resolution (Section 3.6.3)

2 Normally, the high T_{ann} (67 °C) used in the GW method strongly represses the amplification of adaptor-adaptor fragments by the suppression PCR effect

(*Figure 5*). However, in most cases, such a high T_{ann} cannot be used in the FamilyWalker protocol, so amplification of adaptor-adaptor fragments can be problematic. This can be elegantly solved by labelling only the degenerate primer, which prevents the adaptor-adaptor fragments becoming visible.

Protocol 4

Radioactive labelling of degenerate primers

Equipment and reagents

- Degenerate primer (10 μM)
- T4 polynucleotide kinase (PNK)
- 10× T4 PNK buffer (500 mM Tris.HCl pH 7.6, 100 mM $MgCl_2$, 50 mM DTT, 1 mM spermidine-HCL, 1 mM EDTA; Eurogentec)
- (γ-^{33}P)-ATP (10 mCi/ml)
- Two heat blocks

Method

1. For 10 PCR reactions, mix together in a 1.5-ml microfuge tube: 4 μl degenerate primer (10 μM), 4 μl (γ-^{33}P)-ATP, 1 μl T4 PNK 10× buffer, 1 unit T4 PNK, and H_2O to a final volume of 10 μl.
2. Incubate the mixture for 40 min at 37 °C.
3. Inactivate the enzyme by incubating the tubes for 10 min at 80 °C.
4. Centrifuge the mixture briefly to collect any condensate that has formed on the top of the tube.
5. Use the primers immediately or store them at −20 °C[a].

[a] When stored at −20 °C, ^{33}P-labelled primers may be used for 1–2 weeks.

Protocol 5

Secondary PCR for the FamilyWalker protocol

Equipment and reagents

- Template obtained from the primary PCR (see *Protocol 3*)
- Advantage *Tth* Polymerase Mix (50×; Clontech)
- A thermal cycler that can perform complex programmes (e.g. Perkin Elmer GeneAmp PCR System 9600)
- Thin-walled PCR tubes
- dNTPs (10mM)
- dH_2O
- ^{33}P-labelled degenerate primer (see *Protocol 4*)
- Formamide loading dye (98% formamide, 10 mM EDTA, 0.05% bromo-phenol blue, 0.05% xylene cyanol)

Protocol 5 continued

Method

1. Dilute[a] 2 μl of each primary PCR product (see *Protocol 3*) into 18 μl dH_2O.
2. Add 1μl of each diluted primary PCR product to a PCR tube.
3. Combine for one PCR reaction the following reagents[b] in a 1.5 ml microfuge tube: 2 μl 10× *Tth* PCR reaction buffer, 0.88 μl $Mg(OAc)_2$ (25 mM), 0.4 μl dNTPs (10 mM), 1 μl ^{33}P-labelled degenerate primer, 0.4 μl primer AP2 (10 μM), 0.4 μl Advantage *Tth* Polymerase Mix (50×), and 13.92 μl dH_2O.
4. Mix the tube well and briefly spin the tube in a microcentrifuge to collect the contents at the bottom.
5. Add 19 μl of the PCR master mix to the PCR tubes and mix very carefully by pipetting the mixture up and down two or three times.
6. Commence cycling in a Perkin-Elmer GeneAmp PCR system 9600 using the following parameters[c]:

   20 cycles [94 °C for 2 s, (T_{annmin} + 2 °C) for 4 min 30 s];

   10 cycles [94 °C for 2 s, (T_{annmin} + 2 °C) for 4 min 30s and declining 0.5 °C/cycle + adding 20 s extension time/cycle until (T_{annmin} − 3 °C) is reached];[d]

   seven cycles [94 °C for 2 s, (T_{annmin} − 3 °C) for 30 s, 67 °C for 3 min];[e]

   final extension of 67 °C for 10 min.
7. Add an equal volume (20 μl) of formamide loading dye to the PCR products and mix the samples thoroughly.
8. Heat the resulting mixtures at 94 °C for 5 min and immediately cool the samples on ice.
9. Load 2–3 μl of each PCR product on a 3% sequencing gel (see Section 3.6.3), while keeping the samples on ice.

[a] In the original GW protocol, primary PCR products are diluted 50 times.

[b] From here on use filter tips to prevent contamination of your pipette with ^{33}P.

[c] See note b in *Protocol 3*.

[d] This additional TD step is used to obtain a final increase in yield.

[e] For the final cycles, a three-step PCR is used because elongation speed at (T_{annmin} − 3 °C) would be too slow. If (T_{annmin} − 3 °C) is still higher than 60 °C, use a two-step PCR.

3.6.3 Gel analysis

The reaction products are analysed on 3% denaturing polyacrylamide gels. These gels are in principle normal sequencing gels, with the exception that a lower percentage of polyacrylamide is used (see *Protocol 6*). We use a BioRad sequencing gel system (50 × 38 × 0.4 cm, ± 100 ml gel solution), but other sequencing gel systems can work equally well. After electrophoresis, gels are dried on Whatman 3MM paper on a standard slab gel drier for sequence gels. Generally, overnight exposure to Kodak BioMax films gives good results.

Protocol 6

Polyacrylamide gel electrophoresis

Equipment and reagents

- BioRad sequencing gel system (50 × 38 × 0.4 cm) or equivalent system
- 150 ml syringe
- Magnetic stirrer with heating block
- 0.2 μm filter
- Acrylamide stock solution (40%, acrylamide: methylenebisacrylamide, 19:1)
- Urea
- dH_2O
- Amberlite MB-1A mixed bed exchanger (Sigma)
- 10× MAXAM (1 M Tris, 1 M boric acid)
- EDTA (0.5 M, pH 8.0)
- TEMED (tetramethyl-ethylenediamine, Pharmacia Biotech)
- Rain-X (Penzoil Products)[a]
- 10 % ammonium persulfate (APS). Can be stored at 4 °C for up to 3 weeks.

Methods

A. Preparation of 1 l gel solution (3 % (w/v) gel)

1. Mix together 450 g urea, 75 ml acrylamide stock solution, a small spoonful of the ion exchanger amberlite, and add dH_2O to 900 ml.
2. Place the mixture on a magnetic stirrer with heating block (do not heat above 60 °C) and stir until the solution becomes clear.
3. Filter the solution through a 0.2 μm filter to remove the Amberlite and other impurities.
4. Add 100 ml 10 × MAXAM and 4 ml EDTA 0.5 M pH 8.0, mix, and store[b] the solution at 4 °C.

B. Assembly of the gel system

1. Clean both glass plates with soap and rinse the plates well.
2. Wipe the plate with the buffer tank first with ethanol and then with Rain-x.
3. Wipe the other glass plate first with ethanol and then with acetone.
4. Assemble the gel system by putting spacers and the two plates together.

C. Pouring of the gel

1. Put the gel system horizontally (plate with the buffer tank up).
2. Add together 100 ml gel solution, 500 μl 10% APS and 100 μl TEMED.
3. Mix and use a 150 ml syringe to fill the gel (avoid air bubbles in the tube).

Protocol 6 continued

4 Put the comb with the flat side towards the gel in between the glass plates and let the gel polymerize for at least 1 h before use.

D. Gel running

1 Prepare 2 l of 1 × TBE (1 × MAXAM, 2 mM EDTA) and put 400 ml in the lower buffer tank and heat the remaining 1 × TBE in a microwave for 5 min at maximum power.

2 Place the gel in the buffer tank and pour the heated 1 × TBE in the upper buffer tank (fill completely).

3 Take out the comb carefully and clean the slot with a syringe (suck up buffer to do this).

4 Put the comb in place and load the samples.

5 Run the gel at 100 Watts (± 50 mA, ± 2 KV, set power limited) for 4 h (for fragments above 500 bp).

6 After running, open the gel system, lift the gel with Whatman 3MM paper and dry on a standard slab gel drier for sequence gels.

[a] Rain-x is a hydrophobic product which is normally applied on glass or plastic surfaces (such as helmets) as an anti raindrop solution. By putting Rain-x on one of the glass plates of the gel system, the gel system can be opened easily after running the gel (Rain-x prevents the gel sticking to both plates during disassembly). It is important that Rain-x is applied on only one of the glass plates and that it is always the same one.

[b] The gel solution remains stable for at least 3 months when stored at 4 °C.

3.6.4 Isolation, cloning and characterization of the generated fragments

Inherent to the FamilyWalker approach is that a large number of fragments have to be characterized. We have therefore optimized methods to allow a quick characterization of a large number of fragments simultaneously. Firstly, the fragments are cut out from the gel and re-amplified (see *Protocol 7*). Secondly, all the fragments are sequenced using a direct sequencing method with the degenerate primer as a sequencing primer (see *Protocol 8*). In our hands, this method gives easy-to-read sequences for the majority of the fragments. An overview of the results obtained with the degenerate MBG primer described in *Figure 7* is given in *Figure 9*. We sometimes obtained sequence patterns that were the result of a mixture of 2 or more fragments present in the template (e.g. sequencing reaction 1 in *Figure 9*). In such cases, we cloned the fragments into pGEM-T (Promega) and then identified the colonies containing the correct insert according to the method described in *Protocol 9*. In this method, colonies are screened by PCR with a labelled degenerate primer and a reverse primer. PCR products are analysed on a 3% polyacrylamide gel together with the labelled secondary PCR products (see *Protocol 3*), which were used as cloning material. This is necessary because the difference in size between the undesired and the right inserts can be very small. An example of this method is given in *Figure 10*.

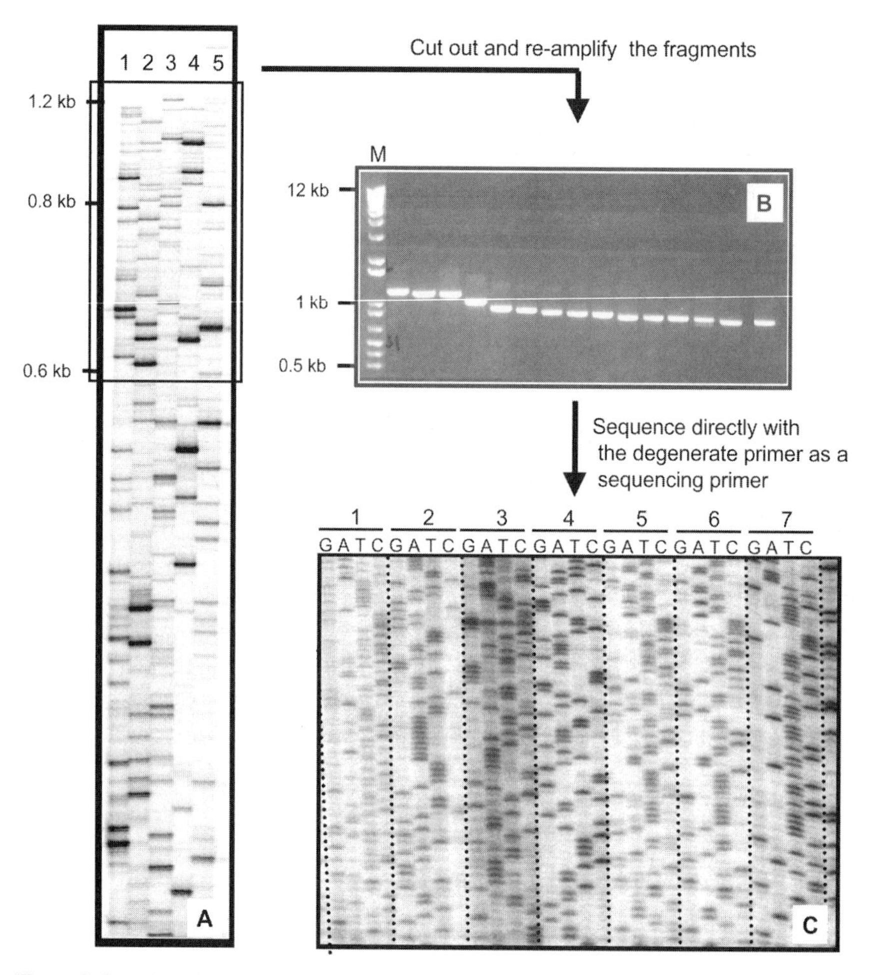

Figure 9 Overview of results obtained with the degenerate MBG primer described in *Figure 7*. (A) Secondary FamilyWalker PCR products run on a 3% polyacrylamide gel. Lanes 1–5: 2 μl of secondary PCR products amplified from *Sca* I, *Dra* I, *Eco* RV, *Pvu* II, and *Stu* I primary PCR products, respectively. (B) Re-amplification products of fragments cut out from the boxed part of the gel presented in picture A and run on a 1 % agarose gel, as described in *Protocol 6*. (C) Representation of a part of a sequencing gel. Sequencing reactions 1–7 are obtained by the direct sequencing method described in *Protocol 7* on re-amplified fragments from picture B.

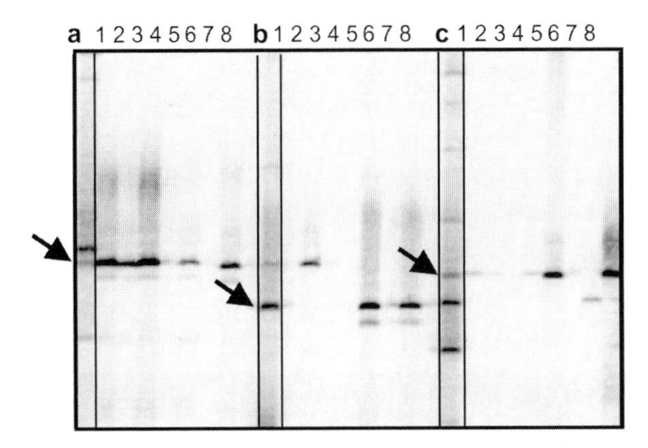

Figure 10 Direct PCR products from colonies and secondary FamilyWalker PCR products run on a 3% polyacrylamide gel. Lanes a–c: Secondary PCR products derived from *Sca* I, *Pvu* II, and *Dra* I libraries, respectively. Lanes 1–8: direct PCR products from colonies obtained from subcloning these fragments. Arrows indicate for which band the colonies were selected in each series.

Protocol 7

Isolation and re-amplification of bands from a polyacrylamide gel

Equipment and reagents

- Kodak BioMax films (Kodak)
- A film cassette
- A sharp razor blade
- Platinum *Taq* DNA polymerase (GibcoBRL)
- dNTPs 10 mM
- Deionized H_20 (dH_2O)
- 10× PCR reaction buffer (with 15 mM $MgCl_2$)
- Forward and reverse primers (a degenerate primer and the AP2 primer if the FamilyWalker method was used for obtaining the fragments)
- A thermocycler

Method

1. Expose the dried gel to a film.[a]
2. Cut the bands from the gel.[b]
3. Put the piece of gel (including the Whatmann paper) in 50 μl dH_2O.
4. Allow the DNA to elute for about 1 h at room temperature and vortex occasionally.
5. Centrifuge the tubes for 1 min at maximum speed to spin down the paper and gel particles. The water phase contains the template DNA.
6. Use 2 μl template DNA in a standard 50 μl PCR reaction to re-amplify the band.[c]

Protocol 7 continued

7 Check the purity of the PCR products on an agarose gel. Single bands should be visible without any background (see *Figure 9*).

[a] Put the film on the gel in such a way that you can exactly position the film on the gel later on.

[b] Expose the gel again to check if the bands have been properly cut out.

[c] Use the same PCR profile that was used to obtain the bands.

Protocol 8

Direct sequencing of bands isolated from polyacrylamide gels

Equipment and reagents

- Thermo sequenase cycle sequencing kit (Amersham Life Science) or any other thermo-cycling system
- Hybaid recovery DNA purification kit II (Hybaid) or any other PCR purification kit
- ^{33}P-labelled primer (see *Protocol 4*)

Methods

A. Purification of the re-amplified PCR products

1 Use 20 μl of the re-amplified PCR product (see *Protocol 7*) for purification on Hybaid spin columns according to the instructions of the manufacturer.

2 In the last step of the Hybaid protocol, elute the DNA in 20 μl H_2O.

B. Sequencing

1 Use 5 μl of the purified PCR product as a template and 2 μl of labelled primer to set up the sequencing reactions as described in the radiolabelled primer cycle sequencing protocol accompanying the Thermo sequenase cycle sequencing kit.

2 Start the following cycling program: 50 cycles (95 °C for 30 s, 55 °C[a] for 30 s, 72 °C for 60 s)

3 Add 4 μl stop solution (supplied with the kit) to the sequencing reactions and heat the samples to 75 °C for 2 min.

4 Analyse the sequencing reactions on a denaturing 6% polyacrylamide gel (adapt *Protocol 6*).

[a] Take the same T_{ann} as used in *protocol 7* to re-amplify the fragments.

Protocol 9

Direct PCR on colonies

Equipment and reagents

- 96-well PCR microtitre plates
- LB (10 g NaCl, 5 g yeast extract, 10 g bactotryptone, 1.5 g agar per 100 ml water) plates containing 100 μg/ml ampicillin
- 10× PCR reaction buffer (with 15 mM $MgCl_2$)
- dNTPs (10 mM each)
- Labelled degenerate primer (see *protocol* 4)
- AP2 primer
- *Taq* DNA polymerase

Method

1. Fill each well of a 96-well PCR plate with 10 μl dH_2O.
2. Pick individual colonies[a] with a tooth pick. Dip the tooth pick first into the 10 μl dH_2O in the appropriate well for about 1 min and then transfer the colony onto a new LB plate (with the same tooth pick) or inoculate a 1.5 ml culture with it. Grow the colonies overnight.
3. Prepare $n + 1$ times the following PCR mix for n colonies: 2 μl 10× PCR reaction buffer, 0.4 μl dNTPs (10 mM each), 0.25 μl labelled degenerate primer, 0.04 μl AP2 primer, 0.1 μl *Taq* DNA polymerase, and 7.26 μl dH_2O.
4. Add 10 μl of the PCR mix to each of the wells.
5. Run the following cycling programme: 94 °C for 5 min, 40 cycles (94 °C for 30 s, (T_{annmin} −3 °C) for 30 s, 72 °C for 1 min/kb).
6. Add an equal volume (20 μl) of formamide loading dye to the PCR products and mix the samples thoroughly.
7. Heat the resulting mixtures at 94 °C for 5 min and then quickly cool the samples on ice.
8. Load 2–3 μl of each PCR product[b] on a 3% sequencing gel (see *Protocol 6*), while keeping the samples on ice (see *Figure 10*).

[a] We normally analyse 8–10 colonies per transformation.

[b] Secondary PCR products of the FamilyWalker protocol should also be loaded on the gel to allow identification of the positive colonies.

Acknowledgements

Our work on the MADS-box gene family was funded by BIOTECH programme grant number Bio4-CT97–2217 from the European Commission.

References

1. Liang, P. and Pardee, A. B. (1992). *Science*, **257**, 967.
2. Bachem, C. W., van der Hoeven, R. S., de Bruijn, S. M., Vreugdenhil, D., Zabeau, M., and Visser, R. G. (1996). *Plant J.*, **9**, 745.

3. Welsh, J., Chada, K., Dalal, S. S., Cheng, R., Ralph, D., and McClelland, M. (1992). *Nucl. Acids Res.*, **20**, 4965.
4. Vos, P., Hogers, R., Bleeker, M., Reijans, M., van de Lee, T., Hornes, M., Frijters, A., Pot, J., Peleman, J., Kuiper, M., and Zabeau, M. (1995). *Nucl. Acids Res.*, **23**, 4407.
5. McClelland, M., Mathieu-Daude, F., and Welsh, J. (1995). *Trends Genet.*, **11**, 242.
6. Goormachtig, S., Valerio-Lepiniec, M., Szczyglowski, K., Van Montagu, M., Holsters, M., and de Bruijn, F. J. (1995). *Mol. Plant-Microbe Interact.*, **8**, 816.
7. Frohman, M. A., Dush, M. K., and Martin, G. R. (1988). *Proc. Natl Acad. Sci. USA*, **85**, 8998.
8. Ohara, O., Dorit, R. L., and Gilbert, W. (1989). *Proc. Natl Acad. Sci. USA*, **86**, 5673.
9. Bertioli, D. (1997). In *Methods in molecular biology* (ed. B. A. White). Vol. 67. Humana Press, Totowa.
10. Bouchez, D. and Hofte, H. (1998). *Plant Physiol.*, **118**, 725.
11. Lee, C. C., Wu, X. , Gibbs, R. A., Cook, R. G., Munzy, D. M., and Caskey, C. T. (1988). *Science*, **239**, 1288.
12. Edwards, J. D. M., Ravassard, P., Icard-Liepkalns, C., and Mallet, J. (1995). In *PCR 2: a practical approach* (ed. M. J. Mc Pherson, B. D. Hames, and G. R. Taylor). IRL Press, Oxford.
13. Preston, M. (1997). In: *Methods in molecular biology* (ed. B. A. White), Vol. 67. Humana Press, Totowa.
14. Bartl, S. (1997). In *Methods in molecular biology* (ed. B. A. White). Vol. 67. Humana Press, Totowa.
15. Siebert, P. D., Chenchik, A., Kellogg, D. E., Lukyanov, K. A., and Lukyanov, S. A. (1995). *Nucl. Acids Res.*, **23**, 1078.
16. Altschul, S. F., Boguski, M. S., Gish, W., and Wootton, J. C. (1994). *Nature Genet.*, **6**, 119.
17. Altschul, S. F., Madden, T. L., Schaffer, A. A., Zhang, J., Zhang, Z., Miller, W., and Lipman, D. J. (1997). *Nucl. Acids Res.*, **25**, 3389.
18. Thompson, J. D., Higgens, D. G., and Gibson, T. J. (1994). *Nucl. Acids Res.*, **22**, 4673.
19. Henikoff, S., Henikoff, J. G., Alford, W. J., and Pietrokovski, S. (1995). *Gene*, **163**, 17.
20. Wada, K. N., Wada, Y., Ishibashi, F., Gojobori, T., and Ikemura, T. (1992). *Nucl. Acids Res.*, **20**, 2111.
21. Smith, T. F., Waterman, M. S., and Sadler, J. R. (1983). *Nucl. Acids Res.*, **11**, 2205.
22. Don, R. H., Cox, P. T., Wainwright, B. J., Baker, K., and Mattick, J. S. (1991). *Nucl. Acids Res.*, **19**, 4008.
23. Hecker, K. H. and Roux, K. H. (1996). *BioTechniques*, **20**, 478.
24. Knoth, K., Roberds, S., Poteet, C., and Tamkun, M. (1988). *Nucl. Acids Res.*, **16**, 10932.
25. Rose, T. M., Schultz, E. R., Henikoff, J. G., Pietrokovski, S., McCallum, C. M., and Henikoff, S. (1998). *Nucl. Acids Res.*, **26**, 1628.
26. D'Aquila, R. T., Bechtel, L. J., Viteler, J. A., Eron, J. J., Gorczyca, P., and Kaplin, J. C. (1991). *Nucl. Acids Res.*, **19**, 3749.
27. Rychlik, W. and Spencer, W. J. (1989). *Nucl. Acids Res.*, **17**, 8543.
28. Rychlik, W. (1994). *J. NIH Res.*, **6**, 78.
29. Schwarz-Sommer, Z., Huijser, P., Nacken, W., Saedler, H., and Sommer, H. (1990). *Science,* **250**, 931.
30. Theissen, G., Kim, J.T., and Saedler, H. (1996). *J. Mol. Evol.*, **43**, 484.

Chapter 10
Western and south-western Library Screens

Rhonda C. Foley
Agriculture Western Australia, CCMAR, Private Bag, #5, Wembley WA 6913, Australia

Karam B. Singh
CSIRO Plant Industry, CCMAR, Private Bag, #5, Wembley WA 6913, Australia

1 Introduction

The screening of cDNA libraries is a staple technique used in most plant molecular biology laboratories at one time or another. Nucleic acid hybridization is the most common screening approach to isolate cDNA clones for a gene of interest, and has the advantages of being relatively straightforward and reliable. However, there are situations when other types of cDNA library screening are required, two of which are covered in this chapter. Western library screening identifies cDNA clones that express proteins that are recognized by specific antibodies and south-western library screening identifies cDNA clones that express proteins that can specifically bind to DNA recognition sites. Unlike conventional cDNA library screening with a nucleic acid probe, both techniques require the cDNA library to be cloned into an expression vector. One disadvantage compared to conventional cDNA library screening is that both western and south-western library screening will only detect cDNA clones that are in the correct reading frame and orientation. This means that more cDNA clones need to be screened in a given experiment and for those target genes that are expressed at low levels, the cDNA expression library must be large ($>10^6$ pfu).

Western library screening, which is related to the western blotting technique described in Volume 2, Chapters 9 and 10, identifies proteins that have antigenic motifs identical or similar to a protein one already has antibodies against. For example, this could be to clone the cDNA for a protein that has been partially purified or, as is more commonly the case these days, to screen for related protein(s) in the same or different plant species. It is in the isolation of related proteins that western library screening can have advantages over conventional screening since as a result of codon degeneracy, conservation at the protein level is typically higher than at the nucleic acid level. While the need for western library screening in model organisms such as Arabidopsis and rice is likely to decrease as their genomes are fully sequenced, western library screening is still likely to be frequently used by researchers working on other plants, for example, those researchers who wish to isolate proteins related to a gene of interest that was first isolated in one of the model organisms.

South-western library screening, which shares a number of steps with western library screening, is used to isolate nucleic acid binding proteins, primarily transcription factors. The last 10 years have seen dramatic progress in the isolation and characterization of plant transcription factors, in many cases as a result of south-western screening experiments. The south-western screening approach developed by Singh *et al.* (1) and Vinson *et al.* (2) is relatively rapid, and simply requires a suitable cDNA library constructed in an appropriate expression vector and a well characterized DNA recognition site probe. However, the south-western screening approach is not suitable for all types of transcription factors, since some factors will not bind to their DNA recognition site when expressed in *Escherichia coli* and cannot be cloned by this approach. These transcription factors may require specific post-translational modifications or one or more additional proteins to bind to the DNA recognition site.

Other methods used to isolate plant transcription factors include biochemical approaches, genetic approaches and the yeast one-hybrid screen. A major advantage of the south-western screening approach over biochemical approaches is that there is no requirement for the purification of a sequence-specific DNA-binding protein and it is well suited for isolating proteins that are only expressed at low levels, as is often the case for transcription factors. However, the south-western screening approach cannot be used if the protein functions in a complex with other proteins or if it requires eukaryotic post-translational modifications in order to bind to DNA.

In the following section we describe the western and south-western library screening approaches including the detailed protocols and critical parameters involved. As the initial steps of plating out the library and adding nitrocellulose filters are similar for both types of screening, we have listed these methods under the same protocol (*Protocol 1* and *2*). Both the western and south-western screening approaches have been described elsewhere (for example, see 3, 4, 5). In addition, some of the information listed here regarding the south-western screening approach has been written in a modified form previously (6).

2 Working with cDNA expression libraries

The construction of cDNA libraries is beyond the scope of this chapter and readers requiring this information should refer to a standard molecular biology technique textbook such as Sambrook *et al.* (3). The choice of cDNA library is important and should represent the mRNA population of the tissue/organ/cell line that has the greatest activity for the particular protein. While it is possible to screen cDNA libraries made in a plasmid expression vector, λ cDNA expression libraries constructed in λgt11 (Promega) or λZAP (Stratagene) and their derivatives are far more commonly used and will be the focus of this chapter. The cDNA inserts are cloned into the λ vectors so that fusion proteins with β-galactosidase are produced. An inducible expression system is used to express the fusion proteins in case the encoded protein is toxic in *E. coli*. We have had success with both λgt11 and λZAP vectors but prefer λgt11 because (1) the size of the λgt11 plaques are smaller allowing more phage to be screened on a single plate, (2) the size of the β-galactosidase fusion protein is larger and this normally provides greater stability. The advantage of the λZAP vector is the ease of subcloning cDNA.

Protocol 1

Plating the cDNA library for western and south-western library screening

Equipment and reagents

- LB medium: 1% w/v bacto-tryptone, 0.5% yeast extract, 1% NaCl
- LB plates: LB medium supplemented with 1.5% agar
- LB top agarose: LB medium supplemented with 0.7% agarose
- Phage dilution buffer: 0.1 M NaCl, 10 mM Tris.HCl pH 8, 10 mM $MgSO_4$
- Bacteria storage solution: 10 mM $MgCl_2$, 0.2% maltose
- Bacteria stock[a]
- cDNA expression library
- 42 °C and 37 °C air incubators

Method

1. Grow *E. coli* to saturation in LB medium containing 0.2% maltose and appropriate antibiotics at 37 °C. Centrifuge the bacteria at 5,000 g for 5 min and resuspend in 0.5 volumes of bacteria storage solution.[b]
2. For each 150 mm plate, mix 300 μl of the *E. coli* culture with 100 μl of phage dilution buffer containing 3–4 $\times 10^4$ pfu from the λ cDNA expression library. Incubate the tubes for 15–20 min at 37 °C to adsorb phage to cells.
3. Melt the top agarose and allow the solution to cool to 47–55 °C in a water bath.
4. Add 10 ml top agarose to each tube. Mix by inverting a few times and plate each suspension onto a 150-mm LB plate. Work fast to ensure that the LB agarose does not set prior to plating. The plates should be freshly made, dry and pre-warmed to room temperature.
5. Once the top agarose has set, incubate the plates for approximately 3 h at 42 °C until tiny plaques are visible.[c]

[a] The choice of the host bacteria for the phage library is important in order to minimize the effects of toxic proteins and to be compatible with the induction chemical (e.g. IPTG). Often it is best to use the bacteria recommended by the supplier of the cDNA vector. For example, with λgt11 and λZAP, use Y1090 and XL1-Blue, respectively.

[b] The bacteria should remain viable for 2 weeks at 4 °C.

[c] The high temperature of 42 °C allows faster growth and promotes lysis.

3 Preparation of nitrocellulose filters

In this section we describe the standard method of preparing nitrocellulose filters using native proteins. While this method is used for western library screening, an alternative method involving denaturation/renaturation steps is sometimes used for south-western library and is described in *Protocol 8*. Unlike south-western screening, where the use of duplicate filters may be important because of weak positive signals over background, it is generally not necessary to use duplicate filters for antibody screening. This is because the positive plaques are normally very obvious, and it is easier to pick and rescreen any putative positives found on a single filter. The advantages and disadvantages of duplicate filters for the south-western screening is discussed in Section 5.2.

Protocol 2

Preparation of nitrocellulose filters for western blot screening and south-western screening with native proteins

Equipment and reagents

- Nitrocellulose filters
- 10 mM IPTG (isopropyl β-D-thiogalactoside)
- Blotting paper

Method

1. While the LB plates containing the phage library (*Protocol 1*) are incubating at 42 °C prepare 132-mm nitrocellulose filters by briefly soaking in 10 mM IPTG.[a,b]
2. Drain the filters and allow to air dry on blotting paper for 30 min.
3. Monitor the phage plates in the 42 °C incubator. When the first pinpricks representing plaques appear, transfer the plates to 37 °C for about 3–4 h. Do not allow plates to cool below 37 °C.
4. Overlay each plate with a dry nitrocellulose filter impregnated with 10 mM IPTG and continue incubation for 5–6 h at 37 °C. Using a needle, dip into Indian ink and place approximately five dots around the edge of the filter, making sure you penetrate into the agar. This procedure serves to orientate the nylon filter to the plate. If duplicate filters are not required proceed to step 6.
5. For duplicate filters the following steps are required. The primary filters are prepared as described in steps one and two above. After 5–6 h (*Protocol 2* step 4) the positions of the first filter is marked on the plate, but the plate is not pre-cooled. After the first filter has been removed the second filter is placed on the plate, returned immediately to 37 °C and incubated for 2 h to overnight.[c]

Protocol 2 continued

6 Cool plates for 20 min at 4 °C.[d]

[a] The nitrocellulose filters should always be handled with gloves and blunt forceps (for example Millipore forceps).

[b] When working with protein it is important to use nitrocellulose, rather than nylon filters.

[c] If duplicate filters are being used it is best if a 37 °C room is available. If duplicate filters are not required the primary filter can be left on overnight. We recommend leaving the duplicate filter on overnight.

[d] Plates are cooled so as to avoid lifting the top agarose with the filter and it is better to use top agarose, rather than top agar for this reason.

4 Western cDNA library screening

If possible, it is a good idea to first perform a Western blot to determine the stringency conditions and the antibody dilution for optimum screening. Generally, there is more protein in a λ plaque obtained with cDNA library screening (*Protocol 3*) than a band on a gel obtained with a western blot. Occasionally, the antibody will cross-react with *E. coli* proteins, which can be rectified by pre-absorbing the antibody against an *E .coli* extract.

4.1 Polyclonal and monoclonal antibodies

Either polyclonal or monoclonal antibodies can be used, although in general polyclonal antibodies give more consistent results as they recognize a larger range of epitopes on the protein. Polyclonal antibodies also have the advantage of being cheaper to produce. However, monoclonal antibodies are more specific and can often distinguish between closely related proteins. When using monoclonal antibodies, a mixture of monoclonal antibodies may give better results. Further details on antibody techniques can be found in Volume 2, Chapter 10.

Protocol 3

Screening cDNA libraries with antibodies

Equipment and reagents

- PBST (130 mM NaCl, 7 mM Na_2HPO_4, 3 mM NaH_2PO_4 + 0.1–0.5% Tween)[a]
- Antibody solution in blocking reagent
- Blocking reagent (3% skim milk powder in PBST)[b]

Method

1 Lift filters carefully starting from one side and place directly into the blocking solution for 30 min at room temperature with gentle shaking.[c]

Protocol 3 continued

2 Using blunt-end tweezers, place filters in hybridization bottles.[d]

3 To the bottles add 20 ml antibody solution to a large bottle (25 cm long) or 10 ml to a small bottle (15 cm long). Generally, a 1/1000 dilution of antibody, either polyclonal or monoclonal, is a good starting point.

4 Incubate overnight in hybridization bottles on a rotary machine at room temperature (or at 4 °C). However, 2 h at room temperature is often sufficient.

5 Discard the antibody and wash the filters with approximately 100 ml PBST for 15 min at room temperature.[e]

6 Repeat the wash with fresh PBST for a further 15 min.

[a] The amount of Tween can range from 0.1% to 0.5% depending on the stringency (0.5% Tween results in higher stringency and less background).

[b] Skim milk powder (also known as non-fat milk powder) is used because it is a cheap source of protein. The proteins in the skim milk will bind non-specifically to proteins on the nitrocellulose filter and prevent subsequent non-specific binding of the antibody.

[c] Do not allow the filters to dry.

[d] Alternatively, plastic bags and a heat sealer can be used. If using a heat sealer to seal the bag, try to avoid trapping any air bubbles in the bag.

[e] If you have limited amounts of the antibody, it is possible to collect and store the antibody at 4 °C with azide (to prevent contamination) and reuse it later.

4.2 Detection systems used in western library screening

Two common detection methods used in western library screening are the colorimetric detection system (*Protocol 4*) and the chemiluminesent system (*Protocol 5*). With the colorimetric system the reaction occurs directly on the filter and, therefore, it is relatively easy to align the original plaque to the putative positive clone. The advantage of the chemiluminesent system is that it is more sensitive, and with the appropriate equipment and software, the images can be exposed to optimum levels. An alternative chemiluminescent detection system, which is not described here, is the CDP-star system that utilizes the alkaline phosphatase secondary antibody (7).

Protocol 4

Colorimetric detection

Equipment and reagents

- PBST (see *Protocol 3*)
- Secondary antibody with conjugated detection system (e.g. alkaline phosphatase) in blocking reagent (see *Protocol 3*)
- Detection buffer: 0.1 M Tris.HCl pH 9, 0.1 M NaCl, 50 mM $MgCl_2$
- Colour detection buffer: detection buffer + 0.15% NBT (nitro blue tetrazolium) + 0.03% BCIP (5-bromo-4-chloro-3-indolyl phosphate)[a]

Protocol 4 continued

Method

1. Remove washing solution and replace with secondary antibody diluted in blocking reagent. The secondary antibody is conjugated with alkaline phosphatase. Use the same volume of secondary antibody that was used for the primary antibody incubation.[b]
2. Incubate at room temperature for 30 min in either a rotary hybridization oven or bench shaker.
3. To remove the unbound antibodies, wash the filters with approximately 100 ml PBST for 15 min at room temperature.
4. Repeat the wash with fresh PBST for a further 15 min.
5. Equilibrate 2 min in detection buffer.
6. Drain the filters and place face up on a hydrophobic membrane such as parafilm.
7. Pour 10 ml of colour detection solution evenly over each of the filters.[c]
8. Once the desired purple colour is achieved, remove the filter and rinse in water to stop the reaction.[d]

[a] Stock solutions of these compounds are stable when stored in the dark at 20 °C: NBT 50 mg/ml in 70% dimethylformamide, BCIP 50 mg/ml in 100% dimethylformamide. Use 33 μl and 16.5 μl of BCIP solution per 5 ml of detection buffer.

[b] Secondary antibodies can be bought commercially and should be diluted to the concentration recommended by the manufacturer. Ensure that the secondary antibody recognizes the primary antibody (for example use anti-mouse for monoclonal antibodies made from mice, and anti-rabbit for polyclonal antibodies produced in rabbit).

[c] The colour solution needs to be made immediately prior to development.

[d] It takes approximately 5 min for the colour to develop, although longer development is possible if the background is low.

When using polyclonal antibodies that are not very specific, all the plaques may give some degree of colour signal and you will need to select the plaque that is significantly darker than the neighbouring plaques. With longer exposure, colour will develop from the bacteria. This colour is due to the low level of alkaline phosphatase endogenously expressed by the bacteria. This background pattern is often advantageous to orientate plaques. It is best to pick positive clones while the filter is still moist, as drying the filter results in fading of the signal and shrinkage of the filter.

Protocol 5

Chemiluminescent detection system using horseradish peroxidase

Equipment and reagents

- Secondary antibody with conjugated detection system (i.e. horseradish peroxidase) in blocking reagent (see *Protocol 3*) using the concentration recommended by the manufacturer [a]
- PBST (see *Protocol 3*)
- Solution A: Add 3 μl of 30% H_2O_2 (just before use) and 27 μl of 90 mM P-coumaric acid (14 mg/ml in DMSO) to 6 ml 100 mM Tris–HCl pH 8.5 (or to a similar ratio)
- Solution B: Add 60 μl of 250 mM 3-Aminophalhydrazin (44 mg/ml in DMSO) to 6 ml 100 mM Tris–HCl pH 8.5
- X-ray film
- X-ray developing facilities

Method

1. Remove washing solution and replace with secondary antibody using the same final volume as used for the primary antibody incubation.
2. Incubate on a shaker at room temperature for 30 min.
3. To remove the unbound antibodies, wash the filters with approximately 100 ml PBST for 15 min at room temperature.
4. Repeat the wash with fresh PBST for a further 15 min.
5. Mix solution A and B at an equal ratio.
6. Drain the filters on a hydrophobic membrane, such as parafilm. Add solution A + B directly on to the nitrocellulose. Incubate for 1 min.
7. Remove the excess solution and wrap the nitrocellulose with plastic membrane (e.g. saran wrap).
8. In the dark place an X-ray film against the plastic covered nitrocellulose membrane. To determine the exposure time, experiment with one filter for different times (e.g. 2 s, 15 s, and 1 min).[b]
9. Develop the film.
10. The filter can be reused by briefly washing the filter with 0.2 M glycine pH 2.5 three times, twice with water, and then rinsing the blot twice with PBST. The filters can be reused up to three times.

[a] Secondary antibodies can be bought commercially and should be diluted to the concentration recommended by the manufacturer. Ensure that the secondary antibody recognizes the primary antibody (for example, use anti-mouse for monoclonal antibodies made from mice and anti-rabbit for polyclonal antibodies produced in rabbit).

[b] Rather than using X-ray film with chemiluminesence, the filters can be analysed directly on a chemiluminescent imager that is equipped to analyse chemiluminescence.

4.3 Characterization of clones isolated by western library screening

It is possible that no positive clones are isolated using the western library screening approach even after repeated efforts. This may be because the protein requires specific post-translational modifications. Alternatively, the antibody may not be able to recognize the protein either due to the poor quality of the antibody or because the target protein is not recognized by this antibody. Normally, these problems can be avoided by performing western blot analysis before initiating a large library screen.

A positive clone needs to be rescreened to ensure validity and if necessary to plaque purify. Further characterization will depend on the nature of the protein in question.

5 South-western screening approach

Before starting the south-western screening approach it is important to have characterized in detail the DNA binding properties of the protein activity under investigation. Ideally, these studies should analyse the protein activity in plant protein extracts using gel retardation, footprinting, and/or south-western blotting experiments, which are described in Volume 2, Chapter 4. These studies also help to establish the optimum binding conditions to be used, including the type and amount of competitor DNA required. South-western blotting also allows one to determine if the DNA binding activity can still function when the protein is bound to nitrocellulose. Non-specific binding proteins are a common problem with this technique and it is therefore important to identify suitable mutant versions of the DNA recognition site in order to distinguish between specific and non-specific DNA binding proteins. An example of the type of results that can be obtained with the south-western library screen using both a wild type and mutant probe is shown in *Figure 1* (taken from ref. 8, with permission).

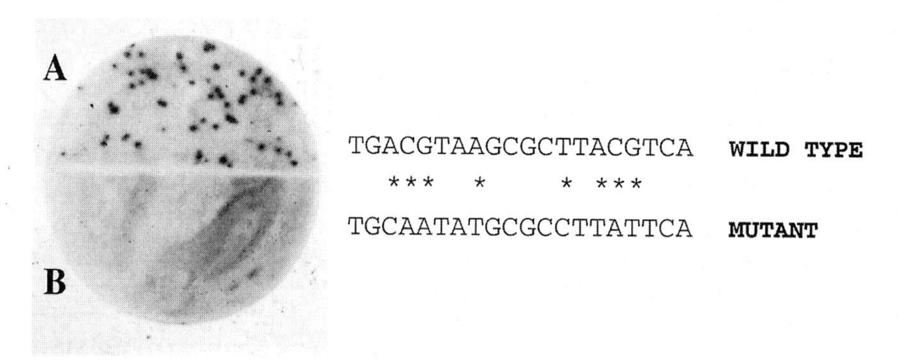

Figure 1 An example of a south-western screen using a wild type (A) and mutant (B) probe to characterize an ocs element binding protein from maize. For further details see Singh *et al.* (8).

Functional DNA binding domains for most types of eukaryotic transcription factors consist of a discrete number of amino acids (approximately 60–200 amino acids) so full length cDNA clones are not necessary. Therefore, the cDNA library should be constructed by random priming in case the DNA binding domain is encoded near the 5′ region of a long mRNA. For the initial steps in the technique, the reader should refer to *Protocol 1* regarding the plating of the library and *Protocol 2* regarding nitrocellulose preparation if the standard non-denaturing procedure is to be used. However, some DNA binding proteins have a higher binding affinity if they are first denatured and renatured, and this procedure is described in *Protocol 8*.

5.1 Preparing the DNA probe for south-western screening

It is critical to try to identify the highest affinity DNA recognition site for the protein of interest, since the signal to background noise ratio can be low in south-western screening, especially during the primary screen. While DNA fragments with a single binding site have been successfully used as probes, fragments with multiple binding sites are likely to perform better. We use double-stranded oligonucleotides from 8 to 25 bp in length, which contain the DNA recognition sequence. The oligonucleotides are then concatermerized by ligation to give an overall length of approximately 200 bp; longer probes yield higher non-specific signals. *Protocol 6* involves the labelling of complementary single stranded oligonucleotides by T4 kinase. The oligonucleotides are designed so that annealing results in the generation of *Bam*HI and *Bgl*II restriction site overhangs at each end. This allows the annealed oligonucleotides to be concatermerized with T4 ligase. DNA recognition probes can also be generated by cloning tandem copies of a synthetic binding-site oligomer in a plasmid vector, gel purifying the fragment and labelling by nick translation, random priming, or end-filling.

Protocol 6

Preparation of the DNA recognition site probe

Equipment and reagents

- Single stranded oligonucleotide binding sites in H_2O at 1 $\mu g/\mu l$
- 10× kinase buffer: 500 mM Tris–HCl pH 7.8, 100 mM $MgCl_2$, 100 mM DTT, 10 mM spermidine
- T4 polynucleotide kinase (10 units/μl), T4 DNA ligase (3 units/μl) and appropriate restriction enzyme(s)
- ^{32}P-γ-ATP approx. >5000 Ci/mmol
- 10 mM ATP
- 100 mM DTT
- Phenol:chloroform:isoamyl alcohol: a mixture of equal parts of equilibrated phenol (containing 0.1% 8-hydroxyquinoline) and chloroform:isoamyl alcohol (24:1).
- 2.5 M NH_4Ac

Protocol 6 continued

- 2.5 M NH_4Ac
- Absolute and 70% ethanol
- G-25 Sephadex columns
- 1 × TBE: 90 mM Tris–HCl pH 8, 90 mM boric acid, 2 mM EDTA
- 10% native polyacrylamide gel, 10% acrylamide:bis acrylamide (29:1), 1 × TBE, 0.01% (w/v) ammonium persulphate, 0.01% (v/v) TEMED
- 85 °C, 65 °C, 37 °C and 12–14 °C water incubator
- Polyacrylamide gel electrophoresis unit
- Scintillation counter
- X-ray film
- X-ray developing facilities

Method

1. For the kinasing reaction add 1 μg of single-stranded oligonucleotide binding site, 2 μl 10 × kinase buffer , 2 μl ^{32}P-γ-ATP approx. >5000 Ci/mmol, 1 μl T4 polynucleotide kinase and H_2O to a final volume of 20 μl. Incubate at 37 °C for 1 h.
2. Anneal the complementary oligonucleotides by mixing 1 μg of each oligonucleotide in a final volume of 40 μl and incubate for 2 min at 85 °C, 15 min at 65 °C, 15 min at 37 °C, 15 min on ice.[a]
3. Set up the ligation by adding to the annealed oligonucleotides: 1.0 μl 10 × kinase buffer, 1.0 μl T4 DNA ligase, 4 μl 10 mM ATP, 4 μl 100 mM DTT. Incubate at 12–14 °C for 1–3 h.[b]
4. Heat to 65 °C for 10 min.
5. Separate the labelled DNA from the unincorporated γATP using centrifugation through G-25 Sephadex columns (Pharmacia medium grade) as described in (3).[c,d]
6. Count an aliquot of the probe in a scintillation counter.
7. Check the size of the probe by running an aliquot (10^6 cpm) on a 10% native polyacrylamide gel in 1 x TBE followed by autoradiography (see Volume 2, Chapter 4).[e, f]

[a] Annealing of the oligonucleotides can also be performed by placing the mixture in a boiling water bath (or use a heating block) for 5 min, then switch off the power, and let the temperature drop slowly to room temperature over a period of at least 30 min.

[b] Ligation time will have to be optimized for each DNA binding site.

[c] Alternatively, the probe can be purified by performing a phenol:chloroform: isoamyl extraction, then precipitating with 2.5 M NH_4Ac and three volumes of ethanol. The pellet is then washed with 70% ethanol and resuspended in water.

[d] We typically obtain 1–5 × 10^8 cpm/μg. A large scale screening of 20 × 150 mm plates will require 10^8 cpm.

[e] We generally transfer the gel onto Whatmann 3MM paper and dry it prior to exposure to obtain sharp bands, although the gel can also be directly exposed without drying. Generally, the gel is exposed for 1–2 h.

[f] If the ligation has progressed too far the probe can be salvaged by digesting with *Bam* HI and/or *Bgl* II. An aliquot of the digested probe should then be checked by gel electrophoresis.

5.2 Preparation of nitrocellulose filters for south-western screening

For the south-western screen it can be helpful to perform the initial primary screen in duplicate, since the background can be high and positive signals often faint, leading to the isolation of false positives. However, the use of duplicate filters significantly increases the workload. We recommend first doing a primary screen with a few plates (three to five) and a single filter. If the background is extensive and/or a large number of potential positives are obtained then the use of duplicate filters will facilitate the identification of true positives.

The preparation of nitrocellulose filters for south-western screening using the non-denaturing screening procedure is the same as western screening and has been described in Section 3 and *Protocol 2*. An alternative protocol involving the denaturation/renaturation of the fusion proteins is described below.

5.3 Using proteins that have been denatured/renatured for south-western screening

In some cases eukaryotic DNA binding proteins have a higher DNA binding affinity when expressed in *E. coli* if they are denatured and then renatured (2), perhaps due to improper folding of the fusion protein in *E. coli*. Alternatively, the high level of expression may lead to the formation of aggregates of the fusion protein. It may be helpful to first perform south-western experiments with plant protein extracts to determine if the binding activity under investigation is improved by the denaturation/renaturation procedure (*Protocol 7*), since the procedure is not always beneficial.

Protocol 7

Preparation of nitrocellulose filters using a denaturation/renaturation cycle for south-western screening

Equipment and reagents

- HEPES binding buffer: 25 mM Hepes, pH 7.9, 25 mM NaCl, 5 mM $MgCl_2$, 0.5 mM DTT
- HEPES binding buffer supplemented with 6 M guanidine hydrochloride
- Blotto: 5% skim milk powder, 50 mM Tris-HCl pH 7.5, 50 mM NaCl, 1 mM EDTA, 1 mM DTT

Method

1 Cool the plates to 4 °C, mark and air dry the filters as described in *Protocol 2*.

2 Carry out all the denaturation/renaturation steps at 4 °C with solutions that have been prechilled to 4 °C. Immerse the filters in 100 ml ice cold HEPES binding buffer supplemented with 6 M guanidine hydrochloride (GuHCl) and incubate with gentle shaking for 5 min.[a,b,c]

Protocol 7 continued

3 Decant the denaturation solution and replace with fresh solution and incubate with gentle shaking for 5 min.[b,c]

4 Decant the denaturation solution into a graduated cylinder and dilute the solution 1:1 with HEPES binding buffer. Incubate the filters with gentle shaking for 5 min.[b,c]

5 Repeat the procedure in step 4 each time diluting the previous solution 1:1 with HEPES binding buffer.[b,c]

6 Wash the filters twice in the HEPES binding buffer for 5 min each time.[b]

7 Block the filters by shaking at 4 °C for 30 min in Blotto.[b]

8 Wash the filters twice for 1–5 min with HEPES binding buffer by shaking at 4 °C.[b]

9 Continue with Step 3, *Protocol 9*.

[a] Although the original procedure recommends that all the steps be carried out at 4 °C, it may be possible to use room temperature and prechilled solutions.

[b] We use 15 cm side, 7 cm high, glass crystallizing dishes (Pyrex) and place up to five 132 mm filters per dish with 100 ml solutions.

[c] The concentrations of guanidine HCl used is 6 M for the first two solutions and 3 M, 1.5 M, 0.75 M, 0.375 M, and 0.187 M for the following solutions.

5.4 Screening the cDNA library for DNA binding proteins

Since high levels of background signals are a common problem with south-western screening, the type and amount of non-specific competitor to be used is important. Gel retardation and/or south-western experiments should be performed with plant protein extracts to test different competitors and to determine the optimum concentration. The two types of non-specific competitors typically used are denatured and sonicated calf thymus DNA (or salmon sperm DNA) or poly(dI-dC).poly(dI-dC). Calf thymus DNA has the advantage of being significantly cheaper.

Protocol 8

Screening the nitrocellulose filters with the DNA recognition site probe

Equipment and reagents

- Binding buffer: 10 mM Tris–HCl pH 7.5, 50 mM NaCl, 1 mM EDTA, 1 mM DTT, 0.25% skim milk powder
- Wash buffer: binding buffer supplemented with 0.1% Triton X-100
- Blotto: (see *Protocol 7*)
- Denatured and sonicated calf thymus DNA (or salmon sperm DNA) or poly(dI-dC).poly(dI-dC)

Protocol 8 continued

Method

1 If using the denaturation/renaturation technique, go to step 3. If using the non-denaturing technique, lift filters from *Protocol* 2, and let air dry, protein surface up, on Whatmann 3MM paper for 15 min.

2 Immerse the filters, protein surface up, in a glass crystallizing dish containing 100 ml of Blotto. Incubate for 60 min at room temperature with gentle swirling on an orbital platform shaker.[a,b,c]

3 Wash each filter twice with binding buffer for 1–5 min at room temperature.[c,d,e] If desired, filters can be stored immersed in a third aliquot of binding buffer for 12–24 h at 4 °C.[f] This is normally done if duplicate filters are being generated. Immerse the filters in binding buffer containing the ^{32}P-labelled DNA binding-site probe (1×10^6 to 2×10^6 cpm/ml) and the appropriate amount of the non-specific competitor DNA [ranging from 2.5–100 μg/ml calf thymus DNA or 5–20 μg/ml poly(dI-dC).poly(dI-dC)]. Use enough solution to just cover all the filters.

4 Incubate at room temperature for 2–4 h with gentle shaking or overnight at 4 °C.[g]

5 Wash the filters three to four times at room temperature with binding buffer, 6 min per wash with 100–200 ml aliquots of wash buffer.

6 Dry the filters on Whatmann 3MM paper, cover with plastic wrap, mark with radioactive ink for orientation and expose to X-ray film with an intensifying screen at −80 °C.[h,i]

[a] In the blocking and all subsequent steps, each filter is incubated protein surface up with gentle swirling. Avoid trapping air bubbles while immersing the filters.

[b] The master plates are sealed with Parafilm or plastic bag sleeves and stored at 4 °C.

[c] Use 100 ml of buffer and up to five 132 mm filters per glass crystallizing dish.

[d] The number and duration of the washes are important parameters and, since the association constants for many DNA binding proteins are low, it is important not to over wash the filters.

[e] The Triton must be completely dissolved in the wash buffer and helps to greatly reduce the spotty background of false positives.

[f] The DNA-binding activities of various recombinant proteins are quite stable under these conditions.

[g] We have got similar results from incubating either at room temperature for 2–4 h or overnight at 4 °C.

[h] Positive signals are often detectable after an overnight exposure although longer exposures particularly with duplicate filters may be necessary.

[i] If high backgrounds are a problem this can be reduced by pre-incubating the probe for 1–2 h with a couple of extra filters prepared the same as the normal filters.

5.5 Selection of positive phage

By screening the recombinant phage in duplicate with both the recognition site probe and a mutant version of this probe it is possible to distinguish between signals resulting from a sequence-specific DNA binding protein versus those due to non-specific DNA-binding proteins.

Protocol 9

Identifying and purifying positive clones obtained with south-western screening

Equipment and reagents

- To pick the plaques, use 1-ml disposable pipette tip with the tips cut off

Method

1. Align the phage plates with the autoradiograms. Pick agarose plugs corresponding to positive signals and use to generate secondary phage stocks as described in Sambrook *et al.* (3).
2. If the primary screen was performed in duplicate it is possible to proceed directly to screening the secondary phage stock with the recognition site probe and mutant probe. If the primary screen was performed with only a single filter the secondary phage stocks should be screened with only the recognition site probe.[a]
3. Plate secondary phage stocks (1000–5000 pfu/90-mm plate) as described previously in *Protocol 1*. For each plate use a 0.1 ml aliquot of the Y1090 culture and 3 ml top agarose. Prepare and screen secondary 82 mm filters as described in *Protocol 2*.
4. Once a phage has been confirmed to be a true positive, screen it with both the recognition site probe and a mutant version of this probe.[b,c]
5. Plaque purify phage that bind specifically to the wild-type probe.

[a] If the primary screen was performed in duplicate then any positive signals are likely to represent true positives. However, if the primary screen was only performed with a single filter then the secondary phage stock must be rescreened.

[b] When screening with the wild type and mutant recognition-site probes we normally use a 100 mm plate and a single 82 mm filter. The filter is then cut in half and each half probed with either the recognition or mutant probe as shown in *Figure 1*.

[c] Ideally, the mutant probe should contain one or a few base changes from the wild type sequence, which prevents binding of the activity present in plant extracts. If such a probe is not available a probe that lacks the recognition site can be used.

6 Characterization of clones isolated by south-western screening

It is possible that no positive clones are isolated using the south-western library screening approach even after repeated efforts. This may because the protein requires specific post-translational modifications or requires other proteins for binding. If this is the case alternative techniques, for example a biochemical approach, need be employed.

Sequence analysis can help to determine if the positive clone is not an artifact, e.g. by showing homology to the DNA binding domain of a previously characterized transcription factor. It is important to further characterize the DNA binding properties of the cloned protein in order to determine if it corresponds to the binding activity in the plant. This is an important issue and one not always possible to resolve, since many plant transcription factors belong to protein families, members of which have considerable overlap in their DNA recognition sequences.

There are a number of ways to express the DNA binding protein. A quick method is to generate a crude extract from a lysogen containing the recombinant phage (1, 3). Another quick method is to express the protein using coupled *in vitro* transcription/translation reactions (9; see also Volume 2, Chapter 7). This method offers the advantage that it is easy to generate and test mutations of the protein. The proteins can also be expressed in *E. coli* or a eukaryotic host such as the baculovirus system and this subject is covered in Volume 2, Chapter 6. The DNA binding properties of the expressed protein can be examined using gel retardation experiments, antibody supershift experiments and DNase I footprinting/methylation interference experiments and these topics are covered in Volume 2, Chapter 4. Antibodies raised against the recombinant protein are another valuable tool to help investigate the relationship between the recombinant protein and the binding activity present in the plant extracts, for example, by performing supershift experiments. By comparing the results of such studies to those obtained with the activity present in plant extracts, it is possible to evaluate if the recombinant protein is a candidate for forming part or all of the plant activity.

Additional approaches to characterize the recombinant DNA binding protein include analysing the corresponding gene expression patterns during plant development and studying the transcriptional properties. For many plant species where the generation of transgenic plants is routine, it is also possible to use powerful reverse genetic approaches to help elucidate the function of the protein. In a model organism like Arabidopsis, these reverse genetic approaches can be coupled to functional genomic techniques to help identify potential target genes regulated by the transcription factor, thereby further helping to understand the role the protein plays in plant growth and development.

References

1. Singh, H., LeBowitz, J. H. , Baldwin, A. S., Jr, and Sharp, P. A. (1988). *Cell*, **52**, 415.
2. Vinson, C. R., LaMarco, K. L., Johnson, P. F., Landschults, W. H., and McKnight, S. L. (1988). *Genes and Development*, **2**, 801.
3. Sambrook, J., Fritsh, E. F., and Maniatis, T. (1989). In *Molecular Cloning: a laboratory manual*, 2nd edn. Cold Springs Harbor Laboratory Press, New York.
4. Hurst, H. C. (1997). In *Methods in Molecular Biology* (ed. I. G. Cowell and C. A. Austin), Vol. 69, pp. 155–59. Humana Press Inc., Totowa.
5. Cowell, I. G. (1997). In *Methods in Molecular Biology* (ed. I. G. Cowell and C. A. Austin), Vol. 69, pp. 161–70. Humana Press Inc., Totowa.
6. Singh, K. B. (1993). In *Plant Molecular Biology Manual* (ed. S. B. Gelvin, R. A. Schilperoot and D. P. S. Verma), Vol. 1, B17, pp. 1–20. Kluwer Academic Publishers, Dordrecht, Belgium.

7. Trayhurn, P., Thomas, M. E., and Duncan, J. S. (1995). *Biochemistry Society Transcripts*, **23**, 495S.
8. Singh, K., Dennis, E. S., Ellis, J. G., Llewellyn, D. J., Tokuhisa, J. G., Wahleithner, J. A. and Peacock, W. J. (1990). *Plant Cell*, **2**, 891.
9. Struhl, K. (1991). In *Methods of Enzymology* (ed. C. Guthrie and G. Fink), Vol. 194, pp. 520–535. Academic Press, Inc., San Diego.

Chapter 11
Complementation cloning

E. Ann Oakenfull and James A. H. Murray
Institute of Biotechnology, University of Cambridge, Tennis Court road, Cambridge CB2 1QT, UK

1 Introduction

Complementation is the restoration to normal of a mutant phenotype by the addition of a wild-type allele, and is a fundamental procedure in genetic analysis. Cloning by complementation makes use of this procedure to isolate genes responsible for restoring normal phenotypes. An essential requirement for a complementation cloning experiment is therefore a mutant phenotype in a suitable screening organism. The most practical screening organisms are prokaryotes, for example, *Escherichia coli*, or simple eukaryotes such as the budding yeast *Saccharomyces cerivisiae*, because they are easily and quickly manipulated. The complementing wild-type allele can be introduced into the mutant-screening organism by transformation with a cDNA library constructed in an appropriate expression vector. Cells that receive the cDNA complementing the wild-type allele of the mutant gene will exhibit a normal phenotype (*Figure 1*). The cDNA responsible for this normal phenotype can then be isolated from the expression vector within the cells.

Originally complementation cloning was only used to clone genes from the screening organism, i.e. *E. coli* mutant strains were complemented with *E. coli* genes, but later experiments revealed that there is sufficient conservation of gene function between organisms for heterologous complementation to be feasible, and plant and mammalian genes have been isolated by complementation screening in mutant strains of *E. coli*, yeast (*S. cerevisiae*) and other relatively simple organisms. The functions of genes isolated by this method have not only been in basic synthetic and catabolic pathways, but also in cell division, cell secretion, and signal transduction. The same principle can be used to confirm the functions of cDNAs isolated using other approaches. As the number of mutant strains of screening organisms increases, it is possible to use complementation cloning to isolate or confirm the functions of an increasing number of genes, and selected recent examples are listed in *Tables 1* and *2*.

An advantage of complementation cloning over more traditional cloning techniques is that it utilizes functional conservation, rather than DNA sequence conservation for recognition of functionally equivalent genes in different species

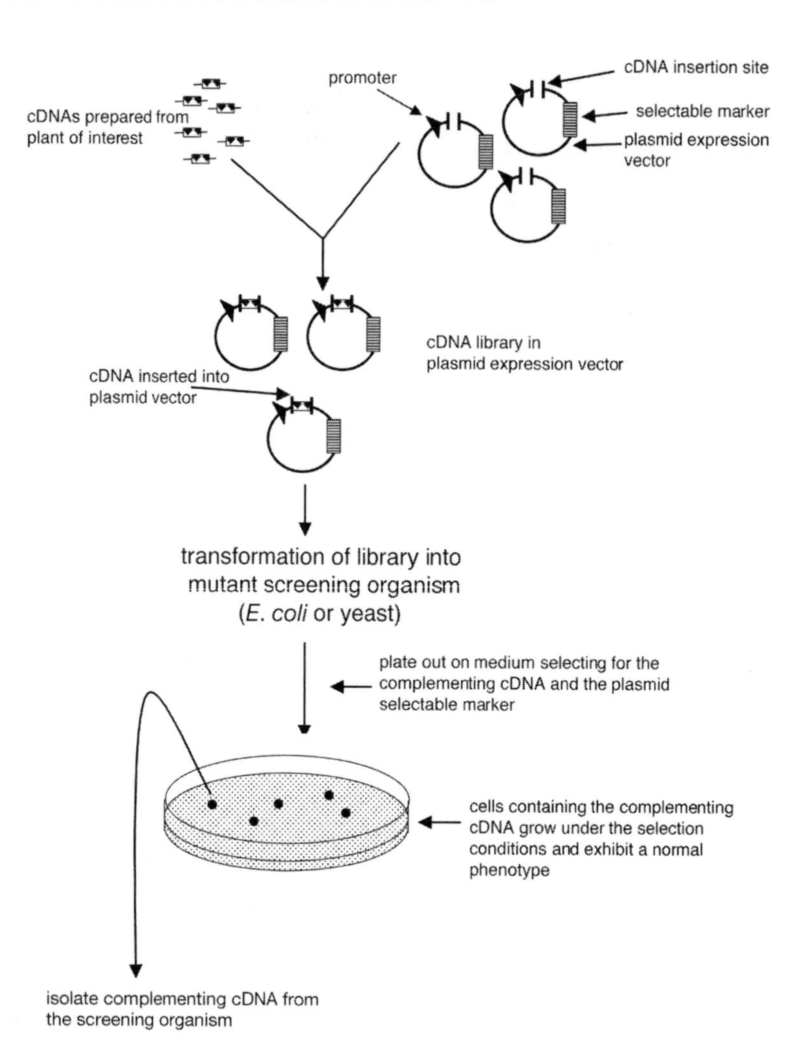

Figure 1 The principle of complementation cloning. cDNA is prepared from the plant tissue in which the gene of interest is expressed, and inserted into a plasmid expression vector containing the appropriate promoter sequence for expression in a screening organism (*E. coli* or yeast), and a selectable marker. The resultant library is used to transform a strain of screening organism that has a mutation in the gene functionally equivalent to the plant gene being cloned. To select for transformants containing a cDNA insert that complements the mutated gene in the screening organism, the transformed cells are plated out onto medium selecting for normal function of the mutated gene. The medium also selects for the plasmid marker to ensure that the cells grow because they contain a plasmid with the appropriate cDNA insert, and not because they contain a mutation that has reverted them to wild type. The complementing cDNA can be isolated from the cells that grow under the selection conditions.

and, therefore, is not dependent on low stringency hybridization. However, as whole genome sequences of many organisms become available to us, database searches will be able to identify this functional conservation by recognizing conservation in the DNA or amino acid sequences in the functional domains of

Table 1 Recent examples of cDNAs isolated by complementation cloning.

Plant	Function encoded	Screening organism	Reference
Arabidopsis thaliana	Chorismate mutase	yeast	(1)
Arabidopsis thaliana	3-phosphoserine phosphatase	*E. coli*	(2)
Arabidopsis thaliana	GABA transporter	yeast	(3)
Arabidopsis thaliana	Gamma-glutamylcysteine synthase	yeast	(4)
Arabidopsis thaliana	Histidine biosynthetic enzymes	*E. coli*	(5)
Arabidopsis thaliana	2 ferrochelatases	yeast	(6)
Arabidopsis thaliana	cDNA corresponding to yeast abc1 gene	yeast	(7)
Arabidopsis thaliana	2 trehalose-6-phosphate phosphatase	yeast	(8)
Arabidopsis thaliana	Translation initiation factors, At.EIF4E1 and At.EIF4E2	yeast	(9)
Arabidopsis thaliana	Assembly of respiratory complexes	yeast	(10)
Arabidopsis thaliana	Galactokinase	yeast	(11)
Arabidopsis thaliana	Serine acetyltransferase	*E. coli*	(12)
Spinacia oleracea	4 phosphoribosyl diphosphate synthase	*E. coli*	(13)
Triticum aestivum	LCT1	yeast	(14)

Table 2 Recent examples of the use of complementation techniques to confirm the function of genes isolated by methods other than complementation cloning.

Plant	cDNA	Method used to isolate gene	Screening organism	Reference
Arabidopsis thaliana	mevalonate diphosphate decarboxylase	Search of EST database for similarity to yeast mevalonate diphosphate decarboxylase, cDNA provided by *A. thaliana* Biological Resource Center (Ohio State University)	yeast	(15)
Arabidopsis thaliana	DNA ligase I	Degenerate PCR used to produce a probe to screen a cDNA library	yeast	(16)
Arabidopsis thaliana	LHT1 (involved in specific amino acid transport)	Search of EST database for similarity to *A. thaliana* NAT2. cDNA provided by *A. thaliana* Biological Resource Center (Ohio State University)	yeast	(17)
Nicotiana tabacum	NAD-dependent isocitrate dehydrogenase	PCR primers designed from equivalent genes in yeast and *A. thaliana* to produce probes to screen a cDNA library	yeast	(18)
Nicotiana tabacum	Sar1 GTPase	PCR primers designed from equivalent genes in *L. esculentum* and *A. thaliana* to produce probes to screen a cDNA library	yeast	(19)

the genes or proteins. Thus, complementation cloning is likely to become less important for isolating genes, but will still be indispensable for testing functional complementation of putative clones. Also, until the whole genome sequences of all organisms are available, there will be a role for complementation cloning in less studied organisms.

This chapter describes techniques for complementation cloning in the two most commonly used screening organisms *E. coli* and the budding yeast *S. cerivisiae* (from here on referred to as yeast). Other screening organisms, such as fission yeast (*S. pombe*), cyanobacteria, and *Chlamydomonas*, can be employed and, although they will not be described here, the principles for their use are the same. More sophisticated developments of complementation cloning are the two-hybrid vector systems used for identifying interacting proteins in a cell. This approach is described in Volume 2, Chapter 9.

2 General requirements for complementation cloning

There are three basic requirements for a complementation cloning experiment:

1. A mutant strain of the screening organism with a recognizable phenotype.
2. A good quality cDNA library constructed in a vector containing the appropriate signals for gene expression in the screening organism and a selectable marker for identifying cells containing the vector.
3. An efficient transformation method for introducing the cDNA library into the screening organism.

The experimental strategy will depend on fulfilling these basic requirements and a more detailed discussion of them can be found in Murray and Smith (20), but the first and perhaps most crucial factor is obtaining an appropriate mutant strain.

2.1 Choice of mutant strain and selection strategy

The level of functional conservation of the screening organism must be considered when choosing the mutant strain, for simple processes, such as nutrient catabolism the functions of the genes involved are likely to be highly conserved, whereas genes that function in, for example, photosynthesis or nitrogen fixation will not be functional in *E. coli* or yeast, and photosynthetic algae or other organisms will be more useful as screening organisms. Another consideration is whether the screening organism can provide the necessary environment for the gene product to become functional. For example, if the product is a eukaryotic protein it may require post-translational modifications or glycosylation that do not occur in prokaryotic bacterial cells. Similarly, some gene products may require location in a specific sub-cellular compartment to be functional or to have access to their substrates.

A further factor in choosing a screening organism is whether a suitable selection scheme for distinguishing mutant and complemented phenotypes is available. There are a variety of selection schemes and the principles of those most commonly used are illustrated below.

2.1.1 Selection schemes for genes involved in growth under certain conditions

Genes coding for enzymes or proteins required for growth under certain conditions are the most easily selected: only mutants with restored phenotypes will be able to grow on the selective medium. For example, to select for a ferrochelatase gene, required for hemin biosynthesis, a strain with a mutated ferrochelatase gene should be transformed and grown on medium lacking hemin; only transformants with a complemented ferrochelatase gene will be able to grow (21). There are many mutant strains available for genes with this type of biosynthetic or catabolic function. However, a large proportion of these are due to single point mutations that have a relatively high frequency of reversion to the normal phenotype, particularly in yeast mutant strains. Using single point mutations in complementation cloning experiments can result in a high level of false positives, which are time-consuming to screen. Although the level of reversion frequency can be tested, ideally non-reverting deletion mutants should be obtained or constructed. Such non-revertant mutants can be constructed in yeast by using a one-step gene disruption strategy, which involves transforming a strain with fragments of DNA containing the gene of interest disrupted by a selectable marker. Due to the identity of the gene sequence on either side of the marker these fragments become integrated into the yeast genome at the correct chromosomal site for the gene. Under the appropriate conditions for the selectable marker, only those cells with the disrupted gene will grow. The fragment containing the disrupted gene can be produced either by enzyme digestion and ligation, or by PCR based methods (22–25). The former requires that the gene has been cloned in the screening organism, which is usually the case in these well-studied organisms and the PCR method requires knowledge of the gene sequence in the screening organism. As the complete genome sequence is now available for both *E. coli* and *S. cerevisiae* design of the primers for these manipulations is straightforward (the yeast genome sequence can be found at http://genome-www.stanford.edu/Saccharomyces/ and that of *E. coli* can be found at http://www.genetics.wisc.edu/). Information on mutant strains already available can be found at http://cgsc.biology.yale.edu for *E. coli* and http://www.atcc.org for yeast, from where deletion strains for many genes can be purchased.

2.1.2 Selection schemes for genes involved in essential cell processes

Selection of genes essential for cell processes, such as cell division or secretion, can be carried out with conditional mutants, i.e. mutants that can only grow under certain conditions (for a detailed discussion of working with essential genes see ref. 26). Four examples of conditional mutants are as follows:

1 Temperature sensitive conditional mutant alleles (*ts*) exist for many essential genes. These alleles are active only at either hot or cold temperatures and, therefore, if the strain is grown at the restrictive temperature then only cells transformed with a complementing gene will grow.

2 One method of constructing conditional alleles is to replace the promoter of the gene to be cloned with a tightly controlled promoter, such as *GAL1* (galactose inducible) in yeast, which is induced only in the presence of galactose. Therefore, the gene will not be activated when the cells are grown on glucose and only cells which have received the complementing cDNA will be able grow under these conditions. Caution should be taken as these promoter systems do not always completely silence the native gene and false positives do occur, therefore the frequency of false positives should always be assessed. If the galactose-controlled construct is placed on a plasmid rather than integrated into the genome, a low copy number expression vector utilizing a centromere (CEN) sequence is preferable over a high copy number plasmid to reduce leaky expression (24).

3 Another method of constructing a conditional phenotype is to replace the wild-type gene in the screening organism by an unstable plasmid containing the gene. The mutated organism is then grown under conditions, which select *against* the unstable plasmid and *for* the transforming plasmid, thus only colonies containing complementary cDNAs will grow (27). This strategy is sometimes termed plasmid shuffling.

4 A further approach is to use the copper-inducible double shut-off procedure as described by Moqtaderi *et al.* (28), where the addition of copper stops the expression of the gene of interest at both the RNA and the protein levels (for practical details see ref. 24).

2.1.3 Selection schemes for non-essential genes

A non-essential gene does not affect the survival of a cell; therefore, straightforward growth properties cannot be used to assess complementation of their mutants and schemes to monitor the gene's activity must be devised. For example, the basis of such a scheme could be a colorimetric assay for the production of a certain metabolite or could involve screening with an antibody for the secretion of a specific protein.

2.1.4 Assessing the reversion frequency of mutant strains to wild-type phenotype

As already mentioned, a serious problem during complementation cloning can be the appearance of false positives and the time taken to disregard them by screening. It is therefore important to assess the reversion frequency in each selection system and determine the conditions which keep it to a minimum. Ideally, these conditions are determined by transforming the screening organism with a control plasmid vector containing a selectable marker, but without a cDNA insert. The transformed cells are plated out both with and without the selection conditions for complementation, combined with conditions to select the plasmid marker. The percentage of the mutant strain that has reverted to wild type, and is thus able to grow under the selection conditions despite its lack of complementing cDNA, can be calculated. The conditions for the assessment are the same for both *E. coli* and *S. cerevisiae* strains and are detailed in *Protocol 1* (for general *E. coli* and yeast manipulation techniques see ref. 24).

The inclusion of a plasmid selectable marker provides an additional precaution against false positive cells because under selection conditions for the marker, only cells containing a plasmid will be able to survive. For *E. coli* plasmids the selectable marker is usually a gene encoding an antibiotic resistance and for *S. cerevisiae* it is often a gene for a biosynthetic process, for example *URA3*, which allows growth of *ura3-* cells on medium without uracil. After consideration of the availability of appropriate mutant strains and selection schemes, there is unlikely to be a choice between screening organisms.

Protocol 1

Assessment of reversion frequency of mutant strains

Method

1 Transform the mutant strain with a control plasmid vector containing a selectable marker.[a]

2 Grow the transformed cells until they reach mid-log phase in medium appropriate for selecting the plasmid marker.

3 Plate the dilutions equivalent to 10^4, 10^5, 10^6, and 10^7 colonies per plate in conditions that normally prevent growth of the mutant, and select for the plasmid with antibiotic or growth medium omission as appropriate.

4 Plate dilutions equivalent to 10^2, 10^3, and 10^4 colonies per plate on the same medium with additional supplement, but not rich medium, to allow growth.

5 Incubate the plates until the colonies are clearly visible, overnight for *E. coli* and approximately 3 days for yeast.

6 Count the number of colonies on the appropriate plates from the selective and non-selective conditions, and calculate the frequency of colony formations due to reversions under complementation selection conditions.[b]

[a] In some *E. coli* strains it is to possible screen bacteriophage libraries of the λYES type (see Section 2.2.2) directly, if this is to be done then the reversion frequency assessment should be carried out using the control bacteriophage vector instead of the plasmid vector.

[b] A maximum reversion frequency of 1 colony per 10^6 transformants is recommended.

However, if there is a choice, then screening in *E. coli* is preferable because higher transformation efficiencies can be achieved which will allow more efficient and comprehensive screening of the cDNA library. In addition, rescue of plasmids from yeast back into *E. coli* for DNA preparation will not be necessary.

2.2 Choice of cDNA library and vector

Some examples of available plant cDNA libraries are shown in *Table 3*. The table is not a comprehensive list and further examples can be found in the references in *Table 1*; however, many of the later complementation cloning experiments used the libraries of Elledge *et al.* (30) and Minet *et al.* (40). Often the most successful method of finding a suitable library is by direct contact with researchers working with the required species. Most of the libraries have been constructed in research laboratories but some have been made commercially. If a suitable pre-made library cannot be found then a new library must be constructed.

2.2.1 cDNA synthesis

We will not go into the details of cDNA library construction here, but the crucial step is the extraction of good quality mRNA (24). This is important since the cDNAs must be full-length, or nearly so, for functional proteins to be expressed in the screening step. Once this has been achieved cDNA can be synthesized using one of the commercially available kits such as from Clontech, Invitrogen, Promega, and Stratagene. When choosing a synthesis kit it is important to consider the compatibility of the cDNA ends with the polylinker sites in the chosen cloning vector.

2.2.2 cDNA library vectors

For a cDNA library to be useful for complementation cloning, its vector must have both the necessary promoter and termination sequences for expression in either *E. coli* or yeast and contain a selectable marker. Either plasmid or bacteriophage vectors can be used for screening in *E. coli*, but only plasmids can be used to transform yeast. *Table 4* illustrates some of the vectors that have been used in previous complementation cloning studies (also see *Table 3*) and suggests some new vectors that could be employed. The simplest vectors for *E. coli* are those based upon pUC or pBluescript, which utilize the *lac* promoter. In yeast vectors there is a greater choice of promoters to stimulate expression at a comparable level to that in wild type yeast; for example, the inducible promoter *GAL1*, the constitutive promoter of *PGK* (phosphoglycerate kinase) that can be induced to a higher level by glucose, or the essentially constitutive promoters of *ADH1* (alcohol dehydrogenase) and *GAPDH* (glyceraldehyde-3-phosphate dehydrogenase) respectively. Additional control over the expression level can be gained by varying the number of plasmids within each yeast cell and, therefore, the relative expression level. Using plasmids with the 2-μ maintenance sequence

Table 3 A selection of plant cDNA libraries.

Plant	Source of mRNA	Vector type	Vector name	Promoter	Selectable marker	Authors/retailer	Examples of further complementation cloning using the same cDNA library
***E. coli* Libraries**							
Arabidopsis thaliana	Leaves and stems	Plasmid	pcDNAII	*lac*	Ampicillin^R	(29)	
Arabidopsis thaliana	Whole plants	Lambda phage/ plasmid	λYES	*lac*	Ampicillin^R	(30)	(30–33, 2)*
Arabidopsis thaliana	Columbia, whole plants	Lambda phage/ phagemid	Unizap	*lac*	Ampicillin^R	Stratagene www.stratagene.com	
Nicotiana plumbaginifolia	Cell suspension	Plasmid	pUC18	*lac*	Ampicillin^R	(34)	
Glycine max (soybean)	21-day-old nodules	Lambda phage/ phagemid	λZAPII/ pBluescript II	*lac*	Ampicillin^R	(35)	(36, 37)
Vigna acontifolia (mothbean)	21-day-old nodules	Plasmid	pcDNAII	*lac*	Ampicillin^R	(38)	(39)
Lycopersicon esculentum (tomato)	Ripening fruit	Lambda phage/ phagemid	λZAPII	*lac*	Ampicillin^R	Stratagene www.stratagene.com	
Nicotiana tabacum (tobacco)		Lambda phage/ phagemid	λZAP	*lac*	Ampicillin^R	Stratagene www.stratagene.com	
Spinacia oleracea (spinach)	Actively growing leaves	Lambda phage/ phagemid	λZAP	*lac*	Ampicillin^R	Stratagene www.stratagene.com	
Hordeum vulgar (barley)	Aleurone tissue	Lambda phage/ phagemid	λZAPII	*lac*	Ampicillin^R	Stratagene www.stratagene.com	
Glycine max (soybean)	Epicotyls	Lambda phage/ phagemid	Unizap	*lac*	Ampicillin^R	Stratagene www.stratagene.com	
Oryza sativa (rice)	Etioloated shoots	Phage	λgt11	*lac*	none	Clontech www.clontech.com	
Yeast (*S. cerevisiae*) Libraries							
Arabidopsis thaliana	Whole plants	Phage/pasmid (CEN)	λYES	*GAL1*	*URA3*	(30)	See ref. 30 in *E. coli* section of this table
Arabidopsis thaliana	Seedlings	Plasmid (2 μ)	Yeast pFL61	*PGK*	*URA3*	(40)	(41, 10, 42, 3, 4, 43, 6,1)*
Spinacia oleracea (spinach)	Leaves	Plasmid (2 μ)	YEp112A1NE	*ADH1*	*TRP1*	(44)	
Solanum tuberosum (potato)	Leaves	Plasmid (2 μ)	YEp112A1NE	*ADH1*	*TRP1*	(45)	(46)
Brassica napus (rape)	Immature seeds	Plasmid (2 μ)	pHR50	*GAPDH*	*URA3*	(47)	

*These references represent a small fraction of the uses of these libraries

Table 4 Examples of vectors suitable for complementation cloning

Screening organism	Vector name	Vector type	Promoter	Selectable marker	Notes	Author/Retailer
E. coli	pUC18	Plasmid	*lac*	Ampicillin^R		(34)
E. coli	pcDNAII	Plasmid	*lac*	Ampicillin^R		(29) or (39)
E. coli	λZAPII	Phagemid	*lac*	Ampicillin^R	Directional cloning possible. Plasmid excised automatically after single-strand replication.	Stratagene www.stratagene.com
E. coli	λtriplexII	Phagemid	*lac*	Ampicillin^R	Directional cloning possible. Plasmid excision by cre-lox system. Expression in all three reading. frames	Clontech www.clontech.com
E. coli and yeast	λYES	Phagemid, the excisable plasmid has CEN sequence	*lac* and *GAL1*	Ampicillin^R and *URA3*	Directional cloning possible. Plasmid excision by cre-lox system.	(30)
Yeast	pYES	2 μ maintenance sequences for yeast	*GAL1*	*URA3* or *TRP1*	Directional cloning possible.	Invitrogen www.invitrogen.com
Yeast	pYC	Plasmid has CEN sequence	*GAL1*	*URA3* or *TRP1*	Directional cloning possible	Invitrogen www.invitrogen.com
Yeast	pESC	Plasmid	*GAL1* and *GAL10*, orientated in opposing directions	*HIS3* or *LEU2* or *TRP1* or *URA3*	Designed for expression and functional analysis of eukaryotic genes in yeast. Directional cloning possible.	Stratagene www.stratagene.com

will produce 10–40 plasmids per cell, while plasmids with the CEN maintenance sequence will be present at only 1–2 copies per cell.

A particularly useful group of vectors are the phagemids; these are bacteriophages with excisable plasmids. Thus, they combine the high cloning efficiency of bacteriophage with the transformation abilities of plasmids. The plasmids are excised either automatically as in λZAP (48), or via the Cre-lox recombination system as in λYES (30), both vectors are illustrated in *Figure 2*. In the former case the plasmid lies between the initiation and termination signals for bacteriophage replication, and a replicating double stranded plasmid is produced after replication. However, the high transformation frequencies now obtainable in *E. coli* with commercially available competent cells has somewhat reduced the advantage of such vectors.

2.3 Transformation efficiency

The transformation efficiency in the screening organism must be great enough to make the amount of labour required for screening the cDNA library realistic. There are approaching 30 000 genes in Arabidopsis, a plant with a relatively small genome. To isolate a low copy mRNA from a cDNA library from this plant it is necessary to screen at least 200 000 and perhaps as many as 10^6 transformed cells. Transformation techniques in both *E. coli* and yeast meet this criteria, whereas transformation methods in plants do not, even though using the available plant mutant strains would avoid the complications of using distantly related organisms for screening (see Section 2.1).

3 Preparation of cDNA libraries and mutant strains for complementation cloning

3.1 Amplification of plasmid cDNA libraries

The amount of plasmid cDNA library DNA required for a complementation cloning experiment depends on the efficiency of the transformation, but will require in the range of 10–100 μg of DNA. To acquire this relatively large amount of cDNA, it will usually be necessary to amplify the library by growth in the *E. coli* host strain. However, libraries may arrive from other laboratories and may not be transformed into a bacterial strain. If the library arrives as plasmid DNA, it must first be transformed into *E. coli*. For yeast complementation; if it arrives as phagemid DNA then the plasmid must be excised by the method appropriate to the vector, for example, the Cre-lox recombination system or automatically after single-strand replication in the bacteriophage, and then transformed into *E. coli*. Once the library is present in the host strain it can be amplified as described in *Protocol 2*. Some bacteriophage libraries designed for expression in *E. coli* can be screened directly as phage, in which case amplification of the library is most efficiently carried out as phage by standard procedures (49, 24).

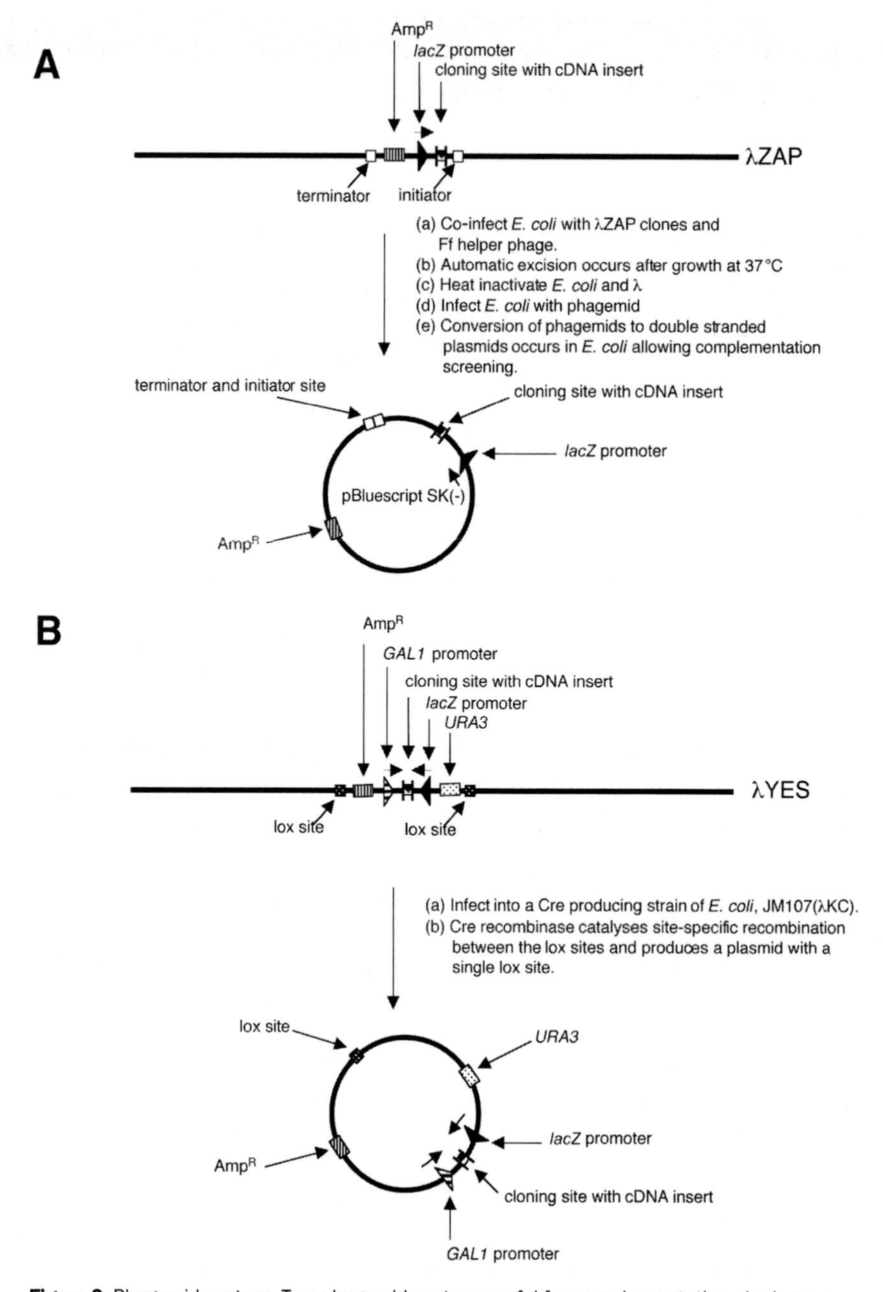

Figure 2 Phagemid vectors. Two phagemid vectors useful for complementation cloning are λZAP (A) and λYES (B). These are both lambda bacteriophages containing unique cloning sites within a linear sequence. After more efficient cloning and replication as bacteriophages, the plasmids can be excised to produce more easily manipulated plasmid vectors for screening cDNA expression libraries in both *E. coli* and yeast. The plasmid in λZAP (46) is excised by co-infecting *E. coli* with the λZAP clones and helper phage, the helper phage recognizes the terminator and initiator domains in the λZAP, and a new DNA strand is produced and the existing strand is displaced. The displaced strand is circularized and packaged as a

Protocol 2

Amplification of plasmid libraries

Reagents

- LB plates and top agarose: 10 g bactotryptone, 5 g bactoyeast extract, 10 g NaCl make up to 1 l with deionized water. Adjust to pH 7.0 with ~0.2 ml 5 M NaOH. For plates add 15 g agar, for top agarose add 7 g agarose
- L-broth: 10 g bactotryptone, 5 g yeast extract, 5 g NaCl make up to 1 l with deionized water. Adjust to pH 7.0 -7.5 with 1-2 pellets of NaOH.
- STE: 0.1 M NaCl, 10 mM Tris-HCl (pH 8.0), 1 mM EDTA (pH 8.0)

Method

1 Make dilutions of the *E. coli* and plate[a] the colonies at 30 000 per 15 cm plate or 10 000 per 9 cm plate.[b]

2 Grow colonies for 12 h or until they are just touching, but with no over-growth.[c]

3 Add the minimal amount of L-broth to the surface of the plates and scrape off the cells with a glass rod. Collect the cells into centrifuge tubes on ice.

4 Centrifuge the suspension for 15 min at 4 °C at 4000 g in a Sorvall GS3 rotor or equivalent.

5 Remove the supernatant and resuspend the pellet in STE. Centrifuge as in step 4.

6 Remove the supernatant and use the pellet for a maxiprep extraction of plasmid DNA using the alkaline lysis method (49),[d] or a reliable proprietary plasmid preparation kit of the appropriate scale.

[a] Amplification usually takes place on solid medium, rather than in liquid to avoid loss of slower growing cells.

[b] To maintain full representation of the library approximately three times the number of colonies in the original library should be plated. For a library with $>10^6$ clones this will require more than thirty 15 cm plates

[c] Growth at 30 °C can be useful to slow down the time taken to reach this point.

[d] If care is taken not to shear the DNA then CsCl gradient purification of the DNA is not necessary. It is also not necessary to remove RNA as it can have beneficial effects acting as a carrier in transformation.

---

filamentous phage by the helper phage proteins, and secreted from the cell. After heat inactivation of the *E. coli* and λ, the packaged phagemids can be infected into *E. coli* where they are converted to double stranded plasmids suitable for complementation screening (A). λYES (28) uses the Cre-lox system to excise the plasmid; the bacteriophage is infected into a Cre producing strain of *E. coli*, for example JM107 (λKC) and Cre recombinase catalyses site specific recombination between the lox sites, at either end of the plasmid sequence, to produce a circular plasmid with a single lox site (B). The plasmids can then be used for complementation screening.

3.2 Optimizing growth conditions of the mutant strain

The best growth conditions for efficient transformation of each strain should be determined before starting complementation cloning experiments (*Protocol 3*). For the most accurate results the tests are carried out with a control plasmid and under the conditions necessary for its maintenance.

Protocol 3

Optimization of transformation growth conditions

Method

1. Inoculate a flask containing 100 ml of the appropriate medium with 100 μl aliquot of a starter culture of the strain.
2. Grow at the appropriate temperature in a shaking incubator with good aeration. Follow the growth of the culture by measuring the OD_{600} of samples taken every 1-2 h for yeast and every 20-30 min for *E. coli*[a].
3. To determine when the optimal time occurs, transformation efficiency should be tested at different stages of growth in log phase.

[a] The optimal time for transformation of a culture is normally when it is growing exponentially. As the number of cells increase the transformation efficiency tends to decrease, and so usually mid-late log phase provides a good compromise between the highest transformation efficiency and a large enough number of cells.

The efficiency of transformations can vary with the size of the culture, particularly during the heat shock stage, therefore to transfer from the test conditions to a full-scale transformation more reproducible results may be obtained if multiple test size transformations are carried out, rather than scaling up to one large culture.

4 Transformation and screening of *E. coli*

4.1 Transformation

cDNA libraries can be transformed into *E. coli* cells by any high efficiency method such as the modified calcium chloride methods (24, 49, 50). However, probably the best method is electroporation (51, 52), because it produces the most linear increase in the number of transformants as the amount of plasmid DNA increases (see Chapter 4 for further details).

4.2 Screening

After transformation the *E. coli* is plated onto medium selecting for both the complementing gene and the plasmid marker. Growing colonies are picked and prepared for DNA extraction and further screening. Under some selection conditions there may only be weak rescue of the normal phenotype by the complementing gene, and in this situation it may be necessary to grow the transformed cells on complete medium containing the appropriate antibiotic for plasmid maintenance, and subsequently replica plating the colonies onto the medium for functional complementation. However, this is more likely to be necessary for yeast strains than bacteria.

After initial functional complementation selection, it is essential to retransform the plasmid DNA into the mutant strain to confirm that it does rescue the normal phenotype. Once functional complementation has been demonstrated in this way the cDNA clone can be analysed in greater detail by restriction mapping and sequencing.

5 Transformation and screening in *S. cerevisiae*

5.1 Transformation

There are several high efficiency transformation methods for yeast such as spheroplast/PEG methods and electroporation, but the most convenient is the lithium salts treatment of whole cells described in *Protocol 4* (53–55). This method has been extensively developed by Gietz and co-workers, and is reported to routinely provide transformation efficiencies of 1×10^6 to 2×10^7 transformants per μg of plasmid DNA (55, 56) and has given more than 10^4 transformants per μg of plasmid DNA from the pFL61 *A. thaliana* cDNA library (21, 57); however, these frequencies are very strain-dependent. Before carrying out the full-scale procedure, it is important to test the transformation conditions to optimize in particular:

1. The concentration ratio of the plasmid DNA compared to the single-stranded carrier DNA (58).
2. The length of the heat shock treatment and the tube volume of the sample for each yeast strain.

As with *E. coli* transformations, greater consistency will be maintained if the transformation of the cDNA library is done as multiple transformations of the same size as the test experiment, rather than scaling up the volumes from the test conditions, because of the difficulty in achieving rapid and homogeneous heat transfer in larger volumes. Alternatively, test scaled-up experiments can be carried out. If low transformation efficiencies are found, it may be preferable to engineer a highly transformable strain such as AB1380, Y190, or CTY10-5D to carry the mutation of interest (56).

Protocol 4

High efficiency yeast transformation (53–56, 58)

Reagents

- YPD/A medium: 10 g yeast extract, 20 g peptone, 20 g dextrose (glucose), 40 mg adenine hemisulphate[a], made up to 1 l in deionized water.
- 1.0 M lithium acetate stock solution: prepare as a 1.0 M stock in deionized water and sterilize by autoclaving. The pH of the solution does not need to be adjusted, but should be in the range of pH 8.4–8.9. Prepare a 100-mM solution by adding 20 ml of the 1.0 M stock to 180 ml of sterile deionized water.
- 50% (w/v) PEG solution: make up PEG (MW 3350, Sigma P3640) with 50% (w/v) of deionized water and sterilize by autoclaving.[b] A convenient way to make up the PEG solution is to dissolve 50 g PEG in 50 ml of deionized water and adjust the volume to 100 ml
- Single-stranded carrier DNA (2 mg/ml): make up a solution of 200 mg of high molecular weight DNA (sodium salt type III from salmon testes, Sigma D1626) in 100 ml of TE buffer (10 mM Tris–HCl (pH 8.0), 1.0 mM EDTA). Disperse the DNA thoroughly with a 10-ml pipette and stir on a magnetic stirrer, preferably overnight at 4 °C. Aliquot the DNA and store at −20 °C. Before use place an aliquot in boiling water for at least 5 min and then cool quickly in ice-water.[c]

Method

1. Prepare an overnight culture of yeast cells by inoculating 5 ml of YPD/A medium with a single yeast colony or 50 μl of a saturated liquid culture. Shake at 30 °C overnight. If the strain already contains a plasmid then selective synthetic complete (SC) medium must be used for this stage only.[d]
2. Use a haemocytometer to determine the titre of the culture by counting the cells in a 10^{-1} dilution, or measure the optical density of a 1:5 dilution.[e] In a 250-ml flask set up a culture with 50 ml of YPD/A warmed to 30 °C and cells at a final concentration of 5×10^6 cells/ml.
3. Incubate the culture at 30 °C, with vigorous shaking (200–300 rpm), until the cell titre is 2×10^7 cells/ml (approximately 3–4 h). This will provide enough cells for 10 standard transformations.
4. Collect the cells by centrifuging at 3000 g for 5 min. Wash the cells with 10 ml of sterile deionized water and pellet the cells again at 3000 g for 4 min.
5. Repeat step 4.
6. Remove the supernatant and resuspend the cells in 10 ml of sterile 0.1 M LiAc and transfer 1 ml to 10 separate 1.5-ml microcentrifuge tubes. Incubate at 30 °C for 10 min. Pellet the cells at full speed for 5 s. Take off the supernatant and add 240 μl PEG solution,[f] 36 μl 1 M LiAc, 52 μl carrier DNA (2 mg/ml), 32 μl plasmid DNA (100 ng to 2 μg) and water to a total volume of 360 μl.

Protocol 4 continued

7. Vortex thoroughly to resuspend the cell pellet in the transformation mix. Incubate the mix at 30 °C for 30 min.
8. Heat shock the transformation reaction at 42 °C for 20 min in a water bath.[g]
9. Pellet the cells at full speed in a microcentrifuge 30 s. Remove the transformation mix.
10. Gently, but thoroughly, resuspend the pellet in 1 ml of sterile deionized water.[h]
11. To recover the transformants place an appropriate[i] amount of the cell suspension on medium to select for the plasmid selectable maker if replica plating, or on medium selecting both for the plasmid and for complementation if using direct selection. Incubate plates for 2–3 days if replica plating, or 2–3 weeks for direct selection. Occasionally, when replica plating very slow growing clones it may be necessary to incubate the plates for 2–3 weeks. For these longer incubations, place the plates in loosely closed polythene bags to prevent dessication. A scaled-up version of this protocol is given in ref. 56.

[a] Adenine hemisulphate can be added before autoclaving, but for maximum effect it is better if it is filter sterilized and added after the rest of the medium has been autoclaved.

[b] Small variations in the concentration of the PEG solution can affect the transformation efficiency and so care should be taken when making up the solution . It should also be stored in a well-sealed tube or bottle that will not allow evaporation.

[c] Carrier DNA is ideally freshly boiled and chilled, but can be frozen after boiling and used 3 or 4 times; however, if transformation efficiencies start to fall then the carrier DNA should be boiled again or exchanged for a new aliquot. If the carrier DNA is of high enough quality, it will not need to be extracted with phenol:chloroform for high transformation efficiencies.

[d] SC medium contains a complete set of supplements, minus the component(s) to be selected (see ref. 24 for preparation of SC).

[e] OD_{600} is 0.1 per 1×10^6 cells for most yeast strains, but should be confirmed for each new strain.

[f] The PEG solution will be very viscous and care should be taken to add the precise volume.

[g] The optimal length of the heat shock can vary from strain to strain from 15 to 25 min and should be optimized for each new strain.

[h] The cells may stick together but forceful mixing can reduce the transformation yield.

[i] The appropriate amount of cell suspension will depend on the amount of plasmid DNA used and the transformation efficiency. A titration to determine the appropriate amount should be carried out before plating out the whole library.

5.2 Screening

Screening of the transformants can be carried out by replica plating or by direct selection. To replica plate the transformed cells they are first plated on medium to select those containing a plasmid, for example, medium without uracil to select for the plasmid *URA3* marker. The resulting colonies are replica plated both onto a medium to select for functional complementation of the mutant strain and a non-selective medium to confirm transfer of the colonies. Replica plating is best carried out when the colonies are pinhead size to prevent too many cells being transferred, which will make it difficult to distinguish new growth on the second and third media plates. Equipment for replica plating can be homemade or obtained commercially, for example, see http://www.biomednet.com/biosupplynet.

By the direct screening method for functional complementation, the transformed cells are plated onto medium that selects for both plasmid uptake and functional complementation at the same time. With this method a small number of transformants must also be plated on medium to select for plasmid uptake alone to determine the efficiency of the transformation. The advantage of the direct screening method is that weakly rescued and therefore slow growing cells can be observed after 1–2 weeks, whereas the slow growth of these cells may not be detected over the relatively large number of cells transferred during replica plating. Replica plating may therefore be preferred for genes where the functional complementation rescue is likely to be stronger, for example, with biosynthetic genes, while direct selection may be better for genes involved in cellular processes, because of their possibly weaker rescue of the mutant strain. However, for cells just recovering from transformation, direct selection can often be too stringent and they may need to recover in liquid rich medium for a few hours before plating onto SC dropout plates, containing all the amino acids except those being used for selection, rather than on to minimal medium with only the essential amino acids added.

The slowest part of the screening process is plating out transformants. If replica plating is required for the screen then each 9-cm plate should have 1000–2000 colonies, with any more there is a risk of confluence and losing individual colonies. About 20 plates are a good number to handle at one time, therefore screening 200 000 transformants will require five to ten plating sessions. For direct selection, higher numbers of colonies (10 000–20 000) per plate can be used and can be plated in a single session.

Yeast mutant strains can have a high rate of reversion to the wild type, therefore after initial screening, it is essential to confirm that the normal phenotype is due to functional complementation and not to reversion. It is also important to be aware that at certain times of the year, in particular the autumn, contaminant colonies can appear during screening. However, contaminant colonies can usually be distinguished from replica-plated colonies because they are smooth, being formed from a single cell, rather than the large irregular colonies, which grow from the multiple cells transferred during replica plating. Confirmation of complementation rather than reversion or contamination can be gained by extraction of plasmid DNA and retransformation of the mutant strain. However, this is a relatively time-consuming process in yeast because obtaining enough plasmid DNA for retransformation involves extracting the plasmid DNA and transforming it into *E. coli* for plasmid preparation. Alternatively, the cDNA region can be amplified via PCR using flanking vector primers and assessed initially by sequencing alone. Another relatively quick method for eliminating the false positives caused by reversion or contamination is segregation analysis (*Protocol 5*). This method can demonstrate co-segregation of the ability of a transformed cell to grow on medium selecting for the plasmid selectable maker and the ability to functionally complement the mutant strain, demonstrating that both properties are conferred by the presence of the transforming plasmid.

Protocol 5
Segregation Analysis

Method

1 Inoculate 10 ml of rich YPD/A (*Protocol 4*) medium with a single colony from a selective plate.

2 Incubate the culture for 24–48 h at the appropriate temperature for the strain, until the culture is in stationary phase.[a]

3 Streak out a small aliquot onto a non-selective plate to produce approximately 1000 colonies. Incubate at 30 °C for 1–2 days.

4 Replica plate the colonies onto one medium to select for the presence of the plasmid vector and another medium to select for the presence of the complementing cDNA.

5 Colonies from cells that contain the complementing cDNA should show co-segregation for the ability to grow on both selective media. Colonies growing on only one selective medium are most likely to be revertants to the wild-type genotype.

[a] This period of growth allows some complemented cells to lose the plasmid, so it can be tested if they display the mutant phenotype again. Genetic revertants will lose the plasmid, but retain a wild-type phenotype, since the phenotype is not dependent on the sequences carried on the plasmid.

In addition to segregation analysis, putative clones can be genotyped as another precaution against unnecessary analysis of contaminants, by checking that they do not grow on plates lacking the nutrient supplement required by the strain. The remaining putative clones can then be isolated from the yeast cells for retransformation into the mutant strain to verify their functional complementation ability (*Protocol 6*). Positive clones can be analysed in more detail by restriction mapping and DNA sequencing. Enough DNA for these studies can be prepared by growth of the plasmid in *E. coli* or alternatively the cDNA inserts can be amplified directly from the yeast colonies (59).

Protocol 6

Isolation of plasmid DNA from yeast

Reagents

- Cracking buffer: 2% (v/v) Triton X-100, 1% (v/v) sodium dodecyl sulfate (SDS), 100 mM NaCl, 10 mM Tris–HCl (pH 8.0), 1 mM EDTA (pH 8.0).
- 0.425–0.600 mm diameter acid-washed glass beads (Sigma G 8772)
- 25:24:1 (v/v/v) phenol/chloroform/isoamyl alcohol
- *E. coli* competent cells HB101 or MH1
- LB plates: 10 g bactotryptone, 5 g bactoyeast extract, 10 g NaCl in 1 l deionized water. Adjust to pH7.0 with ~0.2ml 5 M NaOH. Add 15 g agar before autoclaving.

Method

1. Inoculate 2 ml of SC dropout medium with a single yeast colony containing the plasmid with the cDNA clone of interest. Grow to stationary phase by shaking at 30 °C overnight.
2. Transfer 1.5 ml of the overnight culture into a microcentrifuge tube and spin for 5 s at maximum speed. Pour off the supernatant and vortex briefly to break up the pellet.
3. Resuspend the cells in 200 μl of cracking buffer, add 0.3 g of glass beads (approximately 200 μl volume) and 200 μl of phenol/ chloroform/isoamyl alcohol. Vortex vigorously for about 2 min.[a]
4. Microcentrifuge for 5 min at full speed.
5. Carefully remove the aqueous layer. If the layer appears cloudy, extraction with a further 200 μl of phenol/chloroform/isoamyl alcohol can be carried out.
6. Transform competent *E. coli* with 1–2 μl of the aqueous layer.[b] Plate on LB plates containing the appropriate antibiotic to select for the plasmid.[c]
7. Use standard procedures to extract plasmid DNA from bacterial colonies (49, 24).

[a] The amount of vortexing required will vary between different vortexes. The appropriate amount is the minimum needed to break 80–90% of the cells and can be determined by microscopic examination.

[b] Store the remainder of the aqueous layer at −20 °C for any additional transformations that may be required.

[c] If insufficient colonies are obtained the plasmid DNA may be concentrated by ethanol precipitation.

6 Potential problems

The usefulness of complementation cloning as a technique for isolating genes is best demonstrated by the variety of genes that continue to be isolated by the method (*Table 1*). However, the method is not always successful and probably the most common reason lies in insufficient similarity in function of the gene in the screening organism and the organism from which it is to be cloned. In this situation, rescue of function in the mutant strain is not possible and therefore neither is complementation cloning. Another difficulty that can arise in principle, is when functional complementation occurs, but via an alternative pathway than the one involving the desired gene. Although these alternate pathways may produce interesting avenues for further research, interpretation of the clones can be difficult. Several methods can be used to determine if the clones isolated are homologous to the mutant gene; they are usually recognized from their sequence similarity, they can be assessed by their ability to rescue related mutations in additional strains, and their gene activity can be measured; for instance if the gene is an enzyme, the enzyme activity can be measured in the rescued strain.

Other problems specific to functional complementation experiments are on a more technical level and may be solved by adaptation of the experimental techniques; for example, the cDNA library may not be of high enough quality and the cDNAs may not be of sufficient length to complement the mutant gene or the library may not have a comprehensive representation of the expressed genes in the tissue of origin. In addition, the complementing gene may only be present at a low copy number in the cDNA library and unless sufficient transformants are screened, complementing clones will not be identified. Another variable in complementation experiments is the level of gene expression; this can be affected by the orientation of the cDNA in the plasmid with respect to the promoter and termination sequences, by the amount of upstream sequence included in the cDNA insert and by the number of plasmids per cell. The level of gene expression is likely to affect the success of the mutant rescue, since there may be an optimal level of expression above and below which gene function will be adversely affected. It is also likely that the specificity of this optimal level will vary with the function of the gene, a more precise level being required for some genes than for others. Without knowing the optimal levels of expression for each gene the simplest way to overcome this problem is to screen a larger number of clones, with the expectation that some of them will be expressing at the optimal level. In general, yeast transformants vary considerably in their copy number and their expression level, and different cDNA clones are likely to be transcribed and/or translated at different levels.

7 Conclusions

The intention of this chapter is to provide researchers with the information necessary to plan and execute a strategy for cloning their gene of interest, via a complementation cloning approach. This technique is particularly useful when no molecular information is available for the gene or the gene product, and we envisage that, at least for the next 5 years, there will be a niche for complementation cloning in the many organisms for which the whole genome sequence is not yet available. However, in the post-genomic era it will be increasingly difficult to imagine a scenario where this molecular information is not available, and it seems likely that the future strength of functional complementation techniques will be in confirming the functional properties of genes isolated by computational methods. Finally, there will always be a place for this technique in circumstances where the identity of a gene cannot be predicted from DNA and protein sequence similarity.

Acknowledgements

J. A. H. M. thanks François Lacroute and Alison Smith for starting him along the road of complementation cloning.

References

1. Mobley, E. M., Kunkel, B. N., and Keith, B. (1999). *Gene*, **240**, 115.
2. Ho, C. L., Noji, M., and Saito, K. (1999). *J. Biol. Chem.*, **274**, 11007.
3. Breitkreuz, K. E., Shelp, B. J., Fischer, W. N., Schwacke, R., and Rentsch, D. (1999). *FEBS Lett.*, **450**, 280.
4. May, M. J., Vernoux, T., Sanchez-Fernandez, R., Van Montagu, M., and Inze, D. (1998). *Proc. Natl Acad. Sci. USA*, **95**, 12049.
5. Fujimori, K. and Ohta, D. (1998). *Plant Physiol.*, **118**, 275.
6. Chow, K. S., Singh, D. P., Walker, A. R., and Smith, A. G. (1998). *Plant J.*, **15**, 531.
7. Cardazzo, B., Hamel, P., Sakamoto, W., Wintz, H., and Dujardin, G. (1998). *Gene*, **221**, 117.
8. Vogel, G., Aeschbacher, R. A., Muller, J., Boller, T., and Wiemken, A. (1998). *Plant J.*, **13**, 673.
9. Rodriguez, C. M., Freire, M. A., Camilleri, C., and Robaglia, C. (1998). *Plant J.*, **13**, 465.
10. Hamel, P., Sakamoto, W., Wintz, H., and Dujardin, G. (1997). *Plant J.*, **12**, 1319.
11. Kaplan, C. P., Tugal, H. B., and Baker, A. (1997). *Plant Mol. Biol.*, **34**, 497.
12. Howarth, J. R., Roberts, M. A., and Wray, J. L. (1997). *Biochim. Biophys. Acta*, **1350**, 123.
13. Krath, B. N., Eriksen, T. A., Poulsen, T. S., and Hove-Jensen, B. (1999). *Biochim. Biophys. Acta*, **1430**, 403.
14. Schachtman, D. P., Kumar, R., Schroeder, J. I., and Marsh, E. L. (1997). *Proc. Natl. Acad. Sci. USA*, **94**, 11079.
15. Cordier, H., Karst, F., and Berges, T. (1999). *Plant Mol. Biol.*, **39**, 953.
16. Taylor, R. M., Hamer, M. J., Rosamond, J., and Bray, C. M. (1998). *Plant J.*, **14**, 75.
17. Chen, L. and Bush, D. R. (1997). *Plant Physiol.*, **115**, 1127.
18. Lancien, M., Gadal, P., and Hodges, M. (1998). *Plant J.*, **16**, 325.
19. Takeuchi, M., Tada, M., Saito, C., Yashiroda, H., and Nakano, A. (1998). *Plant Cell Physiol.*, **39**, 590.
20. Murray, J. A. H. and Smith, A. G. (1996). In *Plant Gene Isolation* (ed. G. Foster and D. Twell), p. 177. John Wiley and Sons, Chichester.
21. Smith, A. G., Santana, M. A., Wallace-Cooke, A. D. M., Roper, J. M., and Labbe-Bois, R. (1994). *J. Biol. Chem.*, **269**, 13405.
22. Rothstein, R. (1983). *Methods Enzymol.*, **101**, 202.
23. Smith, J., Zou, H. and Rothstein, R. (1995). *Methods Mol. Cell. Biol.*, **5**, 270.
24. Ausubel, F. M., Kingston, R. E., Seidman, J. G., and Struhl, K. (ed.) (1999). *Current Protocols in Molecular Biology*. John Wiley and Sons, New York.
25. Wach, A., Brachat, A., Rebischung, C., Steiner, S., Pokorni, K., te Heesen, S., and Philippsen, P. (1998) *Methods Microbiol.*, **26**, 67.

26. Stark, M. J. R. (1998) *Methods Microbiol.*, **26**, 83.
27. Elledge, S. J., Bai, C., and Edwards, M. C. (1993). *Methods: Comparison Methods Enzymol.*, **5**, 96.
28. Moqtaderi, Z., Bai, Y., Poon, D., Weil, P. A., and Struhl, K. (1996). *Nature*, **383**, 188.
29. Senecoff, J. F. and Meagher, R. B. (1993). *Plant Physiol.*, **102**, 387.
30. Elledge, S. J., Mulligan, J. T., Ramer, S. W., Spottswood, M., and Davis, R. W. (1991). *Proc. Natl Acad. Sci. USA*, **88**, 1731.
31. Kim, J. and Leustek, T. (1996). *Plant Mol. Biol.*, **32**, 1117.
32. GutierrezMarcos, J. F., Roberts, M. A., Campbell, E. I., and Wray, J. L. (1996). *Proc. Natl Acad. Sci.*, **93**, 13377.
33. Eberhard, J., Ehrler, T. T., Epple, P., Felix, G., Raesecke, H. R., Amrhein, N., and Schmid, J. (1996). *Plant J.*, **10**, 815.
34. Van Camp, W., Bowler, C., Villarroel, R., Tsang, E. W. T., Van Montagu, M., and Inzé, D. (1990). *Proc. Natl Acad. Sci. USA*, **87**, 9903.
35. Delauney, A. J. and Verma, D. P. S. (1990). *Mol. Gen. Genet.*, **221**, 299.
36. Miao, G. H., Hirel, B., Marsolier, M. C., Ridge, R. W., and Verma, D. P. (1991). *Plant Cell*, **3**, 11.
37. Schnorr, K. M., Laloue, M., and Hirel, B. (1996). *Plant Mol. Biol.*, **32**, 751.
38. Hu, C. A., Delauney, A. J., and Verma, D. P. (1992). *Proc. Natl Acad. Sci. USA*, **89**, 9354.
39. Chapman, K. A., Delauney, A. J., Kim, J. H., and Verma, D. P. S. (1994). *Plant Mol. Biol.*, **24**, 389.
40. Minet, M., Dufour, M-E., and Lacroute, F. (1992). *Plant J.*, **2**, 417.
41. Gachotte, D., Husselstein, T., Bard, M., Lacroute, F., and Benveniste, P. (1996). *Plant J.*, **9**, 391.
42. Avelange-Macherel, M. H. and Joyard, J. (1998). *Plant J.*, **14**, 203.
43. Zhou, L., Lacroute, F., and Thornburg, R. (1998). *Plant Physiol.*, **117**, 245.
44. Riesmeier, J. W., Willmitzer, L., and Frommer, W. B. (1992). *EMBO J.*, **11**, 4705.
45. Riesmeier, J. W., Hirner, B., and Frommer, W. B. (1993). *Plant Cell*, **5**, 1591.
46. Klonus, D., Höfgen, R., Willmitzer, L., and Riesmeier, J. W. (1994). *Plant J.*, **6**, 105.
47. Ellerström, M., Josefsson, G., Rask, L., and Ronne, H. (1992). *Plant Mol. Biol.*, **18**, 557.
48. Short, J. M., Fernandez, J. M., Sorge, J. A., and Huse, W. D. (1988). *Nucl. Acids Res.*, **16**, 7583.
49. Sambrook, J., Fritsch, E. F., and Maniatis, T. (ed.) (1989). *Molecular Cloning—A laboratory Manual.* Cold Spring Harbor Laboratory Press, New York.
50. Hanahan, D. (1985). In *DNA Cloning, Volume 1* (ed. D. M. Glover). p. 109. Oxford University Press, Oxford.
51. Inoue, H., Nojima, H., and Okayama, H. (1990). *Gene*, **96**, 23.
52. Zabarovsky, E. R. and Winberg, G. (1990). *Nucl. Acids Res.*, **18**, 5912.
53. Schiestl, R. H. and Gietz, R. D. (1989). *Curr. Genet.*, **16**, 339.
54. Gietz, D., St Jean, A., Woods, R. A., and Schiestl, R. H. (1992). *Nucl. Acids Res.*, **20**, 1425.
55. Gietz, R. D. and Schiestl, R. H. (1995). *Methods Mol. Cell. Biol.*, **5**, 255.
56. Gietz, R. D. and Woods, R. A. (1998) *Methods Microbiol.*, **26**, 53.
57. Soni, R., Carmichael, J. P., Shah, Z. H., and Murray, J. A. H. (1995). *Plant Cell*, **7**, 85.
58. Gietz, R. D. and Woods R. A. (1994). In *Molecular Genetics of Yeast: a practical approach.* (ed. J. R. Johnston), p. 121. Oxford University Press, Oxford.
59. Chen, H-R., Hsu, M-T., and Cheng, S. C. (1995). *BioTechniques*, **19**, 744.

List of suppliers

Ambersil Limited, Wylds Road, Castlefield Industrial Estate, Bridgwater, Somerset TA6 4DD, UK

Amersham Pharmacia Biotech UK Ltd, Amersham Place, Little Chalfont, Buckinghamshire HP7 9NA, UK (see also Nycomed Amersham Imaging UK; Pharmacia)
Tel: 0800 515313
Fax: 0800 616927
URL: http//www.apbiotech.com/

Anderman and Co. Ltd, 145 London Road, Kingston-upon-Thames, Surrey KT2 6NH, UK
Tel: 0181 5410035
Fax: 0181 5410623

Beckman Coulter Inc., 4300 N. Harbor Boulevard, PO Box 3100, Fullerton, CA 92834-3100, USA
Tel: 001 714 8714848
Fax: 001 714 7738283
URL: http://www.beckman.com/

Beckman Coulter (UK) Ltd, Oakley Court, Kingsmead Business Park, London Road, High Wycombe, Buckinghamshire HP11 1JU, UK
Tel: 01494 441181
Fax: 01494 447558
URL: http://www.beckman.com/

Becton Dickinson and Co., 21 Between Towns Road, Cowley, Oxford OX4 3LY, UK
Tel: 01865 748844
Fax: 01865 781627
URL: http://www.bd.com/

Becton Dickinson and Co., 1 Becton Drive, Franklin Lakes, NJ 07417-1883, USA
Tel: 001 201 8476800
URL: http://www.bd.com/

Bio 101 Inc., c/o Anachem Ltd, Anachem House, 20 Charles Street, Luton, Bedfordshire LU2 0EB, UK
Tel: 01582 456666
Fax: 01582 391768
URL: http://www.anachem.co.uk/

Bio 101 Inc., PO Box 2284, La Jolla, CA 92038-2284, USA
Tel: 001 760 5987299
Fax: 001 760 5980116
URL: http://www.bio101.com/

Bio-Rad Laboratories Ltd, Bio-Rad House, Maylands Avenue, Hemel Hempstead, Hertfordshire HP2 7TD, UK
Tel: 0181 3282000
Fax: 0181 3282550
URL: http://www.bio-rad.com/

Bio-Rad Laboratories Ltd, Division Headquarters, 1000 Alfred Noble Drive, Hercules, CA 94547, USA
Tel: 001 510 7247000
Fax: 001 510 7415817
URL: http://www.bio-rad.com/

Boehringer-Mannheim (see Roche)

Branson Ultrasonics Corp., Calibron Instruments Div., 41 Eagle Rd., Danbury, CT 06813-1961, USA.

Calbiochem-Novabiochem GmbH, Postfach 1167, D-65796 Bad Soden/Ts. Germany

Clontech Laboratories, Inc., 4055 Fabian Way, Palo Alto, CA 94303, USA.

Clontech Laboratories, Inc., 1020 East Meadow Circle, Palo Alto, CA 94303–4230, USA.

Clontech Laboratories UK Ltd, Unit 2, Intec 2, Wade Road, Basingstoke, Hampshire, RG24 8NE, UK.
URL: http//:www.clontech.com/

Costar Corning Inc., Science Products Div., 45 Nagog Park, Acton , MA 01720, USA.

CP Instrument Co. Ltd, PO Box 22, Bishop Stortford, Hertfordshire CM23 3DX, UK
Tel: 01279 757711
Fax: 01279 755785
URL: http//:www.cpinstrument.co.uk/

Difco Laboratories

Duchefa, Haarlem, The Netherlands

Dupont Co. (Biotechnology Systems Division), PO Box 80024, Wilmington, DE 19880–002, USA
Tel: 001 302 7741000 Fax: 001 302 7747321
URL: http://www.dupont.com/

Dupont (UK) Ltd, Industrial Products Division, Wedgwood Way, Stevenage, Hertfordshire SG1 4QN, UK
Tel: 01438 734000
Fax: 01438 734382
URL: http://www.dupont.com/

Dynal, Inc., 5 Delaware Drive, Lake Success, NY 11042, USA.

Dynal (UK) Ltd, 26 Grove Street, New Ferry, Wirral, L62 5AZ, UK.

Eastman Chemical Co., 100 North Eastman Road, PO Box 511, Kingsport, TN 37662–5075, USA
Tel: 001 423 2292000
URL: http//:www.eastman.com/

Fastnacht Laborbedarf GmbH, Bonn, Germany

Fisher Scientific UK Ltd, Bishop Meadow Road, Loughborough, Leicestershire LE11 5RG, UK
Tel: 01509 231166
Fax: 01509 231893
URL: http://www.fisher.co.uk/

Fisher Scientific, Fisher Research, 2761 Walnut Avenue, Tustin, CA 92780, USA
Tel: 001 714 6694600
Fax: 001 714 6691613
URL: http://www.fishersci.com/

Fluka, PO Box 2060, Milwaukee, WI 53201, USA
Tel: 001 414 2735013
Fax: 001 414 2734979
URL: http://www.sigma-aldrich.com/

Fluka Chemical Co. Ltd, PO Box 260, CH-9471, Buchs, Switzerland
Tel: 0041 81 7452828
Fax: 0041 81 7565449
URL: http://www.sigma-aldrich.com/

FMC, Flowgen, Lynn Lane, Shenstone, Staffs, WS14 0EE, UK.

FMC Marine Colloids, Bioproducts Department, 5 Maple Street, Rockland, ME 04841, USA

Geno Technology, Inc. 3047 Bartold Ave., Maplewood, MO 63143, USA.

Gibco-BRL (see Life Technologies)

HT Biotechnology Ltd, Unit 4, 61 Ditton Walk, Cambridge CB5 8QD, UK.

Hybaid Ltd, Action Court, Ashford Road, Ashford, Middlesex TW15 1XB, UK
Tel: 01784 425000
Fax: 01784 248085
URL: http://www.hybaid.com/

Hybaid US, 8 East Forge Parkway, Franklin, MA 02038, USA
Tel: 001 508 5416918
Fax: 001 508 5413041
URL: http://www.hybaid.com/

HyClone Laboratories, 1725 South HyClone Road, Logan, UT 84321, USA
Tel: 001 435 7534584
Fax: 001 435 7534589
URL: http//:www.hyclone.com/

IDEXX Laboratories, Inc., One IDEXX Drive, Westbrook, ME 04092, USA.

Invitrogen BV, PO Box 2312, 9704 CH Groningen, The Netherlands
Tel: 00800 53455345
Fax: 00800 78907890
URL: http://www.invitrogen.com/

Invitrogen Corp., 1600 Faraday Avenue, Carlsbad, CA 92008, USA
Tel: 001 760 6037200
Fax: 001 760 6037201
URL: http://www.invitrogen.com/

Life Technologies Inc., 9800 Medical Center Drive, Rockville, MD 20850, USA
Tel: 001 301 6108000
URL: http://www.lifetech.com/

Life Technologies Ltd, PO Box 35, 3 Free Fountain Drive, Inchinnan Business Park, Paisley PA4 9RF, UK
Tel: 0800 269210
Fax: 0800 243485
URL: http://www.lifetech.com/

Merck Sharp & Dohme Research Laboratories, Neuroscience Research Centre, Terlings Park, Harlow, Essex CM20 2QR, UK
URL: http://www.msd-nrc.co.uk/

MSD Sharp and Dohme GmbH, Lindenplatz 1, D-85540, Haar, Germany
URL: http://www.msd-deutschland.com/

Millipore Corp., 80 Ashby Road, Bedford, MA 01730, USA
Tel: 001 800 6455476
Fax: 001 800 6455439
URL: http://www.millipore.com/

Millipore (UK) Ltd, The Boulevard, Blackmoor Lane, Watford, Hertfordshire WD1 8YW, UK
Tel: 01923 816375
Fax: 01923 818297
URL: http://www.millipore.com/local/UKhtm/

Molecular Dynamics, 928 East Arques Ave., Sunnyvale, CA 94086-4520, USA.

Nalge Nunc International. 75 Panorama Creek Drive, P.O. Box 20365, Rochester, NY 14602-0365, USA.

New England Biolabs, 32 Tozer Road, Beverley, MA 01915-5510, USA
Tel: 001 978 9275054

Nikon Corp., Fuji Building, 2-3, 3-chome, Marunouchi, Chiyoda-ku, Tokyo 100, Japan
Tel: 00813 32145311
Fax: 00813 32015856
URL: http://www.nikon.co.jp/main/index_e.htm/

Nikon Inc., 1300 Walt Whitman Road, Melville, NY 11747-3064, USA
Tel: 001 516 5474200
Fax: 001 516 5470299
URL: http://www.nikonusa.com/

Nycomed Amersham, 101 Carnegie Center, Princeton, NJ 08540, USA
Tel: 001 609 5146000
URL: http://www.amersham.co.uk/

Nycomed Amersham Imaging, Amersham Labs, White Lion Rd, Amersham, Buckinghamshire HP7 9LL, UK
Tel: 0800 558822 (or 01494 544000)
Fax: 0800 669933 (or 01494 542266)
URL: http//:www.amersham.co.uk/

Perkin Elmer Ltd, Post Office Lane, Beaconsfield, Buckinghamshire HP9 1QA, UK
Tel: 01494 676161
URL: http//:www.perkin-elmer.com/

Perkin Elmer (Applied Biosystems Division), Kelvin Close, Birchwood Science Park North, Warrington, Cheshire WA3 7PB, UK

Pierce, 3747 N. Meridian Road, P.O.Box 117, Rockford, IL 61105, USA.

Pharmacia, Davy Avenue, Knowlhill, Milton Keynes, Buckinghamshire MK5 8PH, UK (also see Amersham Pharmacia Biotech)
Tel: 01908 661101
Fax: 01908 690091
URL: http//www.eu.pnu.com/

Promega Corp., 2800 Woods Hollow Road, Madison, WI 53711-5399, USA
Tel: 001 608 2744330
Fax: 001 608 2772516
URL: http://www.promega.com/

Promega UK Ltd, Delta House, Chilworth Research Centre, Southampton SO16 7NS, UK
Tel: 0800 378994
Fax: 0800 181037
URL: http://www.promega.com/

Qiagen Inc., 28159 Avenue Stanford, Valencia, CA 91355, USA
Tel: 001 800 4268157
Fax: 001 800 7182056
URL: http://www.qiagen.com/

Qiagen UK Ltd, Boundary Court, Gatwick Road, Crawley, West Sussex RH10 2AX, UK
Tel: 01293 422911
Fax: 01293 422922
URL: http://www.qiagen.com/

Roche Diagnostics Corp., 9115 Hague Road, PO Box 50457, Indianapolis, IN 46256, USA
Tel: 001 317 8452358
Fax: 001 317 5762126
URL: http://www.roche.com/

Roche Diagnostics GmbH, Sandhoferstrasse 116, 68305 Mannheim, Germany
Tel: 0049 621 7594747
Fax: 0049 621 7594002
URL: http://www.roche.com/

Roche Diagnostics Ltd, Bell Lane, Lewes, East Sussex BN7 1LG, UK
Tel: 0808 1009998 (or 01273 480044)
Fax: 0808 1001920 (01273 480266)
URL: http://www.roche.com/

Schleicher and Schuell Inc., Keene, NH 03431A, USA
Tel: 001 603 3572398

Shandon Scientific Ltd, 93-96 Chadwick Road, Astmoor, Runcorn, Cheshire WA7 1PR, UK
Tel: 01928 566611
URL: http//www.shandon.com/

Sigma-Aldrich Co. Ltd, The Old Brickyard, New Road, Gillingham, Dorset SP8 4XT, UK
Tel: 0800 717181 (or 01747 822211)
Fax: 0800 378538 (or 01747 823779)
URL: http://www.sigma-aldrich.com/

Sigma Chemical Co., PO Box 14508, St Louis, MO 63178, USA
Tel: 001 314 7715765
Fax: 001 314 7715757
URL: http://www.sigma-aldrich.com/

Stratagene Europe, Gebouw California, Hogehilweg 15, 1101 CB Amsterdam Zuidoost, The Netherlands
Tel: 00800 91009100
URL: http://www.stratagene.com/

Stratagene Inc., 11011 North Torrey Pines Road, La Jolla, CA 92037, USA
Tel: 001 858 5355400
URL: http://www.stratagene.com/

United States Biochemical (USB), PO Box 22400, Cleveland, OH 44122, USA
Tel: 001 216 4649277

Yakult Hansha Co., Japan

Index

Page numbers in bold refer to this volume, bracketed page numbers in italics refer to volume 2.